Structural Dynamics and Static Nonlinear Analysis From Theory to Application

Mourad Belgasmia
Setif 1 University, Algeria

A volume in the Advances in Civil and Industrial Engineering (ACIE) Book Series

IGI Global
PUBLISHER of TIMELY KNOWLEDGE

Published in the United States of America by
IGI Global
Engineering Science Reference (an imprint of IGI Global)
701 E. Chocolate Avenue
Hershey PA, USA 17033
Tel: 717-533-8845
Fax: 717-533-8661
E-mail: cust@igi-global.com
Web site: http://www.igi-global.com

Copyright © 2021 by IGI Global. All rights reserved. No part of this publication may be reproduced, stored or distributed in any form or by any means, electronic or mechanical, including photocopying, without written permission from the publisher. Product or company names used in this set are for identification purposes only. Inclusion of the names of the products or companies does not indicate a claim of ownership by IGI Global of the trademark or registered trademark.

Library of Congress Cataloging-in-Publication Data

Names: Belgasmia, Mourad, 1973- author.
Title: Structural dynamics and static nonlinear analysis from theory to application / by Mourad Belgasmia.
Description: Hershey, PA : Engineering Science Reference, an imprint of IGI Global, [2021] | Includes bibliographical references and index. | Summary: "This book examines structural dynamics and static nonlinear analysis from theory to application"-- Provided by publisher.
Identifiers: LCCN 2020006929 (print) | LCCN 2020006930 (ebook) | ISBN 9781799843993 (hardcover) | ISBN 9781799854807 (paperback) | ISBN 9781799844006 (ebook)
Subjects: LCSH: Structural dynamics. | Statics. | Nonlinear theories.
Classification: LCC TA654 .B45 2021 (print) | LCC TA654 (ebook) | DDC 621.1/7--dc23
LC record available at https://lccn.loc.gov/2020006929
LC ebook record available at https://lccn.loc.gov/2020006930

This book is published in the IGI Global book series Advances in Civil and Industrial Engineering (ACIE) (ISSN: 2326-6139; eISSN: 2326-6155)

British Cataloguing in Publication Data
A Cataloguing in Publication record for this book is available from the British Library.

All work contributed to this book is new, previously-unpublished material. The views expressed in this book are those of the authors, but not necessarily of the publisher.

For electronic access to this publication, please contact: eresources@igi-global.com.

Advances in Civil and Industrial Engineering (ACIE) Book Series

Ioan Constantin Dima
University Valahia of Târgovişte, Romania

ISSN:2326-6139
EISSN:2326-6155

Mission

Private and public sector infrastructures begin to age, or require change in the face of developing technologies, the fields of civil and industrial engineering have become increasingly important as a method to mitigate and manage these changes. As governments and the public at large begin to grapple with climate change and growing populations, civil engineering has become more interdisciplinary and the need for publications that discuss the rapid changes and advancements in the field have become more in-demand. Additionally, private corporations and companies are facing similar changes and challenges, with the pressure for new and innovative methods being placed on those involved in industrial engineering.

The **Advances in Civil and Industrial Engineering (ACIE) Book Series** aims to present research and methodology that will provide solutions and discussions to meet such needs. The latest methodologies, applications, tools, and analysis will be published through the books included in **ACIE** in order to keep the available research in civil and industrial engineering as current and timely as possible.

Coverage

- Transportation Engineering
- Quality Engineering
- Production Planning and Control
- Structural Engineering
- Urban Engineering
- Coastal Engineering
- Earthquake engineering
- Optimization Techniques
- Operations Research
- Construction Engineering

IGI Global is currently accepting manuscripts for publication within this series. To submit a proposal for a volume in this series, please contact our Acquisition Editors at acquisitions@igi-global.com or visit: https://www.igi-global.com/publish/.

The Advances in Civil and Industrial Engineering (ACIE) Book Series (ISSN 2326-6139) is published by IGI Global, 701 E. Chocolate Avenue, Hershey, PA 17033-1240, USA, www.igi-global.com. This series is composed of titles available for purchase individually; each title is edited to be contextually exclusive from any other title within the series. For pricing and ordering information please visit https://www.igi-global.com/book-series/advances-civil-industrial-engineering/73673. Postmaster: Send all address changes to above address. © © 2021 IGI Global. All rights, including translation in other languages reserved by the publisher. No part of this series may be reproduced or used in any form or by any means – graphics, electronic, or mechanical, including photocopying, recording, taping, or information and retrieval systems – without written permission from the publisher, except for non commercial, educational use, including classroom teaching purposes. The views expressed in this series are those of the authors, but not necessarily of IGI Global.

Titles in this Series

For a list of additional titles in this series, please visit: www.igi-global.com/book-series

Handbook of Research on Developing Smart Cities Based on Digital Twins
Matteo Del Giudice (Politecnico di Torino, Italy) and Anna Osello (Politecnico di Torino, Italy)
Engineering Science Reference • © 2021 • 400pp • H/C (ISBN: 9781799870913) • US $295.00

Data-Driven Optimization of Manufacturing Processes
Kanak Kalita (Vel Tech Rangarajan Dr. Sagunthala R&D Institute of Science and Technology, India) Ranjan Kumar Ghadai (Sikkim Manipal Institute of Technology, Sikkim Manipal University, Majhitar, India) and Xiao-Zhi Gao (University of Eastern Finland, Finland)
Engineering Science Reference • © 2021 • 295pp • H/C (ISBN: 9781799872061) • US $225.00

Developing and Monitoring Smart Environments for Intelligent Cities
Zaigham Mahmood (University of Northampton, UK & Shijiazhuang Tiedao University, China)
Engineering Science Reference • © 2021 • 367pp • H/C (ISBN: 9781799850625) • US $215.00

Driving Transformational Change in the Digital Built Environment
Jason Underwood (University of Salford, UK) and Mark Shelbourn (University of Salford, UK)
Engineering Science Reference • © 2021 • 340pp • H/C (ISBN: 9781799866008) • US $225.00

Applications and Challenges of Maintenance and Safety Engineering in Industry 4.0
Alberto Martinetti (University of Twente, The Netherlands) Micaela Demichela (Politecnico di Torino, Italy) and Sarbjeet Singh (Luleå University of Technology, Sweden)
Engineering Science Reference • © 2020 • 321pp • H/C (ISBN: 9781799839040) • US $265.00

Reconstructing Urban Ambiance in Smart Public Places
Hisham Abusaada (Housing and Building National Research Center (HBRC), Egypt) Ashraf M. Salama (University of Strathclyde, Glasgow, UK) and Abeer Elshater (Ain Shams University, Egypt)
Engineering Science Reference • © 2020 • 280pp • H/C (ISBN: 9781799838562) • US $225.00

Additive Manufacturing Applications for Metals and Composites
K.R. Balasubramanian (National Institute of Technology, Tiruchirappalli, India) and V. Senthilkumar (National Institute of Technology, Trichy, India)
Engineering Science Reference • © 2020 • 348pp • H/C (ISBN: 9781799840541) • US $205.00

IGI Global
PUBLISHER of TIMELY KNOWLEDGE

701 East Chocolate Avenue, Hershey, PA 17033, USA
Tel: 717-533-8845 x100 • Fax: 717-533-8661
E-Mail: cust@igi-global.com • www.igi-global.com

To my parents, Belgasmia Omar & Chabha
To my wonderful wife, Sabah
To my children, Imene and Amine

Table of Contents

Section 4
Numerical Methods

Section 5
Seismics Analysis and Modal Response Contribution of the Structure

Section 6
Nonlinear Transient and Static Nonlinear Analysis

Preface

This book is concerned with the challenging subject of structural dynamic and hydrodynamic principle as well as nonlinear static methods of analyses for seismic design structures. With single goal is to simplify as much as possible all of its principles previously cited to readers with didactic and real examples. This book is suited for academics, researchers, practicing structural engineers and research students in the fields of civil and/or mechanical engineering

This book has three parts. The first deals with single-degree of freedom (DOF) systems before a rich introduction presentation. The second part concerne systems with multiple degree of freedom (DOF), with which we can create analytical and mathematical models of the most complex structures, passing through the hydrodynamic principle with an application in real cases.

The last part makes readers light on the principle of nonlinear static methods and its application on real case.

Understanding the book's contents is within the reach of students in graduation or post-graduation, it is enough that they have acquired knowledge in mathematics and structural analysis. The book breaks down on eleven chapters as follows.

The Introduction provides a rich information in the field of dynamics as Comparison between static and dynamic analysis, Approach type in dynamic study, Dynamic loads type, the definition of dynamic degree of freedom will be given, Motion equations formulation, using different principle will be detailed with example for each formulation. These principles are Dalambert principle newton's second law, Virtual work principle, and Hamilton principle. At the end, we finish the chapter by some basic principles.

All structures are complex because a structure contains several elements, the simplest element contains two nodes and each node has six degree of freedom, three displacements and three rotations. The goal is to try to minimize the number of these equations. The Introduction is devoted solely to the development of all equations concerning construction of an analytical as well as mathematical model of a free vibration system linear oscillator. Mathematical equations of the problem will be detailed as well as the solution that goes with single degree of freedom oscillator for translational vibration for all cases: Free undamped oscillator, as well as free damped oscillator and torsional free undamped vibration passing by critical, subcritical and over damping system. At the end of the chapter, some examples will be treated.

Chapter 1 aims to give a physical interpretation of some interesting case of a forced one degree of freedom system. Passing by all components of the general solution of an undamped forced system, which are transient solution. A large part of this chapter is devoted to one type of loading, which is the harmonic load; the end of the chapter will be devoted to another type of loading and to the construction of an analytical as well as mathematical model of a forced vibration system linear oscillator.

The second part of the book starts from Chapter 2 because a system with one degree of freedom is far from reality, because we do not take into account all the degree of freedom. In order to be close to the reality, it is necessary to use a system with several degrees of freedom. In more complicated structures, it is necessary to use a system with several degrees of freedom, to obtain a satisfactory dynamic model. A real system, generally comprises several masses, connected together by elements of the spring and damper type, in general, there is six degree of freedom for each node three displacements and three rotations. Chapter 2 is devoted solely to the development of all equations concerning construction of an analytical as well as mathematical model of free, forced damped and undamped multi-degrees of freedom vibration linear system linear at the end the hydrodynamic study is detailed using Hozner method.

In practice, all structures, meets resistance during their movements. Consequently, the amplitude of the movement does not remain unchanged; but it will diminish gradually with time, in the case of free vibrations; until the motion stops. The damping forces must depend on the vibratory system itself, as well as external factors surrounding the system. In a system, damping occurs due to the internal friction forces of the materials of the structure and the external effects in which the structure is immersed.

Damping is the most difficult structure characteristic to evaluate, because it depends on several parameters, who are in the structure itself; like support type, internal friction of materials of a studied structure. In addition, parameters surrounding the structure like soil type who has a direct effect on damping. Things that it makes it difficult to evaluate properly. The core of Chapter 3 is the evaluation of all damping types namely Viscous equivalent, Rayleigh, Cauchey damping.

As we know civil structure are usually designed to resist a moderate to mild seismic intensity; these structures may sustain significant damage and deformation during their functionality. In order to limit seismic induced structural displacement and peak accelerations to acceptable levels many rehabilitation techniques are used to extend their lifetime and ensure the functionality. Two main options are available for engineers to reduce seismic effects; enhance the building's lateral stiffness either by increasing the structural size and material quantity, or by increasing the level of damping by means of supplemental devices.

Fluid Viscous Dampers (FVD) are energy dissipation systems, their use for civil applications have emerged after their proven efficiency in various military branches. Chapter 4 is focused on the effect of Fluid Viscous Dampers (FVD); which is a case of passive dampers; in the dynamic behavior of an RC building. With an optimal choice of the linear FVD parameter i.e. damping coefficient, using a simplified and effective approach. It was found that the maximum inter-story drift of the analyzed retrofitted structures can be significantly reduced compared to the original ones.

Chapter 5 describes the different methods of extracting eigenvalues and eigenvectors as well as their classifications and their characteristics. Calculating the eigenvalues and the eigenvectors of matrices is one of the most important problems in linear numerical analysis. Because in the process of solving dynamics problems at a given stage one needs to have the own dynamic characteristic of the structure, in order to be able to push the calculation (resolution) more further these characteristics are the eigenvalues and eigenvectors. Extracting eigenvalues and eigenvectors techniques requiring knowledge of the spectrum of matrices are used in fields as varied as quantum mechanics, structural analysis.

Chapter 6 allows us to describe the different numerical resolution method of dynamic equilibrium equation, which all have the same principle, which is the conversion of a dynamic equation into a static one. Because once the number of degrees of freedom exceeds a certain number, it would be impossible to solve the dynamic equilibrium equation manually, hence the need to switch to a numerical resolution. In Chapter 6 we pass over various types of algorithm namely, a "one-step" algorithm, like that of Newmark and "multi-step" algorithms such as that of Houbolt.

The structural support motion represents another form of solicitation; in which the study constitutes the subject of Chapter 7, which allows us to shed light on the so-called seismic study. This chapter allow us to find out more about the earth structure as well as, earthquakes nature therefore to detail the definition and construction of a spectrum, a presentation of earthquake analysis is given. In order to become familiar with this analysis, two applications are presented at the end with a detailed solution. The study of the response of structures to earthquakes is the most widely used application of structural dynamics in civil engineering. The main reason is that earthquakes generate large forces of inertia for the vast majority of buildings and civil engineering structures. Earthquakes remain the dominant dynamic excitation mode.

Chapter 8 is imperative because it allows us to enlighten reader on the understanding of optimum calculation to better approach the actual behavior of the structure so it is inconceivable to miss the understanding of the modal response contribution and truncation error.

Hence Chapter 8 details all development of certain key points namely, determination of the elastic forces modal contribution, modal participation factors, truncation error, and static correction procedure.

The third part of the book starts from Chapter 9, the interest of researchers and practitioners for the "nonlinear" has increased considerably, because numerical methods coupled with electronic programs make it possible to effectively solve problems of amazing complexity, and this in all domain: structure, flow, diffusion, heat, fluid, dynamics, soils, plasticity, etc. Chapter 10 gives an overview of non-linearity namely; geometric and material source, change in boundary condition, the last presented source is pretensions. Then we go to the physical understanding of non-linear behavior, by presenting the different phases of hysteresis curve sequence of a reinforced concrete structure. In chapter 10 we pass over various numerical formulation, which allow us to deal with non-linearity namely Lagrange and Euler formulation, total Lagrangian formulation, Piola- kirchhoff 2, and corotational formulation. Some examples are exposed at the end of the chapter.

Chapter 11 allows enlightening the reader, on an elegant method by its graphic power and the speed of diagnosing the existing or newly constructed structure called non-linear static method. By detailing all different steps, which allows the correct accomplishment of this method; and compare these results to those of time history analysis on a real structure example. This chapter is devoted to the presentation of nonlinear static methods analyses for seismic design of structures considered by Eurocode 8. The first method is the nonlinear pushover procedure, which is based on the N2 method. The second method is the classical nonlinear time history analysis. The first method is studied in more detail, because the second method is a well-established procedure whose only drawback is the time necessary for the analyses. Nonlinear solvers and procedure in program Z_Soil are described. After a simple nonlinear SDOF application, a test-bed application consisting of an existing two-storey reinforced concrete building in Bone fro, Italy is used to compare the two nonlinear procedures. The selected building is representative of typical residential building construction in Italy in the 1970's and 1980's. The aim of this chapter section is to compare 2D and 3D procedures implemented in Z_Soil software. The second example is a fourteen-story reinforced concrete building designed according to the Algerian code using Sap2000 software.

Mourad Belgasmia
Setif 1 University, Algeria

Introduction

The majority of structures are subject to time-varying charges; no load can really be considered static. Nevertheless, many cases of load where their variations are too long, something which makes it possible to treat the behavior of the structure like static.

The main reason for using static or pseudo-static analysis is the simplicity of the design and the analysis itself. Many structures such as buildings, bridges, dams, ships, airplanes, etc. are studied by a dynamic analysis, which is a more complicated, and time-consuming analysis compared to a static one; such structures studied in this way are safer and their behavior is closer to reality.

Thanks to the important evolution of computer science, numerical methods as well as to mathematical models, we are nowadays boldly confronting the analysis of the most complex structures with huge dimensions, all this in a few hours in order to have an exact behavior of these structures, more and closer to reality.

COMPARISON BETWEEN STATIC AND DYNAMIC ANALYSIS

The main difference between a static and dynamic analysis lies in the interpretation of the load to be considered; dynamic means variation in time, so time is an additional parameter.

Dynamic loads, which unlike static loads, vary over time. These dynamic loads generate displacements, internal forces, reactions and stresses that depend on time.

So there is not a single solution as for a static problem, it is necessary, in a dynamic problem to establish the successive solutions of displacements in time, called dynamic response, before identifying the maximum values of the forces, the reactions or the stresses, which will be used in the dimensioning, or the verification of the structures.

The dynamic problem is distinguished from the static problem by two things:

1) In dynamics the load depends on the time and there are several successive solutions in the time (displacements or dynamic response) but in static, there is only one solution.
2) The main difference between dynamics and a static problem lies in the importance of the forces of inertia (mass multiplied by the acceleration) which are taken into account in the dynamic case.

The primary goal of the structural dynamics is to determine the variations in time of the stresses and deformations generated by any dynamic loading. In this context, static analysis of structures represents a special case of dynamic analysis

APPROACH TYPE IN DYNAMIC STUDY

Deterministic Approach

A dynamic load is a load whose magnitude, direction or application point varies over time. If the loading is a known function of time, it is said that the loading is dynamic prescribed. The analysis of a given structure under the effect of a prescribed load, so called deterministic approach (analysis).

Probabilistic Approach

If the variation of the loading as a function of time is not known, and it can be defined in statistical terms, we say that the loading is dynamic random, a probabilistic analysis is the analysis of the random dynamic loading response.

DYNAMIC LOADS TYPE

Usually all structures are subjected to dynamic loads during their lifetime. Two types of dynamic loads are differentiated: random loads and deterministic loads.

RANDOM LOAD

A random load is always described statistically (mean, standard deviation, frequency content, etc.). Examples of random loads are:

- Vibrations on a dense track.
- Future earthquakes.
- Wind pressures on a building.

DETERMINISTIC LOAD

The amplitude, direction and point of application of a deterministic load are known at all times. There are two classes of deterministic load: periodic loads and non-periodic loads

PERIODIC LOAD

A periodic (or cyclic) load is a repetitive charge that exhibits the same temporal variation over a large number of cycles (Clough & Penzien 1995); case of rotating machinery in a building see figure 1.

The loading can be as periodic, without being harmonic in this case it will be described as anharmonic; case of a propeller force of a ship see figure 2.

Figure 1. Harmonic load

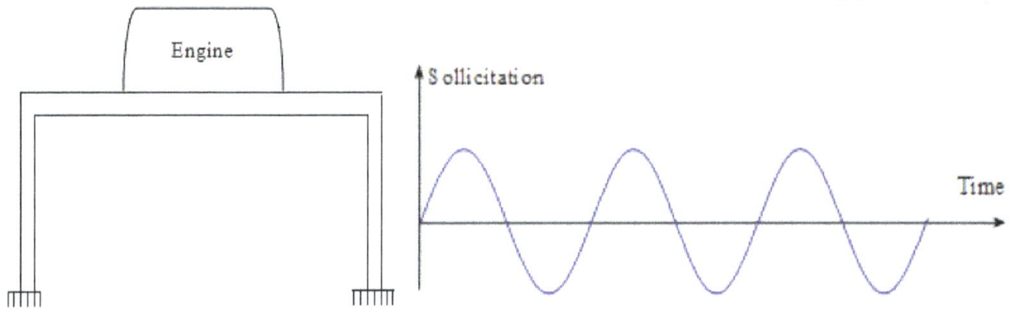

Figure 2. Anharmonic periodic load

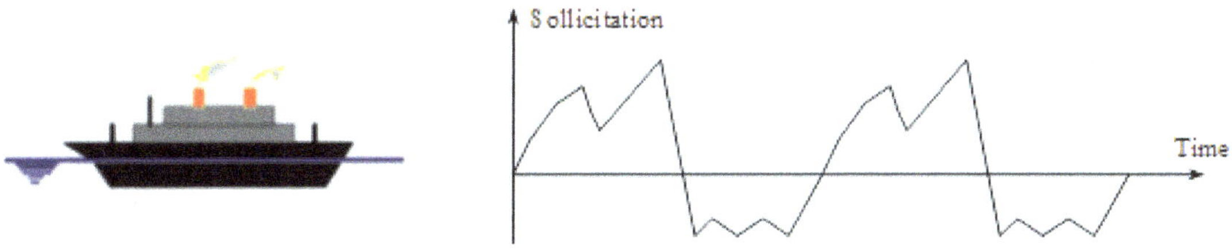

NON-PERIODIC LOAD

Non-periodic load can be of short duration (pulse) or long duration without repetition see figure 3 and figure 4 ; where the first represents the explosion of a bomb near a building (Clough & Penzien 1995). The second represents a building subjected to an acceleration of the ground.

Figure 3. Impulsive load (explosion non-periodic load)

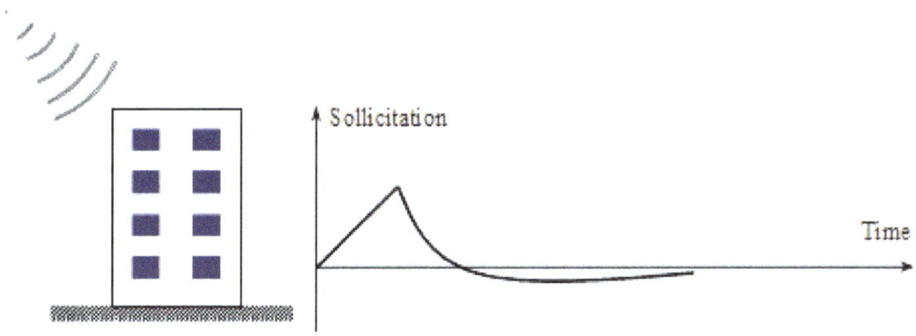

Figure 4. Seismic load (non-periodic load)

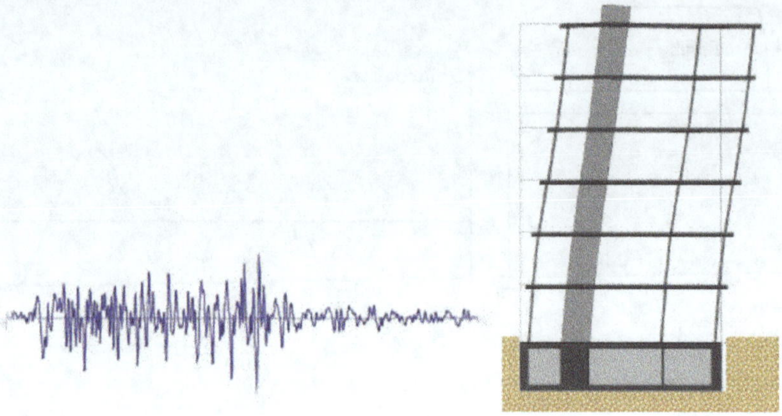

DYNAMIC DEGREE OF FREEDOM (DDF)

Each mass, can undergo six possibilities of displacement in an orthogonal coordinate space system; three translations and three rotations. While in most systems it may be that, they are allowed to move in certain directions, and not free in another directions. Therefore, the number of degrees of freedom in a dynamic system is by definition: the smallest number of coordinates needed to define the position of all particles in the mass system.

In most cases, several components of the motion are negligible in relation to each other, so that a seemingly complicated system can be modeled with only a few degrees of freedom. A crucial step in dynamic analysis lies in the choice of degrees of freedom.

Considering for example, the pendulum represented in figure 5; consisting of a pebble suspended to a ball joint by means of a bar. The system has an infinite number of dynamic degrees of freedom (DDF), taking into account all possible deformations of the bar and the pebble. On the other hand, one observes the following facts, which can be enumerated as follows:

- The bar is much lighter than the pebble (its mass is negligible)
- The dimensions of the pebble are small compared to the length of the bar (point or concentrated mass), thus the system can be reduced to a single DDL which is the rotation of the bar relative to the support.

MOTION EQUATIONS FORMULATION

The mathematical solution expression representing the equation of motion (differential equation), that defines the dynamic displacement of a structural system, provides a complete description of the response of the system, which is a function of time. The way to derive this equation of motion gives rise to three principles:

- DALEMBERT's principle (dynamic equilibrium, Newton's second law)

Figure 5. Pendulum dynamic system

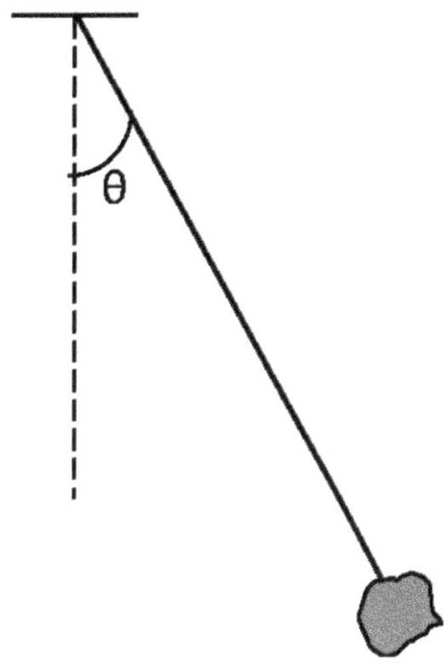

- Virtual work principle
- Hamilton principle

For a structural dynamics problem one must determine

$U(t)$: Displacement as a function of time ;
$\dot{U}(t)$: Velocity as a function of time;
$\ddot{U}(t)$: Acceleration as a function of time.

In order, to know them one must write motion equations system, using three principles:

DALEMBERT PRINCIPLE (NEWTON'S SECOND LAW)

The following example (see figure 6) is considered as an equilibrium problem. The equilibrium state is referred to a dynamic equilibrium, the adopted principle to this solution So-called DALAMBERT's principle.

In a dynamic problem resolution, the inertial force, which is the mass multiplied by the acceleration (its direction is in the opposite direction of the motion), must be incorporated in the free body in addition to the other forces imposed. This is known as the DALAMBERT principle.

In this law, we introduce the concept of the motion body quantity; The product of the mass m, and of its velocity, $\dot{U}$, is supposed as a principle that if a force F acts on a body, the rate of change of its motion amount is equal to the applied force.

$$F(t) = \frac{d}{dt}(m\frac{du}{dt}) = \frac{d}{dt}(mv)$$

$$F(t) - m\ddot{U}(t) = 0 \tag{1}$$

DYNAMIC EQUILIBRIUM (DALAMBERT PRINCIPLE) EXAMPLE 1

Assuming a mass-spring system in a horizontal position as shown in Figure 6, the goal is to write the motion equation system using the D'ALAMBERT's Principe (Patrick 2005).

Figure 6. (a) Mass-spring system (b) Isolated mass

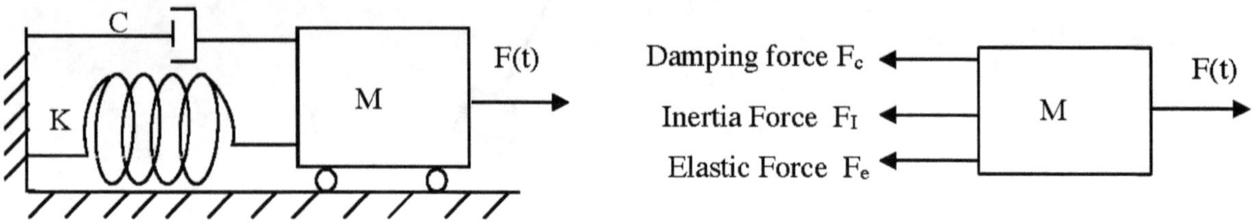

$F_e = KU(t)$; Elastic force (spring restoring force).
$F_C(t) = C\dot{U}(t)$; Damping force that is proportional to the velocity.
$F_I(t) = M\ddot{U}(t)$; Inertia force that is proportional to the acceleration.

The balance of forces shown in Figure 6 b gives:

$$M\ddot{U}(t) + C\dot{U}(t) + KU(t) = F(t) \tag{2}$$

Equation (2) represent the dynamic equation of a system with one dynamic degree of freedom.

DYNAMIC EQUILIBRIUM (ALAMBERT PRINCIPLE) EXAMPLE 2

In this example, the oscillator is supposed to be in a vertical position in order to take into account the effect of the weight. We will notice that this latter (weight) has no effect in the dynamic equation since weight is vanishes with the static displacement that it caused.

$$U(t) = U_{sta} + U_{dyn}(t)$$

$$\dot{U} = \dot{U}_{dyn}$$

As U_{sta} = constant thus $\dot{U}_{sta} = 0$

Figure 7. (a) System mass spring just at the moment of application of the mass (b) System mass spring just after application of the mass (c) System mass spring just after applying the load (d) Isolated mass

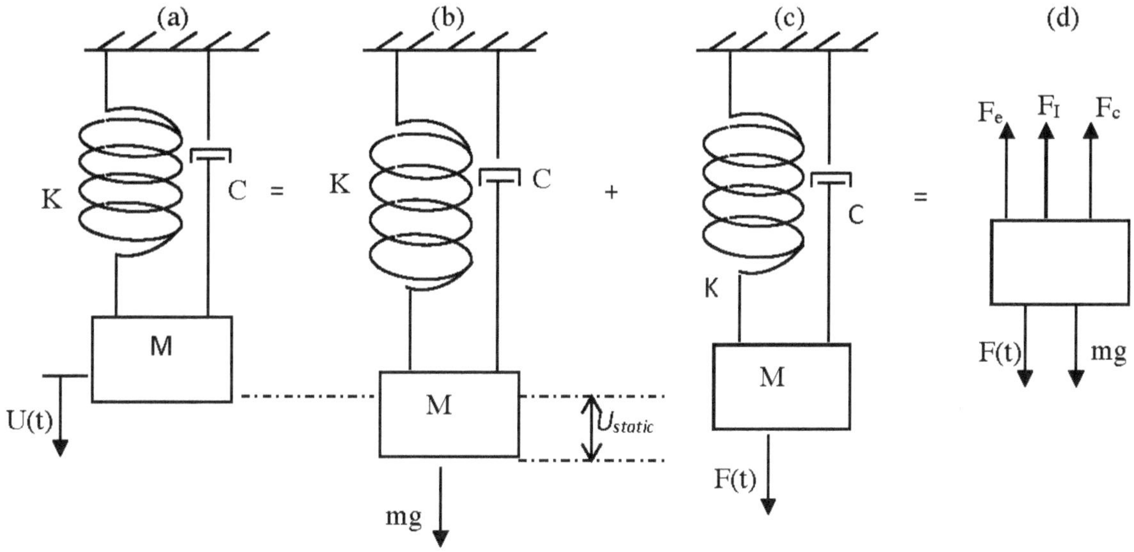

$$\ddot{U}(t) = \ddot{U}_{dyn}(t)$$

As U_{sta} = constant thus $\ddot{U}_{sta} = 0$

Applying the balance forces, we find

$$M\ddot{U}_{dyn}(t) + C\dot{U}_{dyn}(t) + K(U_{sta} + U_{dyn}(t)) = F(t) + mg$$

Knowing that $KU(t)_{sta}=mg$ the final equation becomes

$$M\ddot{U}_{dyn}(t) + C\dot{U}_{dyn}(t) + KU_{dyn}(t) = F(t) \quad (3)$$

VIRTUAL WORK PRINCIPLE

If a system is in equilibrium under the action of a set of forces, this system is subjected to a virtual displacement, thus the work done by all the forces is equal to zero.

VIRTUAL WORK PRINCIPLE EXAMPLE

Assuming a spring mass system as shown in figures 6a and figure 6b, the goal is to write the motion equation of the model using virtual working principle. With dU is a virtual displacement that is not equal to zero.

$\delta WI + \delta WC_{+} \delta We + \delta WF =_{0}$

$-FI(t_{,}\delta U - FC(t)_{\delta}U - Fe(t)\delta_{U} + F(t)\delta U = 0$

$FI(t) + FC(t) + Fe(t) = F(t)$

Motion equation becomes

$$M\ddot{U}(t) + C\dot{U}(t) + KU(t) = F(t) \tag{4}$$

With δWI, $_{\delta}WC$, $\delta_{W}e$, δW_{F} is t_{h}e work of inertia, damping, elastic and external force respectively.

HAMILTON PRINCIPLE (LAGRANGE EQUATION)

In this formulation, only scalar quantities in variational form are involved. Hamilton's principle states that for any time interval [t1, t_2] ($_{P}$aultre 2005).

$$\int_{t1}^{t2} \delta(T - V)\, dt + \int_{t1}^{t2} \delta W\, dt = 0 \tag{5}$$

T: Total kinetic energy of the system.
V: Potential energy of the system comprising both the strain energy and the potential of the external forces.
W: Work done by non-conservative forces (damping and all external charges).
d: Arbitrary variation undergone during the considered time interval.

HAMILTON PRINCIPLE (LAGRANGE EQUATION) EXAMPLE

Assuming a mass system spring as shown in Figures 6a and figure 6b we want to write the motion equation of the model using the Hamilton principle. The Lagrange equations, which represent another well-known form system equation, can be deduced from Hamilton's principle, which have the following expression:

$$+\frac{\partial}{\partial t}\left(\frac{\partial T}{\partial \dot{q}_i}\right) - \frac{\partial T}{\partial q_i} + \frac{\partial V}{\partial q_i} = F_i \tag{6}$$

The kinetic energy and the potential energy is respectively a function of the generalized coordinates as shown by the following expression:

$$T = T(q_i, \dot{q}_i),\ V = V(q_i)$$

Where $\dot{q}_i$ denotes the derivative with respect to the time of q_i.

The work done by non-conservative forces is equal to the work of dissipative forces and applied forces.

$$\delta w = \sum_i Q_i \delta q_i - C\dot{q}_i \delta q_i$$

Hamilton's principle (see equation 5) allows us to write:

$$\int_{t1}^{t2} \sum_i \left(\frac{\partial T}{\partial q_i} \delta q_i + \frac{\partial T}{\partial \dot{q}_i} \delta \dot{q}_i - \frac{\partial V}{\partial q_i} \delta q_i + Q_i \delta q_i - \mathrm{C}\, \dot{q}_i \delta q_i \right) dt = 0$$

Integrating by part the following term

$$\int_{t1}^{t2} \frac{\partial T}{\partial \dot{q}_i} \delta \dot{q}_i \, dt = \left[\frac{\partial T}{\partial \dot{q}_i} \delta q_i \right]_{t_1}^{t_2} - \int_{t1}^{t2} \frac{d}{dt} \left(\frac{\partial T}{\partial \dot{q}_i} \right) \delta q_i \, dt$$

And recognizing that the first term of the right member ($\left[\frac{\partial T}{\partial \dot{q}_i} \delta q_i \right]_{t_1}^{t_2}$) is equal to zero it comes

$$\int_{t1}^{t2} \sum_i \left[\frac{d}{dt} \left(\frac{\partial T}{\partial \dot{q}_i} \right) - \frac{\partial T}{\partial q_i} - \frac{\partial V}{\partial q_i} + Q_i - C\dot{q}_i \right] \delta q_i \, dt = 0 \qquad (7)$$

Equation (7) must be valid for any variation dq_i, from this equation Lagrange equation comes from as follows:

$$+\frac{d}{dt} \left(\frac{\partial T}{\partial \dot{q}_i} \right) + \frac{\partial T}{\partial q_i} - \frac{\partial V}{\partial q_i} = Q_i - C\dot{q}_i = F_i \qquad (8)$$

In our example the only variable is U, the previous equation becomes as follows:

$$\int_{t1}^{t2} \left(-\frac{d}{dt} \left(\frac{\partial T}{\partial \dot{U}} \right) - \frac{\partial V}{\partial U} + \frac{\partial T}{\partial U} + Q_i - C\dot{U} \right) dt = 0$$

Knowing that kinetic energy of the system shown in Figures 6a & 6b is as follows:

$$T = \frac{1}{2} M \dot{U}^2$$

Its potential energy in the case of a linear system is:

$$V = \frac{1}{2} K U^2$$

$$\frac{\partial T}{\partial \dot{U}} = M\dot{U}, \ \frac{d}{dt}(\frac{\partial T}{\partial \dot{U}}) = M\ddot{U}, \ \frac{\partial V}{\partial U} = KU$$

which is the conservative force

We know That $\frac{\partial T}{\partial U} = 0$

Hence the following equation:

$$\int_{t1}^{t2} (-M\ddot{U} - C\dot{U} - KU + Q_i)\, dt = 0$$

The following equation follows immediately:

$$M\ddot{U}(t) + C\dot{U}(t) + KU(t) = f(t) \tag{9}$$

Equations (9, 4 and 2) are identical using the three principles mentioned above in.

SOME BASIC PRINCIPLES

When an elastic system is subjected to any disturbance, it will oscillate back and forth. A vibratory system includes means for storing potential energy (spring), means for storing kinetic energy (mass or inertia), and means by which the energy is progressively lost (damping). The vibration of a system involves the transfer of alternative energy between its potential and kinetic forms. In a damped system, energy is dissipated at each vibration cycle and must be replaced by an external source if constant vibration is to be maintained.

MASS

Dynamically, mass is the property that describes how much the material is present. Mass is the property that describes how a free body resists the application of an external force. In simple terms, when the mass is big, the force necessary to accelerate it will be greater. Mass is obtained by dividing the weight of a body by the acceleration of gravity, g

STIFFNESS

Stiffness is a spring-like property that describes the level of strength of resistance that results when a body undergoes a change in length, the unit of stiffness is forces / length (N / m)

DAMPING

Damping is a means of reducing speed through resistance to movement, particularly by forcing an object through a liquid or a gas, or along another body.

The damping unit is often proposed as force * length / time (N.m / s). In practice, all structures, encounter resistance during their motion. Consequently, his motion amplitude does not, remain unchanged; but it will diminish gradually with time.

STRUCTURAL MODEL

A model is a mathematical representation of a structure, it can be discrete or continuous, it is interesting to make a comparison between the two models.

$$\frac{\partial^2}{\partial x^2}\left(EI\frac{\partial^2 U}{\partial x^2}\right)+m(x)\frac{\partial^2 U}{\partial t^2}=f(x,t)$$

$$M\ddot{U}+KU=f(t)$$

SPRING CONNECTION

Sometimes more than one spring act in a vibration system. The spring, which is considered as an elastic element of constant stiffness, can take many forms in practice, for example, it can be a beam. Combined spring elements, can be replaced in the analysis by a single spring of equivalent stiffness as follows.

SERIAL SPRING CONNECTION

The three-spring system shown in Figure 9 a can be replaced by an equivalent spring see Figure 9b

The equivalent stiffness in this case is calculated as follows:

$$\frac{1}{k_{eq}}=\frac{1}{k_1}+\frac{1}{k_2}+\frac{1}{k_3}$$

Figure 8. Continuous and discrete structural model

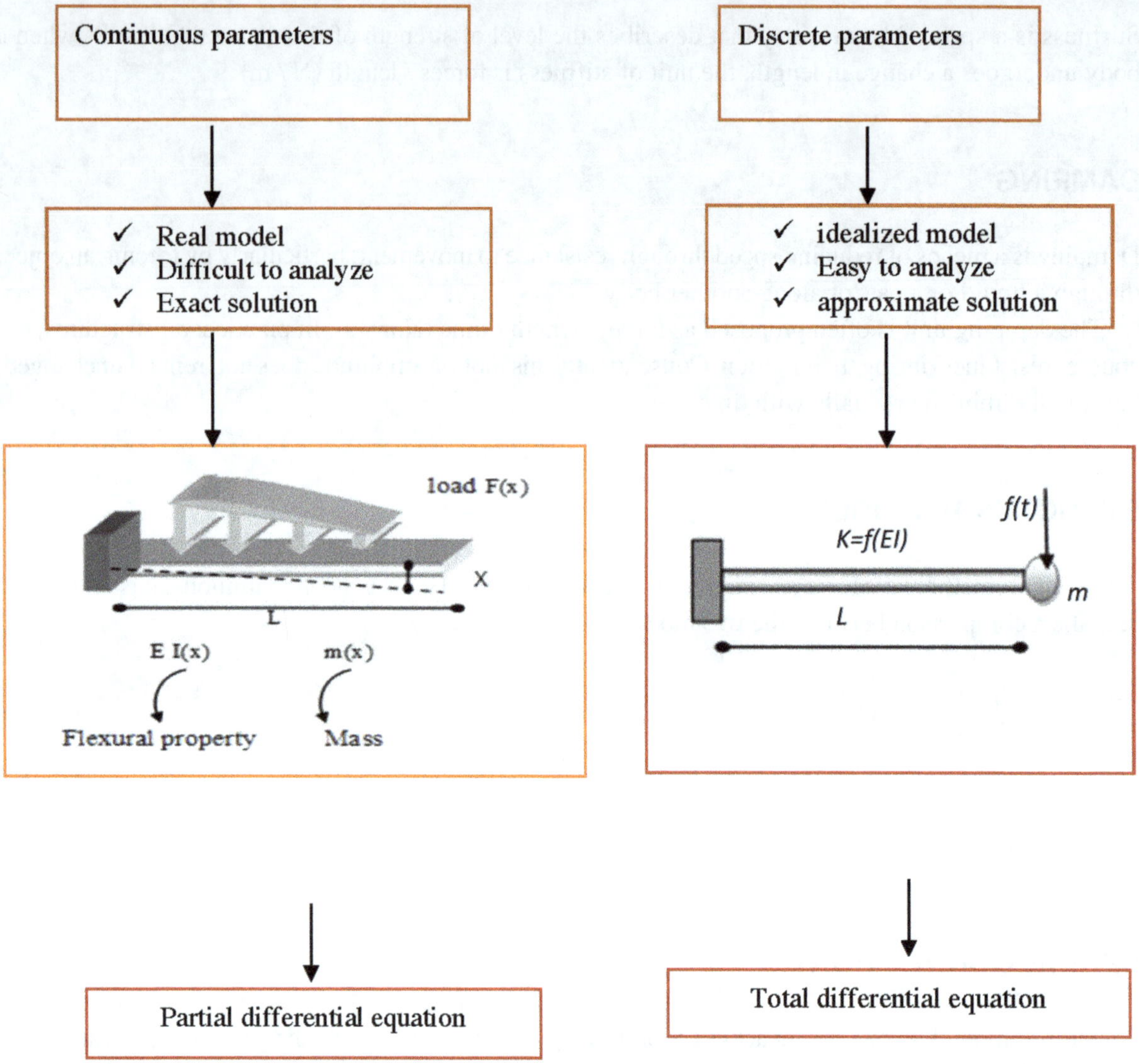

Because the three element bar represented by stiffness k_1 k_2 k_3 in figure 10a does not deform at the same time; the element bar number bearing the stiffness k_3 is deformed first then the second (k_2) finally the third (k_3).

PARALLEL SPRING CONNECTION

The three spring system shown in figure 10a can be replaced by an equivalent spring see figure 10b

The equivalent rigidity in this case is calculated as follows:

$$k_{eq} = k_1 + k_2 + k_3$$

Figure 9. (a) Serial spring System (b) Equivalent Spring System

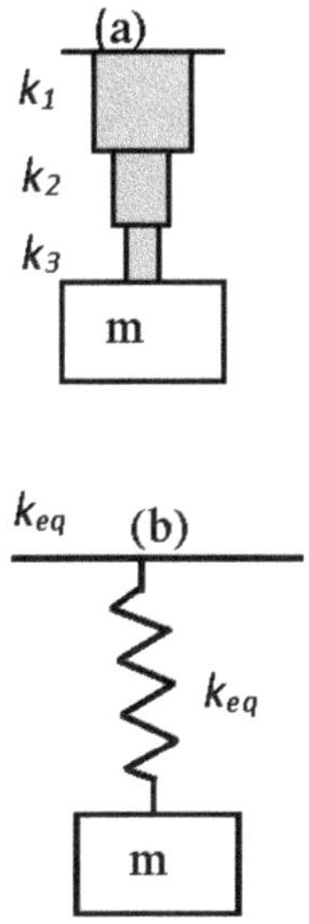

Figure 10. (a) Parallel spring System (b) Equivalent Spring System

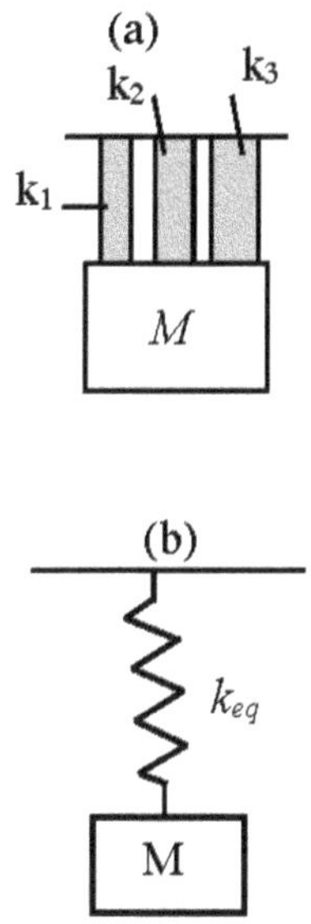

Because in this case the three element bar represented by stiffness $k_{1,}$ k_2, k_3 in figure 10a deform at the same time.

REFERENCES

Clough, R.W. & Penzien, J. (1995). *Dynamic of structures.* Compute and Structure, Inc.

Paultre, P. (2005). *Dynamic of structures.* John Wiley & Sons, Inc.

Section 1
Single Degree of Freedom System Study

Chapter 1
Free Vibration of Single Degree of Freedom Linear Oscillator

ABSTRACT

This chapter is dedicated to understanding and studying a didactic case represented by a free vibration of a linear oscillator with a single degree of freedom. Mathematical equations of the problem will be detailed as well as the solution that goes with single degree of freedom oscillator for translational vibration for all cases: free undamped oscillator, as well as free damped oscillator, and torsional free undamped vibration passing by critical, subcritical, and over damping system. At the end of the chapter, some examples will be treated.

INTRODUCTION

This chapter is devoted solely to the development of all equations concerning construction of an analytical as well as mathematical model of a free vibration system linear oscillator.

All structures are complex because a structure contains several elements, the simplest element contains two nodes and each node has six degree of freedom (dof), three displacements and three rotations. The goal is to try to minimize the number of these equations.

To reduce the number of equations, we will make a simple mathematical model that will reflect as much as possible the real structure shown in Figure 1.

In this case, a column with the same physical and mechanical characteristics replaces the metallic structure, and the mass M replaces the platform (see Figure 2).

DOI: 10.4018/978-1-7998-4399-3.ch001

Copyright © 2021, IGI Global. Copying or distributing in print or electronic forms without written permission of IGI Global is prohibited.

Figure 1. Complex structure

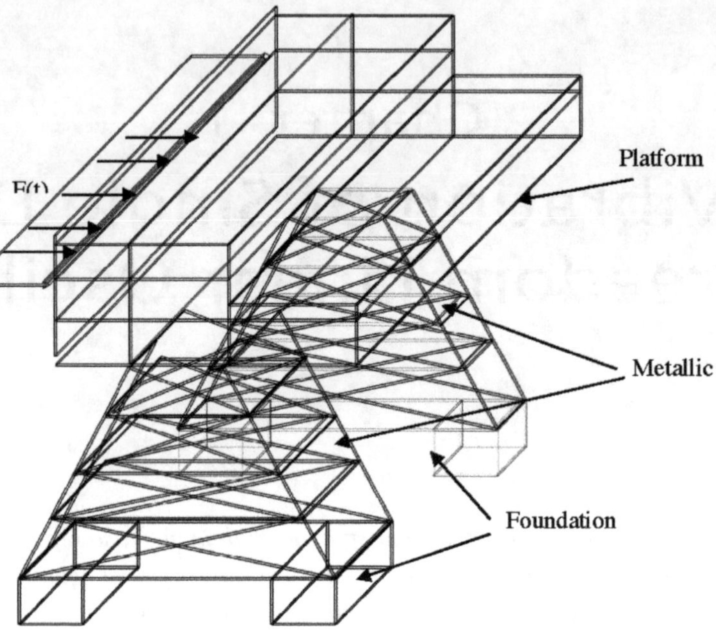

DYNAMIC MOTION EQUATION OF A FREE UNDAMPED OSCILLATOR

The simplest vibrating mechanical system is constituted by a point mass m, subject to frictionless displacement on an axis Ox. The spring force of a stiffness K and a neglected mass is the only force exerted on this mass (see figure 2). This system has a single degree of freedom because the position of the mass m is represented by a single variable U(t) which is the variation of m with respect to its equilibrium position.

Figure 2. Simplest model (mass stiffness)

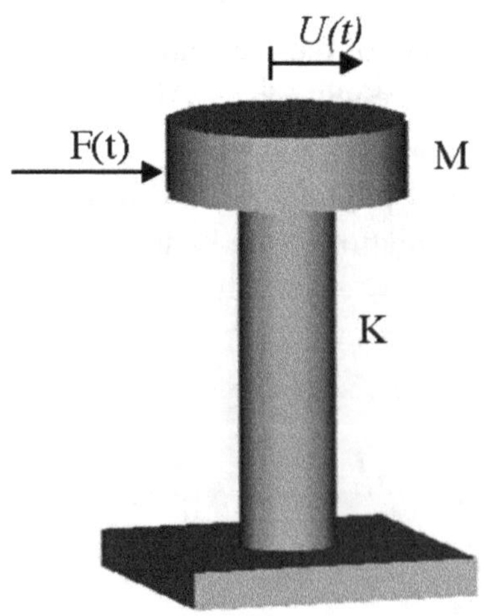

Figure 3. (a) Mass-spring system; (b) Isolated mass

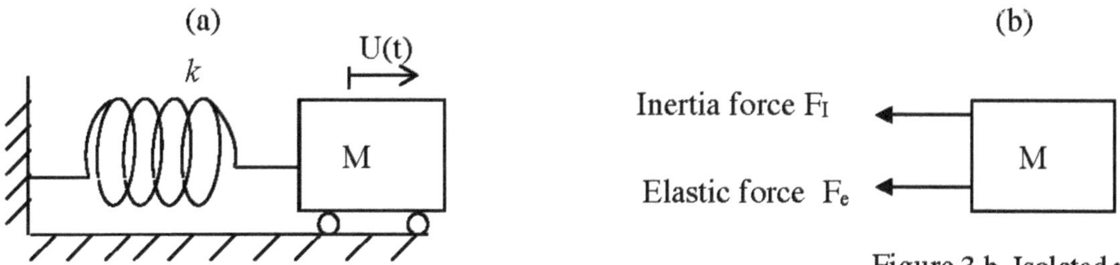

Figure 3 b. Isolated mass

FREE UNDAMPED OSCILLATOR

In this case, the external load is zero as well as the damping, so the equation is as follows (Clough & Penzien 1995):

$$M\ddot{U} + KU = 0$$

$$n = \xi\omega = \frac{C}{2M} = 0 \text{ car } C = 0$$

Dividing the equation above by M, we will have:

$$\ddot{U} + \frac{K}{M}U = 0$$

We have $\omega^2 = \frac{K}{M}$ thus

$$\ddot{U} + \omega^2 U = 0 \qquad (1)$$

The solution of equation (1) assumed as follows:

$$U(t) = Ae^{\lambda t}, \ddot{U}(t) = A\lambda^2 e^{\lambda t}$$

replacing $U(t)$ and $\ddot{U}(t)$ in the equation(1) we have *:*

$$A\lambda^2 e^{\lambda t} + \omega^2 A e^{\lambda t} = 0$$

Knowing that $A \neq 0$ because it's a constant and $e^{\lambda t} \neq 0$ thus $\lambda 2 + \omega 2 = 0$ Thus $\lambda = \pm i\omega$ (i2=-1)
The solution of equation (1) is

$$U(t) = A_1 e^{i\omega t} + A_2 e^{-i\omega t}$$

$$U(t) = A_1(\cos\omega t + i\sin\omega t) + A_2(\cos\omega t - i\sin\omega t)$$

$$U(t) = (A_1 + A_2)\cos\omega t + i(A_1 - A_2)\sin\omega t$$

$$U(t) = A\cos\omega t + B\sin\omega t$$

A and B are real constants. The velocity expressed in this way:

$$\dot{U}(t) = -\omega A\sin\omega t + B\omega\cos\omega t$$

Constants A and B can be determined from initial conditions U(0) and at t=0.

$$A = U(0) \text{ and } B = \frac{\dot{U}(0)}{\omega}$$

Taking into account these expressions the solution of an undamped free oscillator is written as follows:

$$U(t) = U(0)\cos\omega t + \frac{\dot{U}(0)}{\omega}\sin\omega t \quad (2)$$

The motion, is defined by the sum of two simple harmonic functions, that are repeated after a certain time interval this interval (cycle) is called period.

$$\omega T = 2\pi \text{ D'où } T = \frac{2\pi}{\omega} = 2\pi\sqrt{\frac{M}{K}} \quad (3)$$

Equation (3) represents the period of an undamped free vibration. Frequency, however, is the opposite of the period as shown in equation (4).

$$f = \frac{1}{T} = \frac{1}{2\pi}\sqrt{\frac{K}{M}} \quad (4)$$

The first term $U_1(t)$ of equation (2), depends on the initial displacement, the second term $U_2(t)$ depends on the initial velocity.

$$U_1(t) = U(0)\cos\omega t \text{ and } U_2(t) = \frac{\dot{U}(0)}{\omega}\sin\omega t$$

The two terms $U_1(t)$ and $U_2(t)$ are shown separately in Figure 4a and 4b respectively, for Figure 4c, it represents the sum of $U_1(t)$ and $U_2(t)$; in an other words, it represents equation (2).

Figure 4a. Component of the harmonic motion term proportional to the coswt and which depends on U(0)

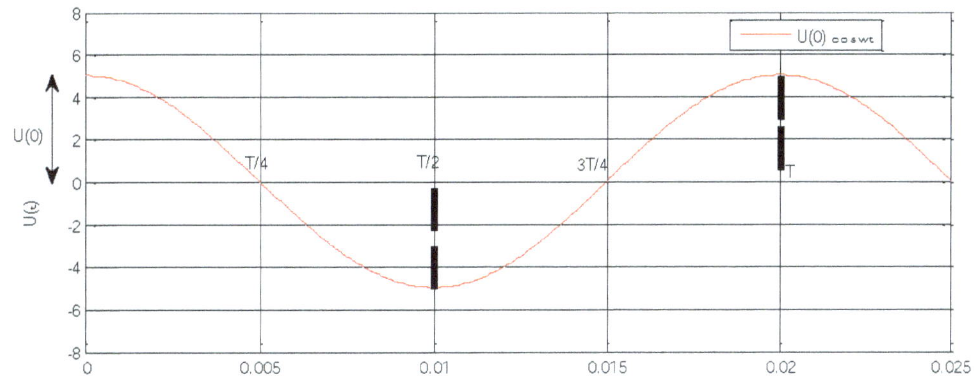

Figure 4b. Component of harmonic motion proportional to sinωt and who depends on $\dot{U}(0)$

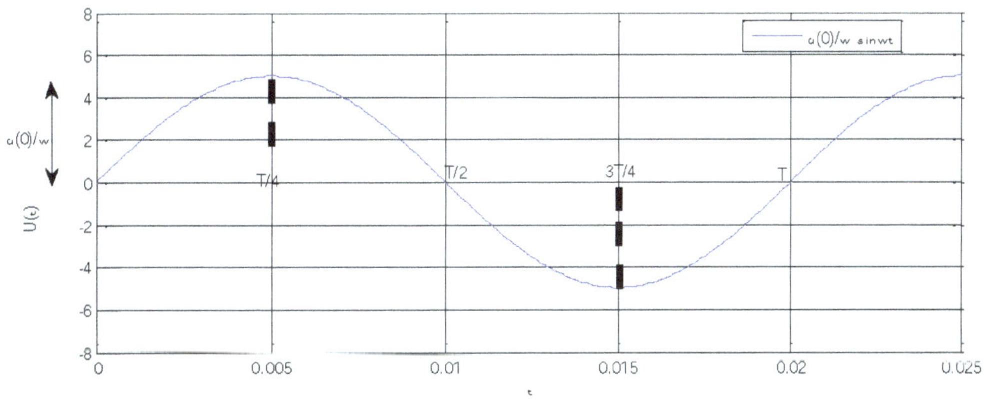

Figure 4c. Harmonic motion resulting from the sum of terms proportional to consωt and sinωt

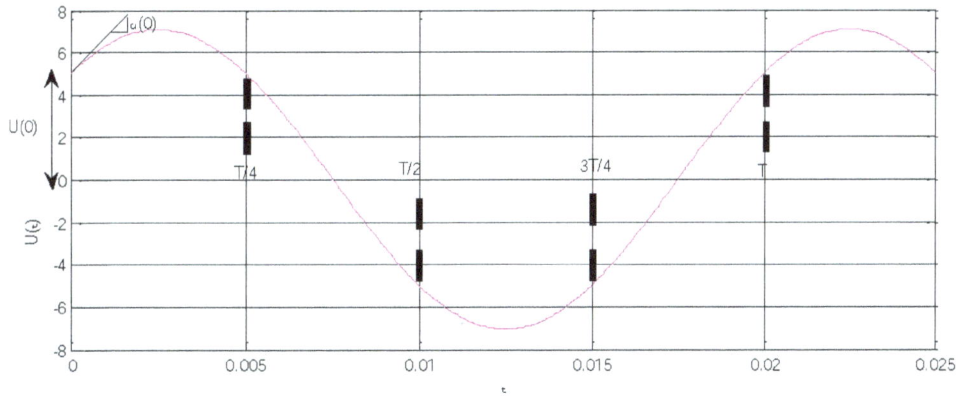

DYNAMIC MOTION EQUATION OF DAMPED FREE OSCILLATOR

In the absence of the external force F(t), the mass oscillates freely. This motion can, for example, be obtained by releasing the oscillator after having removed it from its equilibrium position. The simplest mathematical model is that shown in figure 5a. The damping is represented by the coefficient C, and its role is to reduce the displacement.

Figure 5. (a) Mass-spring system; (b) Isolated Mass

(a)

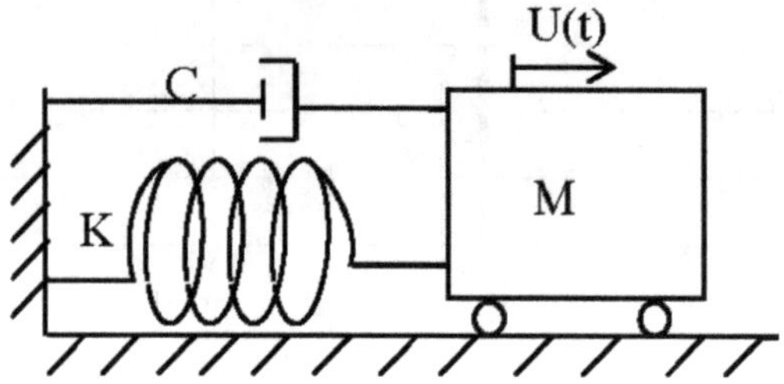

(b)

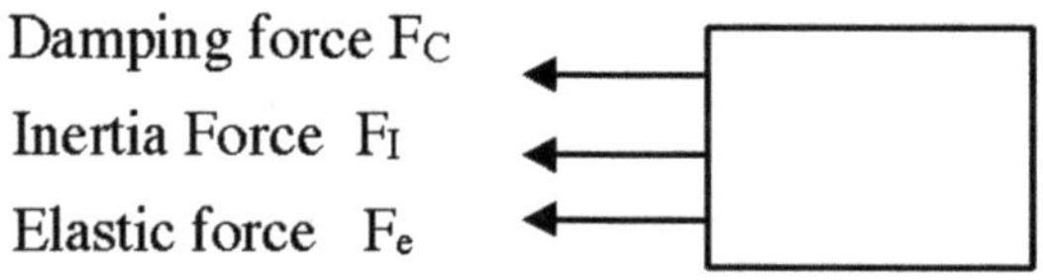

From figure 5a, the equation of equilibrium is as follows:

$$M\ddot{U} + C\dot{U} + KU = 0 \tag{5}$$

Equation (5) represents the dynamic equation of a system with one degree of freedom, in this case the external forces are zero (F(t)=0), and after dividing the equation by M; the dynamic equilibrium equation becomes:

$$\ddot{U}(t) + \frac{C}{M}\dot{U}(t) + \frac{K}{M}U(t) = 0$$

We assume

$$\omega^2 = \frac{K}{M} \text{ and } 2\xi\omega = \frac{C}{M}$$

ξ Is the damping ratio
ω Is the own pulsation

$$\ddot{U}(t)+2\xi\omega\dot{U}(t)+\omega^2 U(t)=0 \quad (6)$$

It is a second-order differential equation without second member as a function of U(t), the solution of this equation is an exponential function

$$U(t)=Ae^{\lambda t},\ \dot{U}(t)=A\lambda e^{\lambda t} \text{ and } \ddot{U}(t)=A\lambda^2 e^{\lambda t}$$

Replacing $U(t)$, $\dot{U}(t)$ and $\ddot{U}(t)$ expressions in the differential equation (6) we have

$$A\lambda^2 e^{\lambda t}+2\xi\omega A\lambda e^{\lambda t}+\omega^2 Ae^{\lambda t}=0$$

We obtain $\lambda 2+ 2\xi\omega\lambda+ \omega 2= 0$ because $A\neq 0$ *is* a constant.
Hence, the equation that follows:

$\lambda 2+ 2^{\xi}\omega\lambda+ \omega 2= 0$

With λ is a solution of the equation
We assume $\xi\omega$= n we w*ill* finally have

$$\lambda 2+ 2n\lambda+ \omega 2=0 \quad (7)$$

I[t] is a second-degree equation as a function of λ the solution is:

$$\lambda=-n\pm\sqrt{n^2-\omega^2} \quad (8)$$

With $n=\xi\omega=\dfrac{C}{2M}$

Note that this solution depends on the value of n an*d* therefore the damping C. Three cases occur according to the term, which is under the radical:

CRITICAL DAMPING SYSTEM (n = ω or ξ = 1)

In this case, the term that is inside the radical n2-*ω2*=0 *f*rom where n=*ω*= *ξ ω so ξ=1, a*nd the characteristic equation (7) admits a double root λ1=λ2 = -ω
The solution is of the form:

$$U(t)=(A_1+A_2 t)e^{-\omega t} \quad (9)$$

DEMONSTRATION

The general solution of any second-order equation is the sum of two linearly independent solutions. A particular solution of equation (6) is of the form $U_1(t) = e^{-\omega t}$ we have to find a second solution because the function $e^{-\lambda_2 t}$ is identically equal to $e^{-\lambda_1 t}$ so let us look for a second solution in form $U_2(t) = y(t)\,e^{-\omega t}$

The first two derivatives give:

$$\dot{U}_2(t) = (\dot{y}(t) - \omega\, y(t))\,e^{-\omega t}$$

$$\ddot{U}_2(t) = (\ddot{y}(t) - 2\omega\,\dot{y}(t) + \omega^2 y(t))\,e^{-\omega t}$$

Substituting expressions of $U_2(t), \dot{U}_2(t)$ and $\ddot{U}_2(t)$ in equation (6) taking into account that $\xi = 1$ we obtain:

$$\ddot{y}(t)\ e^{-\omega t} = 0$$

From where $y(t) = E_1 t + E_2$ we can put E1 = 1 and E2 = 0 so we will have

$$U_2(t) = t\,e^{-\omega t}$$

We immediately check that the two solutions are linearly independent

$$\frac{U_1(t)}{U_2(t)} = \frac{e^{-\omega t}}{t\,e^{-\omega t}} = \frac{1}{t} \neq C^{st}$$

The general solution is written as shown in equation (9).

$$U(t) = (A_1 + A_2 t)\,e^{-\omega t} \text{ and } \dot{U}(t) = (A_2 - \omega A_1 - \omega A_2 t)\,e^{-\omega t}$$

$$\text{At t=0} \begin{cases} U(0) = A_1 \\ \dot{U}(0) = A_2 - \omega U(0) \end{cases} \Rightarrow \begin{cases} A_1 = U(0) \\ A_2 = \dot{U}(0) + \omega U(0) \end{cases}$$

After substitution in equation (9) we will have

$$U(t) = (U(0)(1 + \omega t) + \dot{U}(0)\,t)\,e^{-\omega t}$$

We notice that t = 0 $U(0) = U_0$ and a $t = \infty \Rightarrow U(\infty) \to 0$

Figure 6. Motion of an oscillator under critical damping

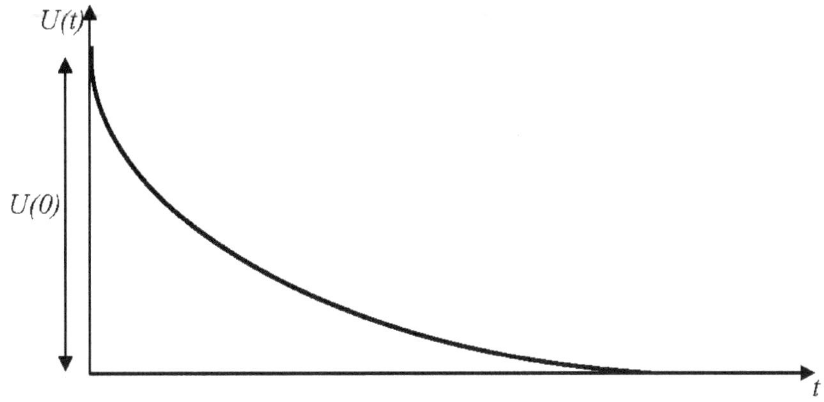

SUBCRITICAL DAMPING SYSTEM (n<ω or ξ <1)

Damping is sub-critical when the damping rate is less than unity (ξ <1), in the case where (n<ω) ($n^2 - \omega 2^{\leq} 0$), the roots of equation (8) are complex and are represented in the following equation:

$$\lambda_{1,2} = -n \pm \sqrt{n^2 - \omega^2}$$

On *aura* $\lambda 1{,}2 = -n \pm i\omega d$

We assume

$$\omega_d = \sqrt{\omega^2 - n^2}$$

with

$$\omega_d^{\;2} = \omega^2 - n^2 \succ 0$$

$$\omega_d^{\;2} = \omega^2 (1 - \frac{n^2}{\omega^2}) = \omega^2 (1 - \xi^2)$$

Thus $\omega_d = \omega \sqrt{(1 - \xi^2)}$

ω_d: is the own pulsation (pseudo angular frequency) of the undamped oscillator

Its natural frequency is

$$f_d = \frac{1}{T_d} = \frac{\omega_d}{2\pi} \leq f \text{ with } T_d = \frac{2\pi}{\omega_d} = \frac{2\pi}{\omega\sqrt{1-\xi^2}}$$

T_d: is the pseudo period
ξ: is the ratio of system damping to critical damping

$$\xi = \frac{C}{C_{cr}} = \frac{C}{2m\omega}$$

The general solution is given by the equation

$$U(t) = A_1 e^{-\xi\omega t + i\omega_d t} + A_2 e^{-\xi\omega t - i\omega_d t}$$

$$U(t) = e^{-\xi\omega t}(A_1 e^{i\omega_d t} + A_2 e^{-i\omega_d t})$$

$$U(t) = e^{-\xi\omega t}(A\cos\omega_d t + B\sin\omega_d t)$$

Where A and B are real

$$\dot{U}(t) = -e^{-\xi\omega t}\left[(\xi\omega A - \omega_d B)\cos\omega_d t + (\xi\omega B + \omega_d A)\sin\omega_d t\right]$$

A t=0 A=U(0) and $B = \dfrac{\xi\omega U(0) + \dot{U}(0)}{\omega_d}$

$$U(t) = e^{-\xi\omega t}\left(U(0)\cos\omega_d t + \frac{\xi\omega U(0) + \dot{U}(0)}{\omega_d}\sin\omega_d t\right) \quad (10)$$

Equation (10) is the response of an underdamped system; its graphical representation is as follows: The natural period T is connected to the pseudo period Td, by means of the following formula:

$$T_d = \frac{2\pi}{\omega_d} = \frac{2\pi}{\omega\sqrt{1-\xi^2}} = \frac{T}{\sqrt{1-\xi^2}} \text{ From where}$$

$$\frac{\omega_d}{\omega} = \frac{T}{T_d} \text{ and } \left(\frac{\omega_d}{\omega}\right)^2 + \xi^2 = 1 \quad (11)$$

Figure 7. Motion of a simple oscillator underdamped

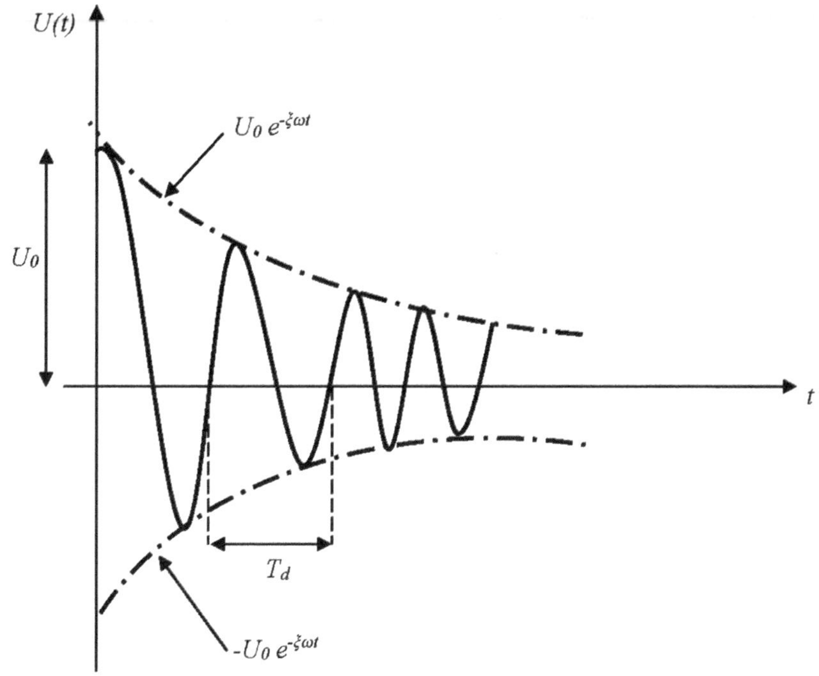

Figure 8. Effect of damping on free vibration (Chopra 2001)

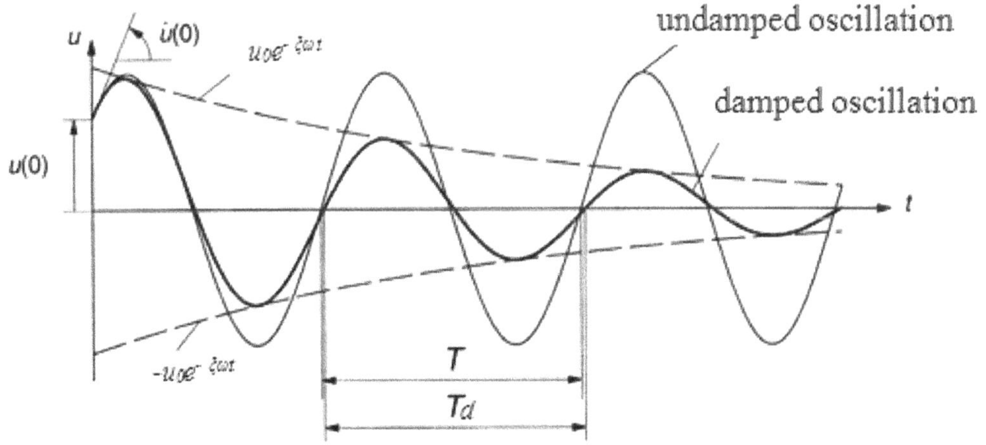

The damping has the effect of reducing the natural frequency ω to ω_d, and increasing the period from T to Td. This effect is negligible for a damping coefficient below 20%, which is the interval that includes all civil engineering structures (figure 9) where the coefficient is ω / ωd = T / Td is trace as a function of ξ.

For most structures, ω_d and T_d are approximately equal to ω and T respectively.

The most important damping effect is on the attenuation rate of the free vibration, which is shown in Figure 10 where the free vibration, due to the initial displacement u(0) is represented for four systems having the same natural period T, but different damping rates ξ = 2.5,10 and 20%.

Figure 9. Damping rate ξ

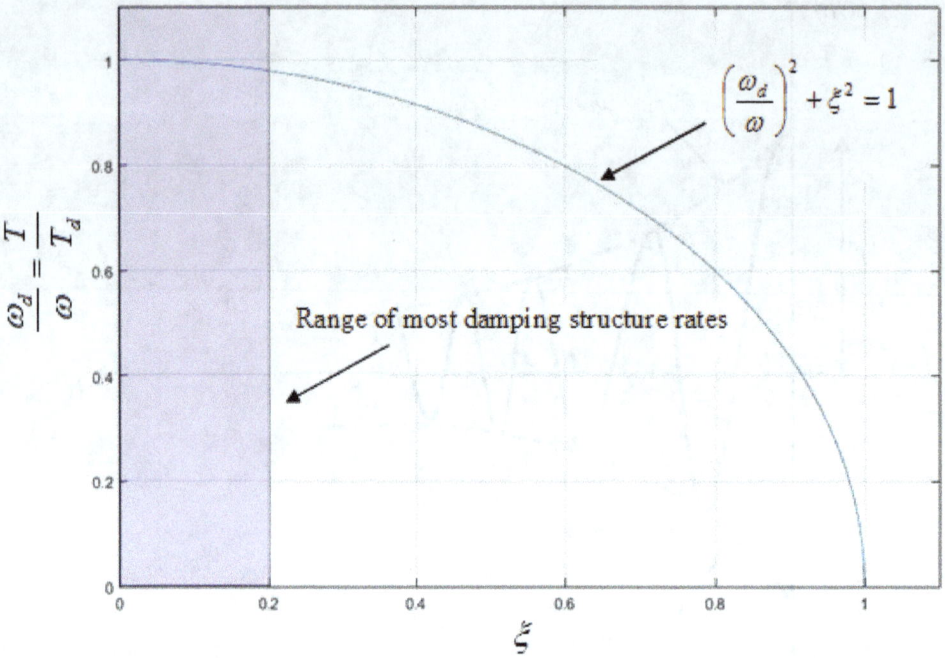

Figure 10. Free vibration of a system with four damping levels ξ = 2.5,10 and 20% (Chopra 2001)

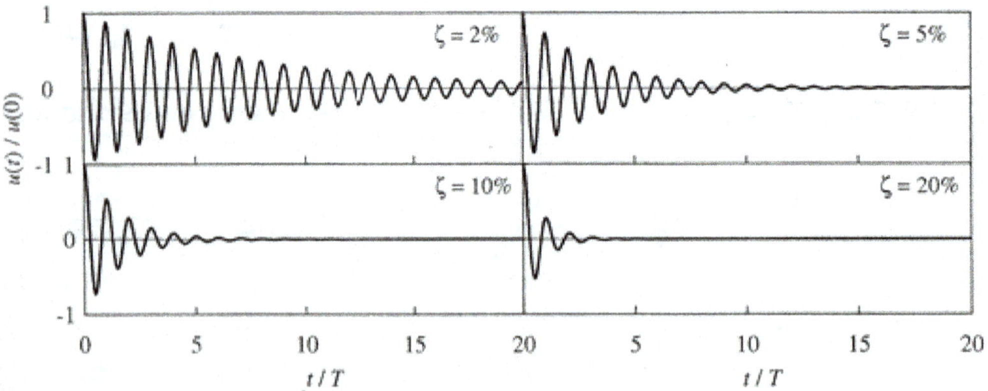

OVER DAMPING SYSTEM (n>ω or ξ >1)

The roots λ_1 and λ_2 of equation (7) are real and negative when n> ω the solution of this equation is: $n^2 > \omega^2$ from where

$$\sqrt{n^2 - \omega^2} \succ 0$$

we have

$$\lambda_{1,2} = -n \pm \sqrt{n^2 - \omega^2}$$

with *(ξ>1) n=ξω*

We assume

$$\lambda_1 = -\xi\omega + \hat{\omega} \text{ and } \lambda_2 = -\xi\omega - \hat{\omega}$$

with

$$\hat{\omega} = \omega\sqrt{\xi^2 - 1}$$

$$U_1(t) = e^{-(\xi\omega - \hat{\omega})t} \quad U_2(t) = e^{-(\xi\omega + \hat{\omega})t}$$

$$U(t) = e^{-\xi\omega t}(Ae^{\hat{\omega}t} + Be^{-\hat{\omega}t}) = e^{-\xi\omega t}(A_1 \operatorname{Cosh}\hat{\omega}t + B_1 \operatorname{Sinh}\hat{\omega}t)$$

Where A_1=A+B and B1=A-B

$A_1 = U(0)$ and $B_1 = \dfrac{\xi\omega U(0) + \dot{U}(0)}{\hat{\omega}}$

$$U(t) = e^{-\xi\omega t}\left(U(0)\operatorname{Cosh}\hat{\omega}t + \frac{\xi\omega U(0) + \dot{U}(0)}{\hat{\omega}}\operatorname{Sinh}\hat{\omega}t\right) \tag{12}$$

Equation (12), represents the response of an over-damped system, it is a case, which is not very interesting in civil engineering.

The damping is so great that the mass returns towards the equilibrium, the equation (12) is an equation similar to equation (11); whose trigonometric functions are replaced by hyperbolic functions, and there is no oscillation with respect to the equilibrium position.

Three cases of damping system can then occur: underdamping, critical damping or over damping. Figure 11 illustrates these three damping levels.

From figure 11, it is clear that the underdamped system is the most dangerous regime for civil engineering structures.

LOGARITHMIC DECREMENT

The damping of structures is often expressed in terms of equivalent viscous damping obtained from free vibration tests (after an initial release). A measure of the damping of an elementary oscillator is given by the reduction of the amplitude after a vibration cycle.

Figure 11. Free vibration of a underdamped, critical damped and overdamped system (Chopra 2001)

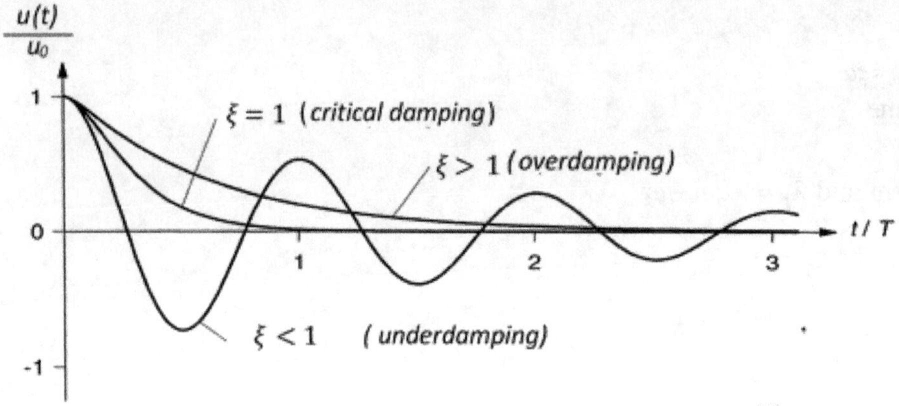

Figure 12. Initial release tests for determination of damping by logarithmic decrement

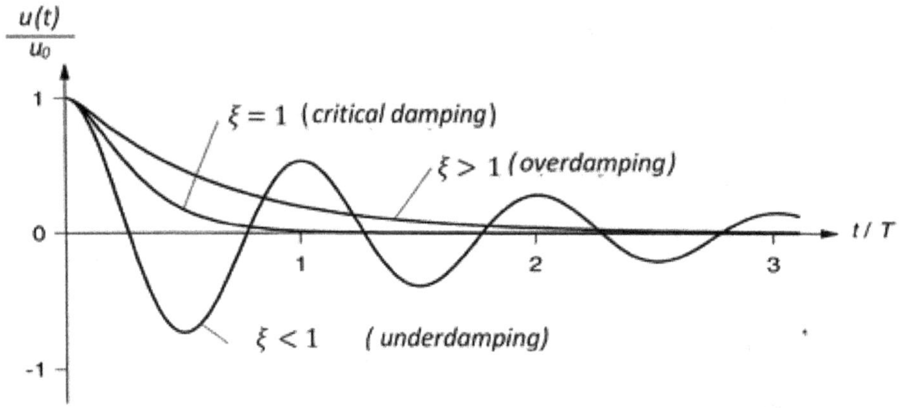

This result is the basis of an experimental method to determine the damping of a structure: the mass is removed from its equilibrium position and then, without initial velocity, it is released and allowed to vibrate; the ratio of the amplitudes U(t) and U(t + T_d) is written as follows:

$$\frac{U(t)}{U(t+Td)} = \frac{U_0 e^{-\xi\omega t}\,\mathrm{Cos}(\omega_d t-\theta)}{U_0 e^{-\xi\omega(t+T_d)}\,\mathrm{Cos}(\omega_d(t+T_d)-\theta)}$$

Knowing that equation (10) can be written in this form

$$U_0 e^{-\xi\omega t}\operatorname{Cos}(\omega_d t-\theta) \text{ with } \begin{cases} U_o=\sqrt{(U(0))^2+\left(\dfrac{\xi\omega U(0)+\dot{U}(0)}{\omega_d}\right)^2} \\ \theta=\operatorname{Tan}^{-1}\left(\dfrac{\xi\omega U(0)+\dot{U}(0)}{\omega_d U(0)}\right) \end{cases}$$

We have:

$$\operatorname{Cos}(\omega_d(t+T_d)-\theta)=\operatorname{Cos}(\omega_d(t-\theta)+\omega_d T_d)=\operatorname{Cos}(\omega_d(t-\theta)+2\pi)=\operatorname{Cos}\omega_d(t-\theta)$$

Thus equation (13) become.

$$\frac{U(t)}{U(t+Td)}=\frac{e^{-\xi\omega t}}{e^{-\xi\omega(t+T_d)}}=e^{-\xi\omega T_d}$$

Take the logarithm of the two members

$$\delta=\ln\frac{U(t)}{U(t+Td)}=-\xi\omega T_d=2\pi\xi\frac{\omega}{\omega_d}=\frac{2\pi\xi}{\sqrt{1-\xi^2}}$$

δ is the logarithmic decrement.

The damping of a system can be calculated by measuring the displacement motion of the system at one interval cycle. Damping is given by the following relationship

$$\xi=\frac{\delta}{\sqrt{(2\pi)^2+\delta^2}} \tag{14}$$

For low damping values $\omega \approx \omega_d$ and $\delta \approx 2\pi\xi$ .This last equation is used to estimate the overall damping of a structure from two successive measurements of the response. In order to improve the accuracy, we measure the maximums at m interval cycles, that is to say U_n and U_{n+m} the amplitudes corresponding to the times t_n and $t_n + mT_d$

$$\frac{U_n}{U_{n+m}}=e^{(\xi\omega T_d)^m}=e^{m\xi\omega T_d}$$

We have $\delta=\xi\omega T_d$

$$\frac{U_n}{U_{n+m}}=e^{m\delta}$$

From where

$$\delta = \frac{1}{m} Ln \frac{U_n}{U_{n+m}}$$

$$\xi \approx \frac{\delta}{2\pi} = \frac{1}{2\pi m} Ln \frac{U_n}{U_{n+m}}$$

With m: is the number of cycles

We can have another formula for determining damping; the demonstration of this formula is as follows.

The logarithm between two successive amplitudes is:

$$U_n = Ae^{-\eta t} \quad \eta=\xi\omega$$

$$U_{n+1} = Ae^{-\eta(t+T_d)}$$

$$Ln\frac{U_n}{U_{n+1}} = Ln\left[\frac{Ae^{-\eta t}}{Ae^{-\eta(t+T_d)}}\right] = Ln\ e^{\eta t_d} = \eta T_d = \xi\omega T_d$$

$$Ln\frac{U_n}{U_{n+1}} = \xi\omega T_d = \xi\omega\frac{2\pi}{\omega\sqrt{1-\xi^2}} = \frac{2\pi\xi}{\sqrt{1-\xi^2}}$$

if $\xi^2 << 1$ $\delta = Ln\frac{U_n}{U_{n+1}} \approx 2\pi\xi$ d'où $\delta=2\pi\xi$

$$\frac{U_n}{U_{n+1}} = e^{\delta} = 1 + 2\pi\xi + \frac{(2\pi\xi)^2}{2!} + \ldots + \frac{(2\pi\xi)^n}{n!}$$

When $\xi<<1$ we can neglect the terms of the second order

$$\frac{U_n}{U_{n+1}} = e^{\delta} = 1 + 2\pi\xi$$

Thus

$$\xi = \frac{1}{2\pi}\left[\frac{U_n - U_{n+1}}{U_{n+1}}\right]$$

We obtain a better precision if we have separate peaks t = mT

$$\xi = \frac{1}{2\pi\, m}\left[\frac{U_n - U_{n+m}}{U_{n+m}}\right] \tag{15}$$

Equation (14) is represented graphically as shown in figure 13.

Figure 13. Curve connecting the logarithmic decrement to the damping rate

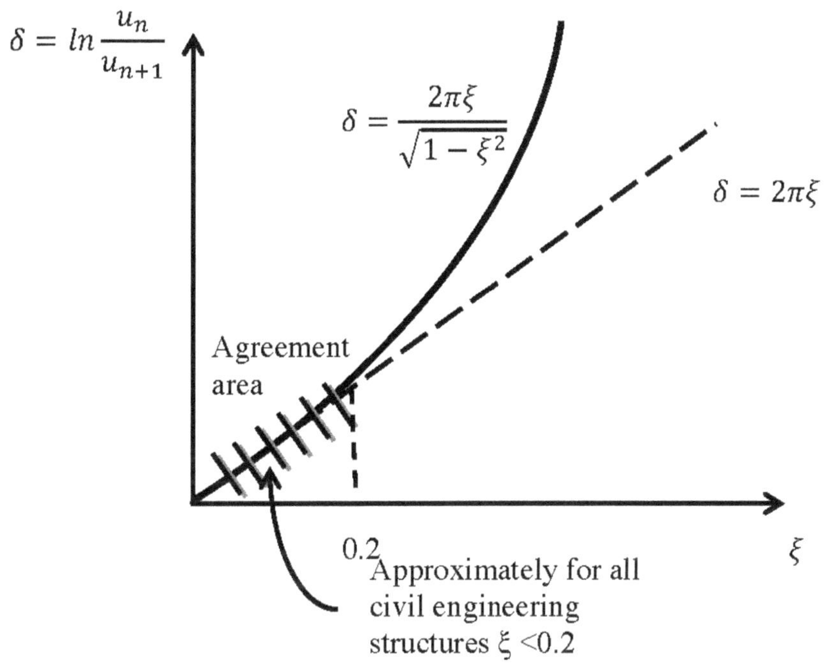

FREE UNDAMPED AND DAMPED SYSTEM VIBRATION ENERGY

When vibrating an undamped free system, the total energy E_t is independent of time and equal to the energy that the system receives; the demonstration is as follows:

Knowing that the expressions of the kinetic energy as well as the potential energy are written as follows respectively:

$$E_{kinetic} = \frac{1}{2} M\dot{U}^2 \quad E_{potential} = \frac{1}{2} KU^2$$

Knowing also that the response of free undamped system is as follows

$$U(t) = U(0)\cos\omega t + \frac{\dot{U}(0)}{\omega}\sin\omega t$$

With

$$\dot{U}(t) = -U(0)\omega\sin\omega t + \frac{\dot{U}(0)}{\omega}\omega\cos\omega t$$

$$\left[\dot{U}(t)\right]^2 = \omega^2\left[-U(0)\sin\omega t + \frac{\dot{U}(0)}{\omega}\cos\omega t\right]^2$$

From where

$$T = \frac{1}{2}m\omega^2\left[-U(0)\sin\omega t + \frac{\dot{U}(0)}{\omega}\cos\omega t\right]^2$$

$$V = \frac{1}{2}k\left[U(0)\cos\omega t + \frac{\dot{U}(0)}{\omega}\sin\omega t\right]^2$$

$$E_t = E_{kinetic} + E_{potential} = \frac{1}{2}m\omega^2\left[-U(0)\sin\omega t + \frac{\dot{U}(0)}{\omega}\cos\omega t\right]^2 + \frac{1}{2}k\left[U(0)\cos\omega t + \frac{\dot{U}(0)}{\omega}\sin\omega t\right]^2$$

$$E_t = \frac{1}{2}m\omega^2\left[U(0)^2\sin\omega t^2 + \left[\frac{\dot{U}(0)}{\omega}\right]^2\cos\omega t^2 - 2\frac{U(0)\dot{U}(0)}{\omega}\sin\omega t\cos\omega t\right] + \frac{1}{2}k\left[U(0)^2\cos\omega t^2 + \left[\frac{\dot{U}(0)}{\omega}\right]^2\sin\omega t^2 + 2\frac{U(0)\dot{U}(0)}{\omega}\sin\omega t\cos\omega t\right]^2$$

$$E_t = \left[\frac{1}{2}m\omega^2U(0)^2 + \frac{1}{2}k\left[\frac{\dot{U}(0)}{\omega}\right]^2\right]\sin\omega t^2 + \left[\frac{1}{2}kU(0)^2 + \frac{1}{2}m\omega^2\left[\frac{\dot{U}(0)}{\omega}\right]^2\right]\cos\omega t^2 + \frac{U(0)\dot{U}(0)}{\omega}\sin\omega t\cos\omega t\left[k - m\omega^2\right]$$

Knowing that:

$$\frac{1}{2}m\omega^2 U(0)^2 = \frac{1}{2}kU(0)^2, \frac{1}{2}k\left[\frac{\dot{U}(0)}{\omega}\right]^2 = \frac{1}{2}m\left[\dot{U}(0)\right]^2 \text{ and } k - m\omega^2 = 0car\ \omega^2 = \frac{k}{m}$$

From where

$$E_t = T + V = \frac{1}{2}k\left[u(0)^2\right] + \frac{1}{2}m\left[\dot{u}(0)^2\right] = \text{constant} \tag{18}$$

From equation (18), it can be deduced that the total energy E_t is independent of time and equal to the energy received by the system (constant) see figure 14.

Figure 14. Vibration energy of free undamped system

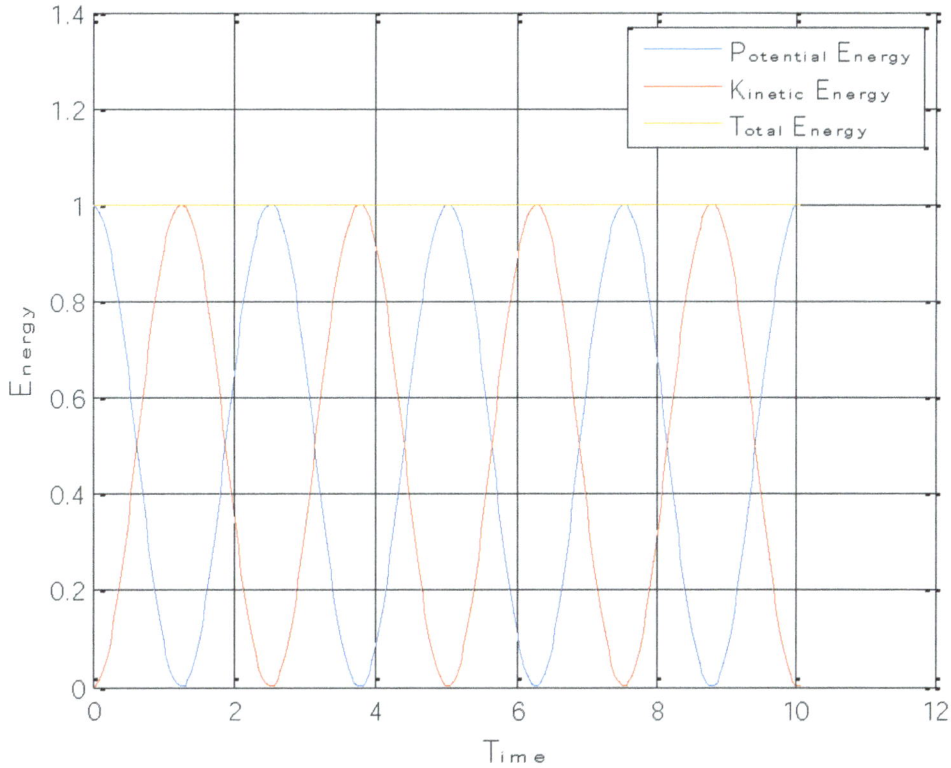

When the mass is in an extremal position with respect to its starting position, its velocity is vanished (maximum of potential energy, minimum of kinetic energy).

Conversely, when the mass returns to the initial position, its velocity is extremal (maximum of kinetic energy, minimum of potential energy).

However, for the case of a damped system, the total energy E_t is not independent of time, because this latter decrease with time, and because of the energy due to damping; which is written as follows:

$$E_d = \int_0^u f_d(t)\, du = \int_0^u C\dot{U}\, du$$

With $dU = \dot{U} dt$. From where

$$E_d = \int_0^t C\dot{U}^2\, dt$$

It is clear that the total energy E_t will depend on time, because of the damping energy E_d (see figure 15).

Figure 15. Vibration energy of free damped system

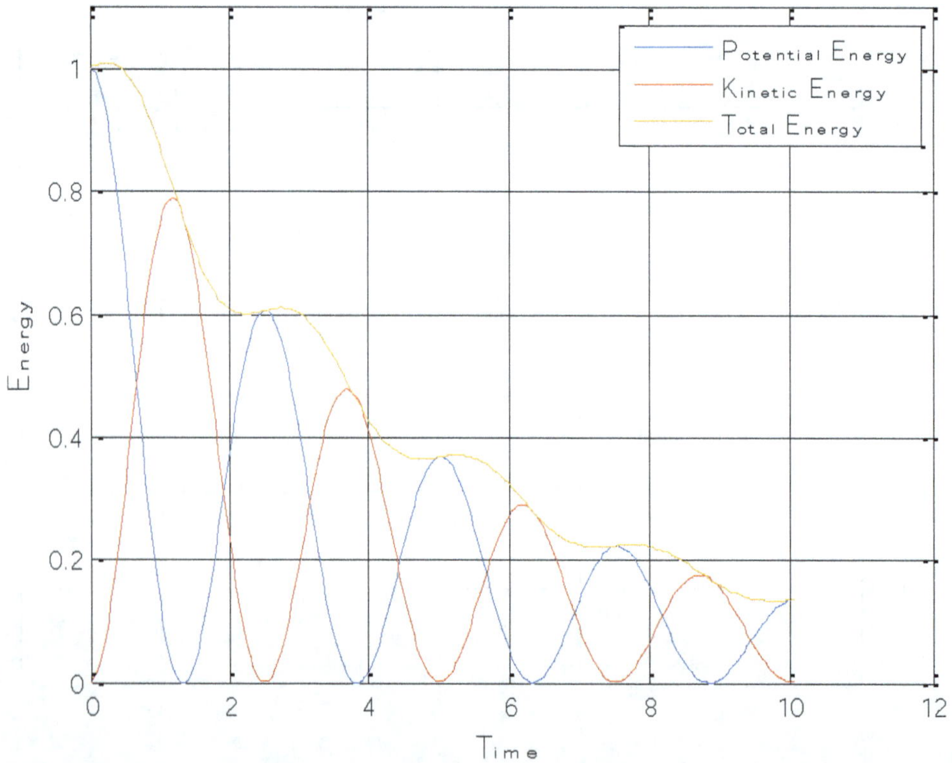

SIGNIFICANT SPRING MASS SYSTEM VIBRATION

In the case the use of springs with a non-negligible mass, can have a considerable effect on the vibration frequency of the structures. Considering a translational system shown in Figure 16.

Figure 16. Single degree of freedom system with significant mass spring

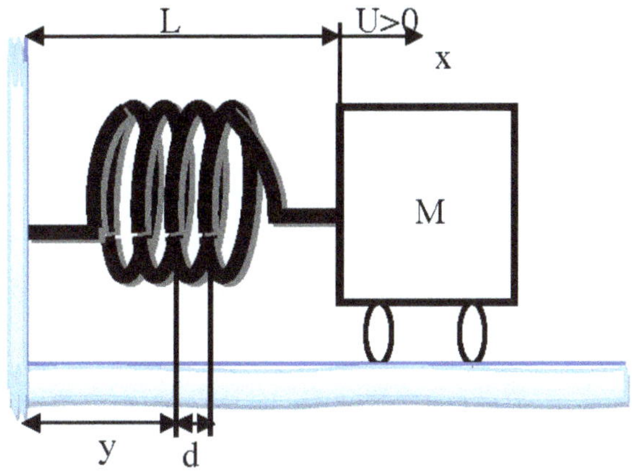

Where a mass M is connected by a spring having a mass m of length L and stiffness k, the mass M moves in the x direction only if the dynamic deflection shape of the spring is assumed to be the same as the static shape, the velocity of the spring element is:

$$\dot{y} = \left(\frac{y}{l}\right)\dot{U}$$

And element mass is

$$\left(\frac{m}{L}\right)dy \left(\frac{m}{l}\right)dy$$

thus

$$T = \frac{1}{2}M\dot{U}^2 + \int_0^L \frac{1}{2}\left(\frac{m}{l}\right)\left[\frac{y}{l}\dot{U}\right]^2 dy$$

$$T = \frac{1}{2}\left(M + \frac{m}{3}\right)\dot{U}^2$$

$$V = \frac{1}{2}KU^2$$

Applying the Lagrange equation, we find:

$$(M+\frac{m}{3})\ddot{U}-KU=0$$

$$\ddot{U}-\frac{K}{M+\frac{m}{3}}U=0$$

$$\ddot{U}-\omega^2U=0$$

With

$$\omega^2=\frac{K}{M+\frac{m}{3}} \quad f=\frac{1}{2\pi}\sqrt{\frac{K}{M+\frac{m}{3}}} \tag{19}$$

What we can conclude is that the period increases when we take into account the effect of mass spring.

SINGLE DEGREE OF FREEDOM TORSIONAL VIBRATION SYSTEM

Figure 17 shows a rod connected to a disk, knowing that: d, L and t are the diameter, the length, the thickness of the rod, on the other hand, I_p, D represents the moment of inertia, and the diameter of the disk respectively, with ρ is the mass density of the disk (Beards 1995).

Figure 17. Torsional vibration model with one degree of freedom

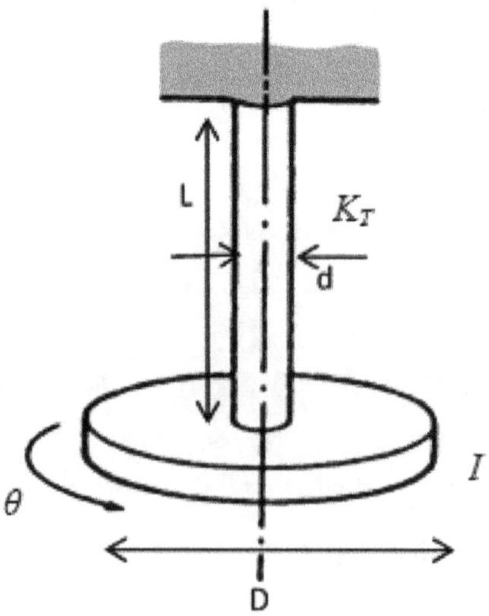

$$\rho = \frac{4W}{\pi D^2 tg} \tag{20}$$

$$I_p = \rho J_D t = \frac{4W}{\pi D^2 tg} \frac{\pi D^4}{32} t = \frac{WD^2}{8g} \tag{21}$$

Where J_D is the polar moment of inertia of the disk.

In torsional free vibrations of the rod, the stiffness is the ratio between the torsional force applied and the length given as follows:

From where $k_T = \frac{J_D G}{L}$

$$k_T = \frac{\pi d^4 G}{32L} \tag{22}$$

With G is the shear modulus of the rod.

If θ is the angular displacement as a function of time, the kinetic energy of the system is

$$T = \frac{1}{2} I_p \dot{\theta}^2 \tag{23}$$

The strain energy U is given:

$$U = \int_0^\theta k\theta \, d\theta = \frac{1}{2} k\theta^2 \tag{24}$$

Substituting equations (23) and (24) in the Lagrange equation, we find:

$$I_p \ddot{\theta} + k_T \theta = 0 \tag{25}$$

Equation (25) represents the free torsional vibration of the stem. Dividing the equation (25) by I_p we find:

$$\ddot{\theta} + \omega^2 \theta = 0 \tag{26}$$

Equation (25) is similar to equation (1) and its solution is:

$$\theta = \theta_0 \cos \omega t + \frac{\dot{\theta}_0}{\omega} \sin \omega t \tag{27}$$

With $\theta 0$ and $\dot{\theta}_0$ are: displacement and initial angular velocity respectively

The expressions of angular frequency and period are:

$$\omega = \sqrt{\frac{k_T}{I_p}} = \sqrt{\frac{\pi d^4 G}{32L}\frac{8g}{WD^2}} = \sqrt{\frac{\pi G g d^4}{4W D^2 L}} \tag{28}$$

$$T = 2\pi \sqrt{\frac{4W L D^2}{\pi G g d^4}} = 4\frac{D}{d^2}\sqrt{\frac{\pi W L}{G g}} \tag{29}$$

APPLICATIONS

Example 1

If we observe that the oscillation frequency of a mass suspended at a spring which have a Stiffness K = 500N / m is equal to 1.59 Hz. Calculate this mass?

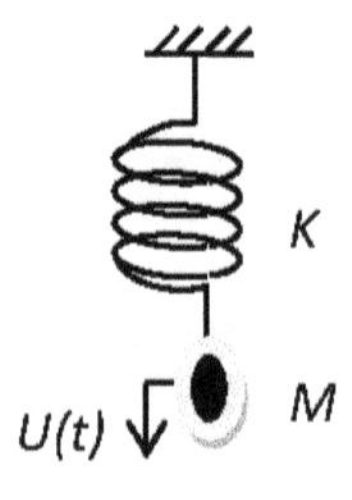

Solution

It is a damped free oscillator

$$M\ddot{U} - KU = 0 \quad \Rightarrow \ddot{U} - \omega^2 U = 0$$

With $\omega^2 = \dfrac{K}{M}$ then $M = \dfrac{K}{\omega^2}$ (1)

Calculation of ω (the own pulsation system)

$$T = \frac{2\pi}{\omega} = \frac{1}{f} \quad \Rightarrow \omega = 2\pi f \tag{2}$$

By replacing equation (2) in (1) we find

$$M = \frac{K}{(2\pi f)^2} \tag{3}$$

$$M = \frac{500}{(2\pi 1.59)^2} = 5Kg$$

Note: we must just convert each given characteristic of the structure in are international unit to have results automatically in international unit the main characteristics of a structure in international unite are

Stiffness K [N/m]
Mass M [Kg]
Damping C [N.m/s]
Period T [s]
Frequency f [Hz]
Pulsation [rad/s]
Displacement U [m]
Velocity $\dot{U}$ [m/s]
Acceleration $\ddot{U}$ [m/s2]

For example for the equation (3) allows the determination of the mass just put the other characteristics in their international unit, namely the rigidity in N/m the frequency in Hz, at that moment, we do not care about the unit of mass m found, it will be automatically in international unit which is Kg

Example 2

It is observed that a mass of 0.5 kg suspended to a spring, the elongation created by this mass is 5 cm. Calculate the spring stiffness ?

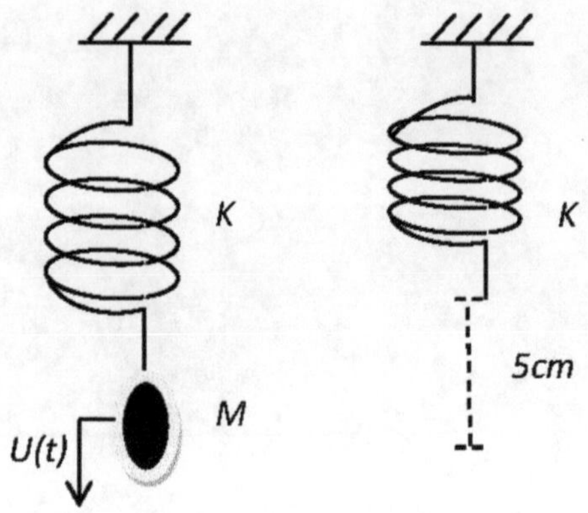

Solution

m=0.5Kg

The elongation is 5cm. Calculation of the stiffness spring K

$$P_{st} = KU_{st} \quad \Rightarrow K = \frac{P_{st}}{U_{st}} = \frac{mg}{Ust}$$

$$K = \frac{P_{st}}{U_{st}} = \frac{mg}{Ust} = \frac{0.5(9.81)}{5.10^{-2}} = 98.1N/m$$

Example 3

Demonstrate that by writing the equation of motion considering the gravitational forces W, the dynamic motion equation is independent of the static equation.

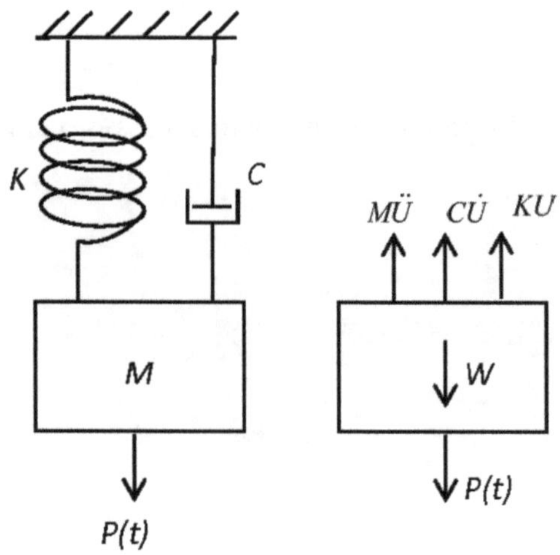

Solution

Considering gravity the equation of dynamic motion is written as follows:

$$M\ddot{U} + C\dot{U} + KU = p(t) + W$$

Knowing $U = U_{st} + U_{dyn}$

$$\dot{U} = \dot{U}_{st} + \dot{U}_{dyn} = \dot{U}_{dyn} \quad \left(\dot{U}_{st} = 0 \quad car \quad U_{st} = cst\right)$$

$$\ddot{U} = \ddot{U}_{st} + \ddot{U}_{dyn} = \ddot{U}_{dyn} \quad \left(\ddot{U}_{st} = 0 \quad car \quad U_{st} = cst\right)$$

By replacing the values of U $\dot{U}$ and $\ddot{U}$ in the equation of motion we find:

$$M\ddot{U}_{dyn} + C\dot{U}_{dun} + K(U_{st} + U_{dyn}) = p(t) + W$$

$$M\ddot{U}_{dyn} + C\dot{U}_{dun} + U_{dyn} + KU_{st} = p(t) + W$$

Knowing That $K\,U_{st} = W = mg$
We have

$$M\ddot{U}_{dyn} + C\dot{U}_{dun} + U_{dyn} = p(t)$$

Example 4

For an undamped free system it is assumed that the mass M is 2.95x105 Kg and K = 8.5×10^6 Nm (C = 0).

If the initial displacement U (0) = 4.5cm for t = 0, there is a displacement U1 = 4.5cm for t = 1.2s as well. In addition, the initial velocity is zero.

Determine the displacement at time t = 2.4 s?

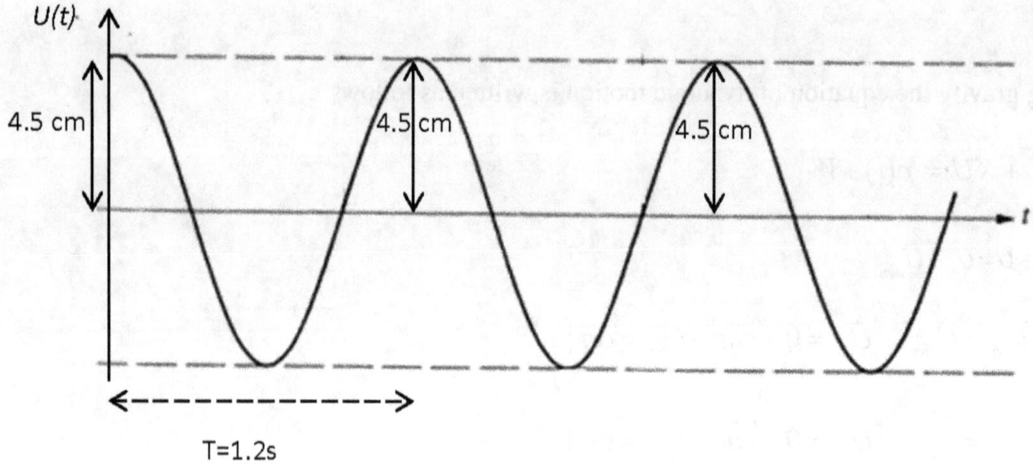

Solution

The equation of motion of an undamped free system is as follows

$$M\ddot{U} + KU = 0$$

The solution of this equation is

$$U(t) = Acons\omega t + B\sin\omega t$$

$$U(t) = U_0 cons\omega t + \frac{\dot{U}_0}{\omega}\sin\omega t$$

Knowing That $U_0 = c^{st}$ et d'où

$U(t) = U_0 cons\omega t$ $\dot{U}_0 = 0$

$$\omega^2 = \frac{K}{M} \Rightarrow \omega = \sqrt{\frac{K}{M}} = \sqrt{\frac{8x10^6}{2.95x10^5}} 5.236 rad/s$$

ω is :

U(t=2s)=4.5cos5.236 (2.4)=4.49cm

U(t=2s)= 4.5cm=0.045m

Example 5 (Commend 2001)

Write the equation of motion of a vertical oscillator at one DDL, subjected to gravity.

1. Determine the theoretical and numerical response when released from its undeformed rest position?
2. Determine the natural frequency, period, displacement, velocity and max acceleration?

Knowing that m = 50 kg, k = 200 N/m

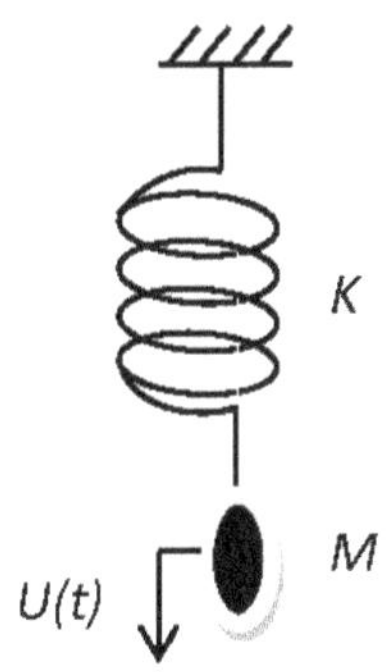

Solution

The equation of motion of a single degree of freedom vertical oscillator, subjected to gravity is:

$m\ddot{u} + ku = f(t)$ with : $f(t)=cst.=mg$

$$\omega = \sqrt{\frac{k}{m}} = 4 rad/s \qquad T = \frac{2\pi}{\omega} = 1.57 s$$

Numerical response with m = 50 kg and k = 200 N/m:

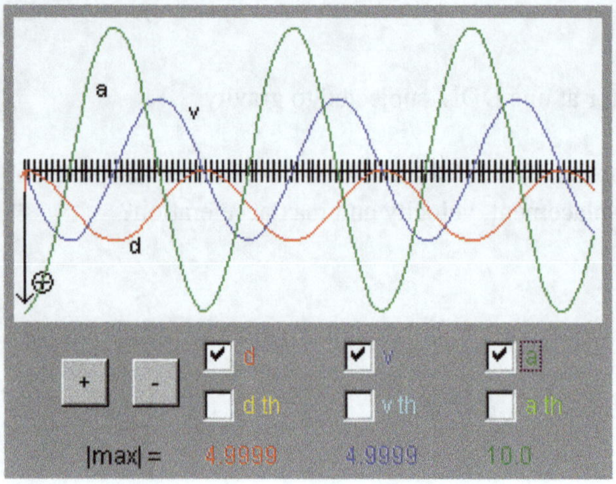

Observation: $u_{\max} = \dfrac{2mg}{k} = 2u_{stat}$

Example 6 (Commend 2001)

Determine the theoretical and numerical response of the same oscillator in previous exercise but this time we apply an instantaneous displacement of the support $\bar{u}_{s0} = 1m$ and we release the oscillator.

Solution

With instantaneous displacement $\bar{u}_{s0} = 1m$:

$m\ddot{u} + ku = f(t)$

with *:*

$f(t) = cst. = mg + k\bar{u}_{s0}$

Numerical response, with m = 50 kg et k = 200 N/m: is

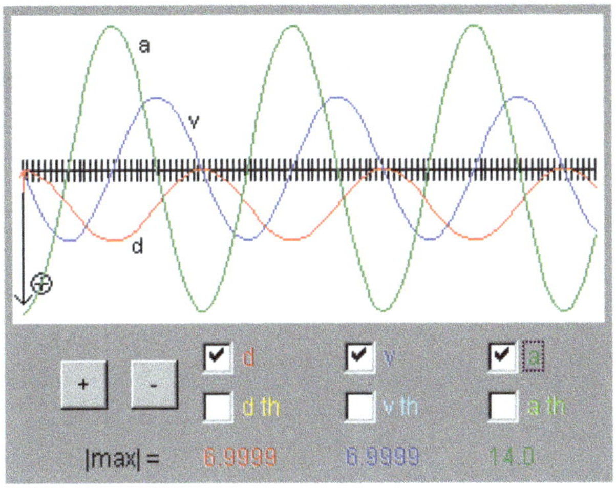

Observation : $u_{\max} = 2\left(\frac{mg}{k} + \bar{u}_{s0}\right)$

Example 7 (Commend 2001)

Determine the theoretical and numerical response of the same oscillator of the previous exercise but in horizontal position, when we apply to it an instantaneous displacement of the support $\bar{u}_{s0} = 1m$.

Solution

Equation of the motion of a single degree of freedom horizontal oscillator (or vertical not subject to gravity), subjected to an instantaneous displacement $\bar{u}_{s0} = 1m$:

$$m\ddot{u} + ku = f(t)$$

With :

$$f(t) = cst. = k\bar{u}_{s0}$$

Numerical response, with m = 50 kg and t k = 200 N/m:

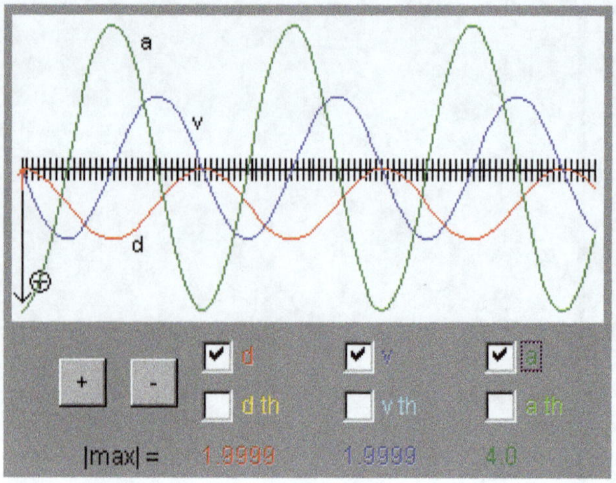

Observation: $u_{\max} = 2\bar{u}_{s0}$

Example 8

The system and his characteristics represented in the following figure

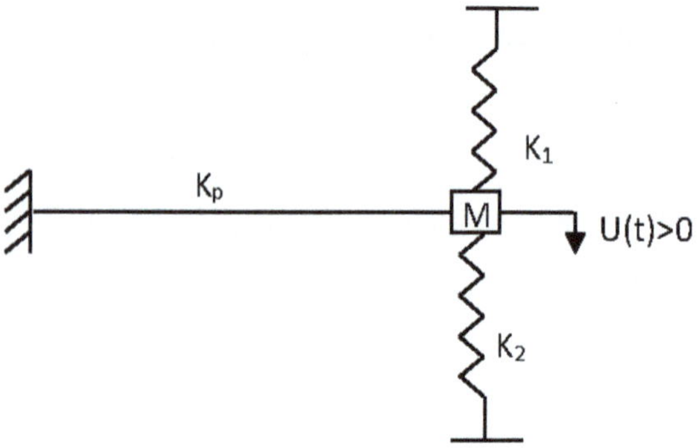

K_1=3,85 Kg/cm

K_2=1,78Kg/cm

K_p =1,27Kg/cm

P=5,28Kgf, g=9,81m/s^2

we have the initial condition as follows U(0)=4cm and $\dot{U}(0) = 0$

1. Determine the own pulsation of the system ?

2. Determine the velocity with which the mass passes through the first equilibrium position?
3. Assuming that the rate of damping is 6% determine the amplitude of the displacement of the first positive peak?

Solution

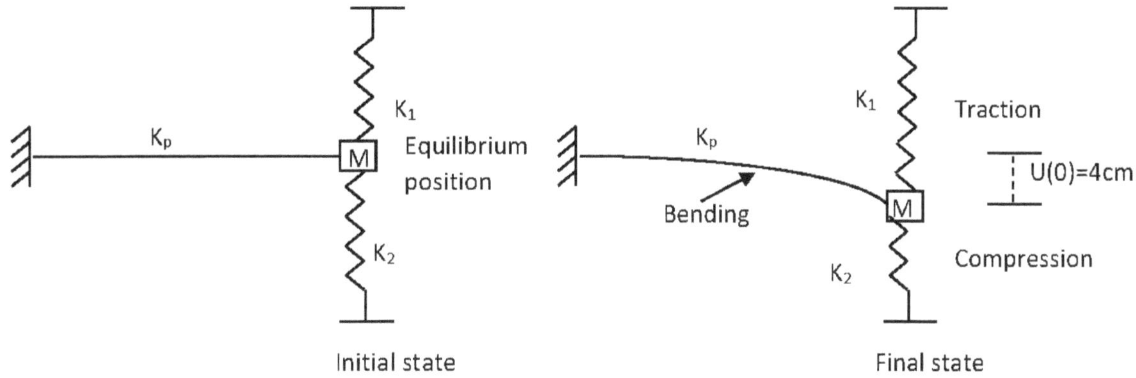

We have an undamped free system his equation of motion is as follows

$$M\ddot{U} + KU = 0$$

The equivalent mathematical model of this structure is as follows:

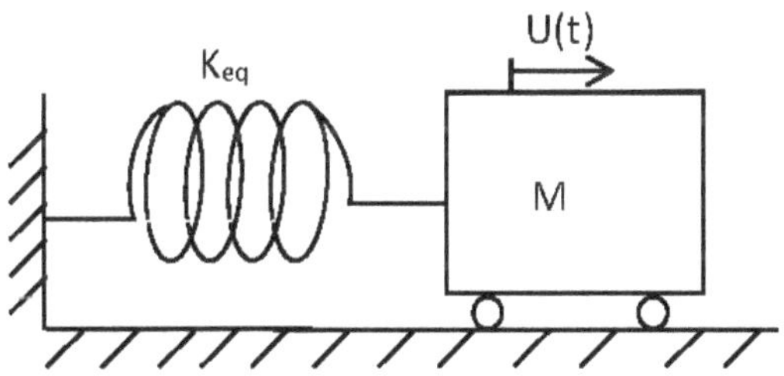

As the components of the system are parallel, that means, that the two springs and the beam deforms at the same time from where

$$K_{eq} = \Sigma K_i \Rightarrow K_{eq} = k_1 + k_2 + k_p$$

K_{eq}=6.9Kg/cm =6.9x10^{3}N/m

$$\omega = \sqrt{\frac{K_{eq}}{M}} = \sqrt{\frac{6.9x10^3}{\frac{P}{g}}} = \sqrt{\frac{6.9x10^3}{\frac{5.28x10}{10}}} = 35.8 rad/s$$

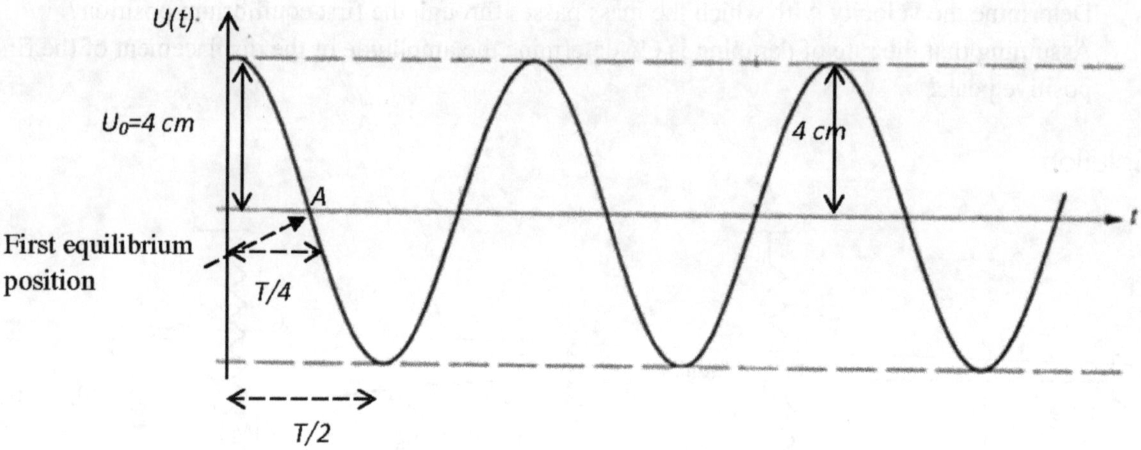

We have an undamped free vibration the solution is as follows:

$U(t)$= A $cons\omega$t+ B sin ωt

The initial conditions are $\begin{cases} U(0) = 4cm \\ \dot{U}(0)0 \end{cases}$

At t=0 U(0)=Ax1+Bx0=A from where A=U(0)

$\dot{U}(t) = -A\omega \sin \omega t + B\omega \cos \omega t \dot{U}(t) = 0\, B\omega$=0⇒B=0 because $\omega \neq 0$

From where U(t)=4×10-^{2c}osωt *and* U(t)= -4×10-2^{si}nωt

Time estimation at the equilibrium position

First Method

We have at the first equilibrium position (point A) U (t) = 0

From where

$$U(t) = 4 \times 10 - 2\cos \omega t = 0 \Rightarrow \cos \omega t = 0 \Rightarrow \omega t = \frac{\pi}{2} \Rightarrow t = \frac{\pi}{2\omega}$$

Second Method

Either time is determined by the graph; at equilibrium position t = T / 4

$$\text{T=}\frac{2\pi}{\omega} \Rightarrow \frac{T}{4} = \frac{\frac{2\pi}{\omega}}{4} = \frac{\pi}{2\omega} \Rightarrow t = \frac{\pi}{2\omega}$$

$$\dot{U}(t=\frac{\pi}{2\omega})=-4\times 10-2\sin\omega\frac{\pi}{2\omega}=-4\omega \quad \dot{U}(t=\frac{\pi}{2\omega})=-4\omega$$

Determination of the first positive peak displacement (the rate damping is 6%)

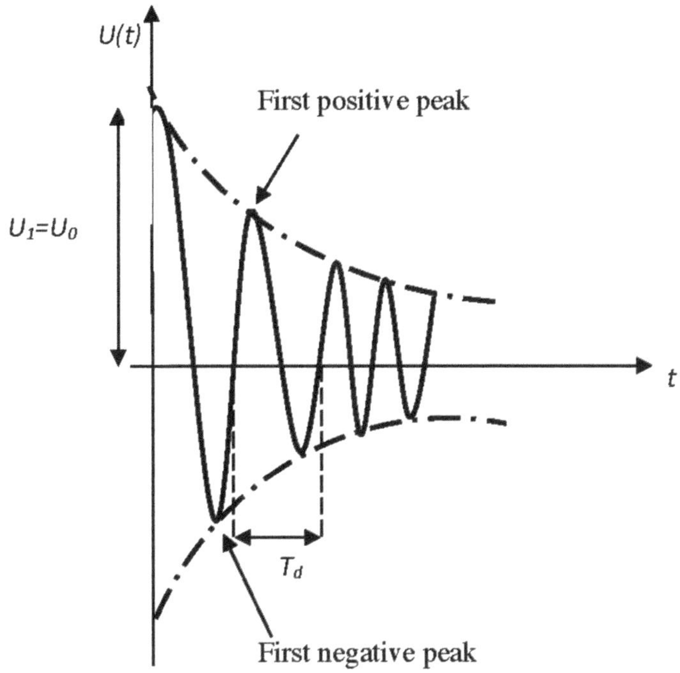

We are in the case of a low damping system there are two methods to determine the first positive peak displacement:

First Method

We have the response of a single degree of freedom low damping system, which is as follows

$$U(t)=e^{-\xi\omega t}\left(A cons\omega_d t+B\sin\omega_d t\right)$$

Damping constant C =2ξω from where ξ=C/2mω

Low damping system $\xi\approx\frac{\delta}{2\pi}=\frac{1}{2\pi m}Ln\frac{U_n}{U_{n+m}}$ m is a number of cycles

For the first peak

$$\xi = \frac{\delta}{2\pi} = \frac{1}{2\pi} Ln \frac{U_0}{U_1} \Rightarrow \delta = Ln \frac{U_0}{U_1} \Rightarrow \frac{U_0}{U_1} = e^{\delta} \Rightarrow U_1 = \frac{U_0}{e^{\delta}} = \frac{4x10^{-2}}{e^{0.377}}$$

U_1=2.74x10^{-2}m

The second method is:
We have the response of a single degree of freedom low damping system

$$U(t)=e^{-\xi\omega t}\left(A cons\omega_d t + B \sin\omega_d t\right)$$

$U(0)$=4x10^{-2}m

$A=U(0)$= 4x10^{-2}m

$$\dot{U}(t)=-\xi\omega e^{-\xi\omega t}\left(A cons\omega_d t + B \sin\omega_d t\right) + e^{-\xi\omega t}\left(-A\omega_d \sin\omega_d t + B\omega_d \cos\omega_d t\right)$$

$$\dot{U}(t)=0 \Rightarrow -\xi\omega(A)+(B\omega_d)=0 \Rightarrow A\xi = B \quad because \quad \omega_d \approx \omega \quad thus \quad B=0.24$$

$$U(t)=e^{-\xi\omega t}\left(4x10^{-2} cons\omega_d t + 0.24x10^{-2} \sin\omega_d t\right)$$

At first positive peak $t=T=2\pi/\omega_d$

$$U(t=T)=e^{-\xi\omega\frac{2\pi}{\omega_d}}\left(4x10^{-2} cons\omega_d \frac{2\pi}{\omega_d} + 0.24x10^{-2} \sin\omega_d \frac{2\pi}{\omega_d}\right)$$

$$U(t=T)=e^{-2\pi\xi}\left(4x10^{-2} cons2\pi + 0.24x10^{-2} \sin 2\pi\right)$$

$$U(t=T)=4x10^{-2} e^{-2\pi 0.06}$$

$U(t=T)$=2.74x10^{-2}m

Example 9

The mass, the spring and the damping are attached to a rigid beam represented in the following figure:

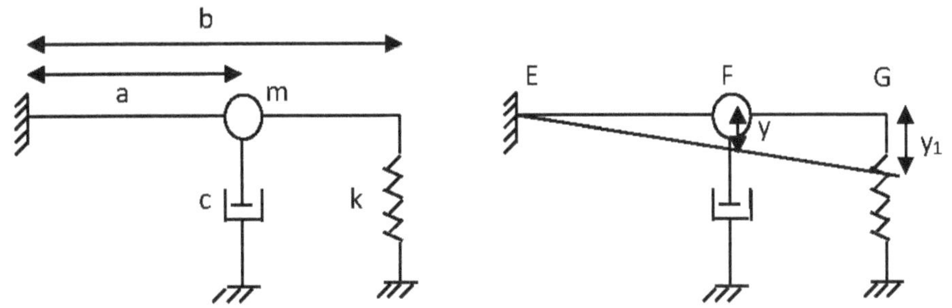

1. Write the equation of motion as a function of y, a, and b by applying the sum of the moments of the forces acting on the beam with respect to the point E (with explanation)?
2. Determine the natural frequency of the free damped system and the critical damping coefficient as a function of k, m, a, and b?

Solution

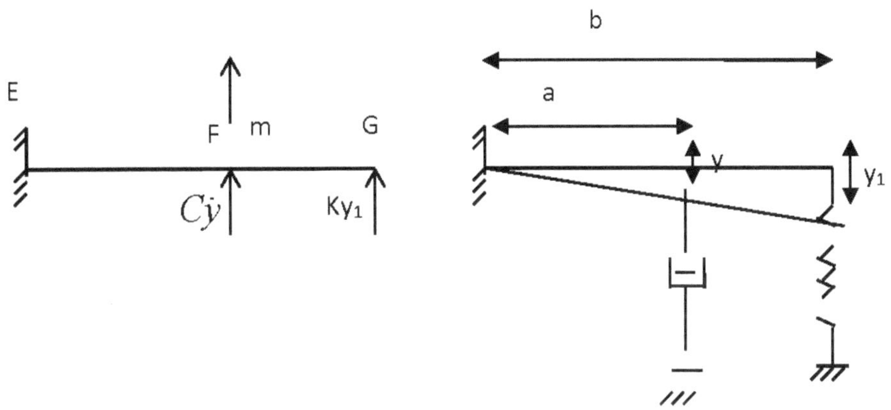

By similar triangles $y_1=(b/a)y$ (a)

$$\sum \text{Force/E} = 0 \Rightarrow \text{Ky}_1 xb + C\dot{y}a + m\ddot{y}a = 0$$

$$m\ddot{y} + C\dot{y} + \text{Ky}_1 x\frac{b}{a} = 0 \quad \text{(b)}$$

By replacing equation (a) in the equation (b)

$$m\ddot{y} + C\dot{y} + \text{Ky}x\left(\frac{b}{a}\right)^2 = 0$$

$$\ddot{y}+2\eta\dot{y}+\omega^2 y=0$$

With

$$m\ddot{y}+C\dot{y}+\omega^2=\frac{K}{m}\left(\frac{b}{a}\right)^2 \qquad \textit{and} \qquad \eta=\xi\omega=\frac{C}{2m}$$

$$\omega_d=\sqrt{\omega^2-\eta^2}=\sqrt{\frac{K}{m}\left(\frac{b}{a}\right)^2-\left(\frac{C}{2m}\right)^2}$$

At the critical damping ω = η so ξ = 1

$$\frac{C}{2m}=\frac{b}{a}\sqrt{\frac{K}{m}} \qquad C=\frac{2b}{a}\sqrt{Km}$$

Example 10

The figure shows a pendulum which consists of a rigid rod of length L, which can pivot about a point fixed at one end and having a mass m fixed at its other end. A stiffness spring k is fixed as illustrated at a distance a from the pivot point.

Knowing that the rod is vertical, and that the spring is horizontal with respect to the position shown and that the spring is not stressed.

Determine free oscillations frequency?

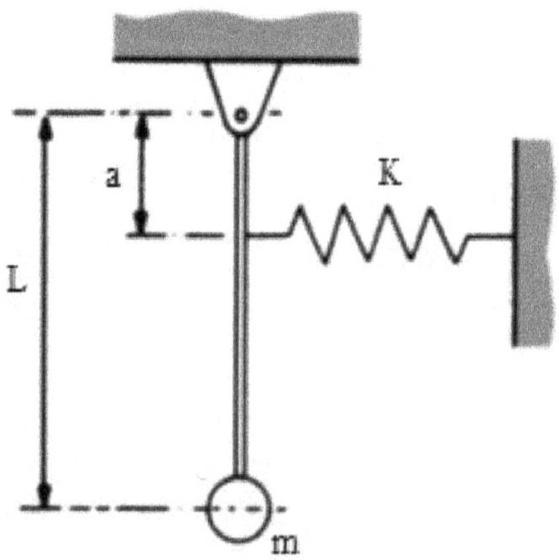

Solution

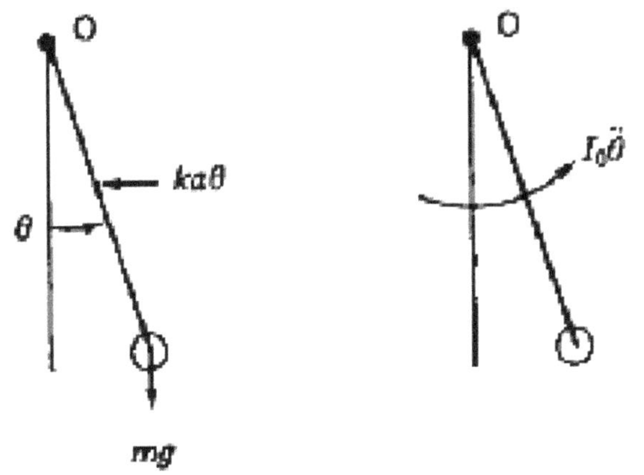

The sum of the moments around the point O is

$$I_0\ddot{\theta} = -mgL\theta - ka^2\theta$$

Chère I_0=mL^2. Thus

$$\ddot{\theta} + \left(\frac{mgL + ka^2}{mL^2}\right)\theta = 0 \quad f = \frac{1}{2\pi}\sqrt{\frac{mgL + ka^2}{mL^2}}$$

REFERENCES

Beards, C.F. (1995). *Engineering Vibration Analysis with Application to Control Systems*. Edward Arnold.

Chopra, A. K. (2001). *Dynamics of Structures: Theory and Applications to Earthquake Engineering* (2nd ed.). Prentice Hall.

Clough, R. W. & Penzien, J. (1995). *Dynamic of structures*. Compute and Structure, Inc.

Commend, S., & Zimmermann, T. (2001). *LSCDYN 1DOF and nDOF: two applications for computer-assisted teaching of structural dynamics. LSC Internal Report*. Swiss Federal Institute of Technology.

KEY TERMS AND DEFINITIONS

Critical Damping: Is a system in which rate damping is equal to 1 ($\xi = 1$).
Logarithmic Decrement: Is the logarithm ration two different amplitudes response of the system.
Over Damping: Is a system in which rate damping is great than 1 ($\xi > 1$).
Subcritical Damping: Is a system in which damping rate is less than unity 1 ($\xi < 1$).
Torsional Vibration: Is a vibration system under moment or rotation effect not displacement one.
Undamped Oscillator: Is a system in which damping is zero.

Chapter 2
Single Degree of Freedom Forced Vibration System

ABSTRACT

This chapter concerns the study of forced vibration of a single degree of freedom system, treating undamped and damped system under harmonic, periodic, and arbitrary loading with different cases and examples. Passing by all components of the general solution of an undamped forced system, which are a transient solution, depends only on initial conditions, transient solution due to the load at the end the stationary solution. In this chapter, a study of the dynamic influence factor depends on the ration between load frequency and structure one is presented.

INTRODUCTION

Time-dependent physical phenomena are generated initially by differential equations. In the simplest case, there is a single quantity, which varies and speaks of a system with one degree of freedom, most of the time governed by a second order differential equation. Natural phenomena arc usually non-linear, but in many cases, the small motion hypothesis provides an excellent approximation provided by a linear differential equation with constant coefficients of second order. The forced oscillator evolves according to a transient regime then a permanent regime. This chapter aims to give a physical interpretation of some interesting case of a forced one degree of freedom system. A large part of this chapter is devoted to one type of loading, which is the harmonic load; the end of the chapter will be devoted to another type of loading and to the construction of an analytical as well as mathematical model of a forced vibration system linear oscillator

HARMONIC LOADING

Among the existing external forces, we have the harmonic forces. Other types exist, and will be studied later at the end of this chapter.

DOI: 10.4018/978-1-7998-4399-3.ch002

Copyright © 2021, IGI Global. Copying or distributing in print or electronic forms without written permission of IGI Global is prohibited.

Undamped System (c = 0)

We will assume a system with a single degree of freedom represented in the following figure.

Figure 1. Undamped single degree of freedom model system

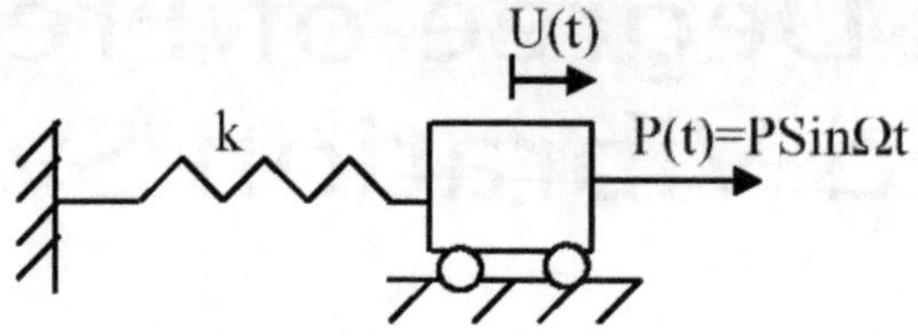

In order to write the motion of equation we use Dalambert principle so we isolate the mass as follows:

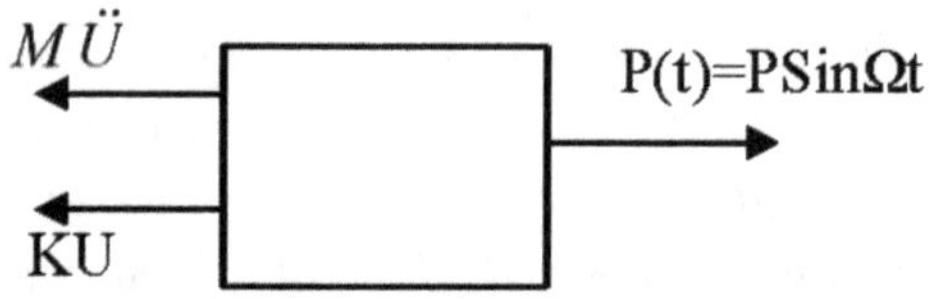

The dynamic equilibrium equation of forced undamped system is:

$$M\ddot{U}(t)+KU(t)=P_0\sin\Omega t$$

$$\ddot{U}(t)+\omega^2 U(t)=\frac{P_0}{M}\sin\Omega t \quad (1)$$

With $\omega^2 = K/M$

Equation (1) is a second-degree differential equation with second member.
The general solution = the particular solution + the homogeneous solution
The dynamic equilibrium of a forced, undamped system is:

$$U(t)=U_p(t)+U_h(t)$$

The homogeneous solution is as shown by the following equation:

$$U_h(t) = A \sin \omega t + B \cos \omega t \quad (2)$$

On the other hand, for the particular solution we can assume that Up (t) has of the same form of the harmonic excitation force

$$U_p(t)=C\sin\Omega t,\ \dot{U}_p(t)=\Omega(C_3\operatorname{Cos}\Omega t-C_4\operatorname{Sin}\Omega t),\ddot{U}_p(t)=-\Omega^2 C\operatorname{Sin}\Omega t=(-\Omega^2)U_p(t)$$

By replacing $U_p(t)$ and $\ddot{U}_p(t)$ in the equation (1) we find:

$$M\,\ddot{U}_p + K\,U_p = P_0 \sin \Omega t$$

$$M((-\Omega^2) + K)U_p = P_0 \sin \Omega t$$

$$U_p(t) = \frac{P_0}{K(1-\beta^2)} \sin \Omega t \quad with \quad \beta = \frac{\Omega}{\omega} \tag{3}$$

$$A_{U_p} = \frac{P_0}{K(1-\beta^2)}$$

Is the amplitude of particular solution.

By drawing a curve, that represents the amplitude of the particular solution A_{U_p} as a function of the frequency ratio load and structure (β) respectively, we find:

Figure 2. (a) Particular solution of forced undamped system; (b) Particular solution amplitude as a function of structure and load frequency

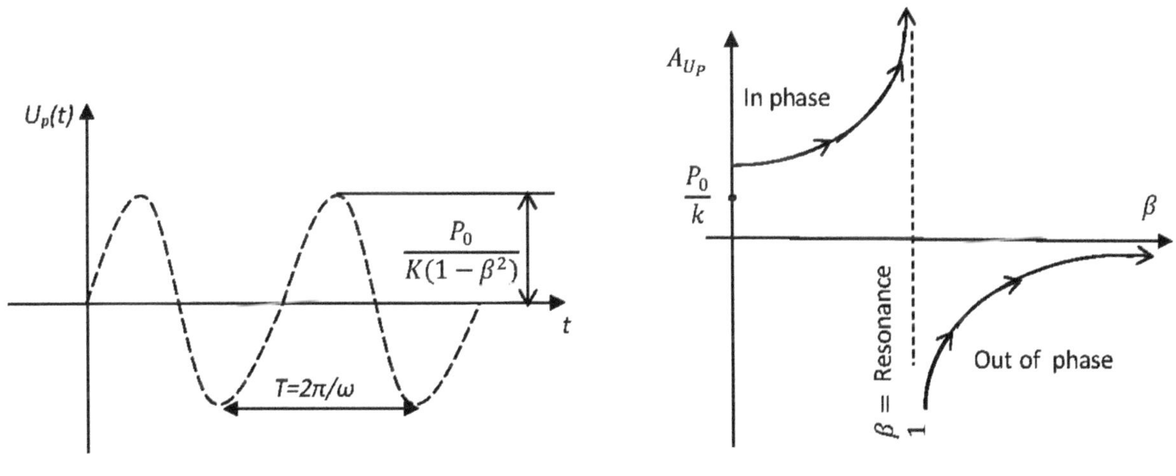

As the general solution = the particular solution + the homogeneous solution

$$U(t) = A\,\text{Sin}\,\omega t + B\,\text{Cos}\,\omega t + \frac{P}{K(1-\beta^2)} \sin \Omega t$$

We need to determine constants A and B

At t=0 U(0)=0+B+0

From where B=U(0)

$$\dot{U}(t) = A\omega \operatorname{Cos}\omega t - U(0)\,\omega \operatorname{Sin}\omega t + \frac{P_0\,\Omega}{K(1-\beta^2)}\cos\Omega t$$

$$\dot{U}(0) = A\omega + \frac{P_0\,\Omega}{K(1-\beta^2)}$$

$$A = \frac{1}{\omega}\dot{U}(0) - \frac{P_0\,\Omega}{\omega\,K(1-\beta^2)}$$

The general solution is:

$$U(t) = \underbrace{U_0 \operatorname{Cos}\omega t + \frac{\dot{U}}{\omega}\operatorname{Sin}\omega t}_{\text{Transient solution that depends on the initial conditions } U_1(t)} - \underbrace{\frac{P_0\beta}{K(1-\beta^2)}\sin\omega t}_{\text{Transient solution due to applied load } U_2(t)} + \underbrace{\frac{P_0}{K(1-\beta^2)}\sin\Omega t}_{\text{Stationary solution } U_p(t)} \quad (4)$$

The figure below gives more details on the general solution represented in equation (4).

Figure 3. Graphical representation of all components of the general solution of an undamped forced system

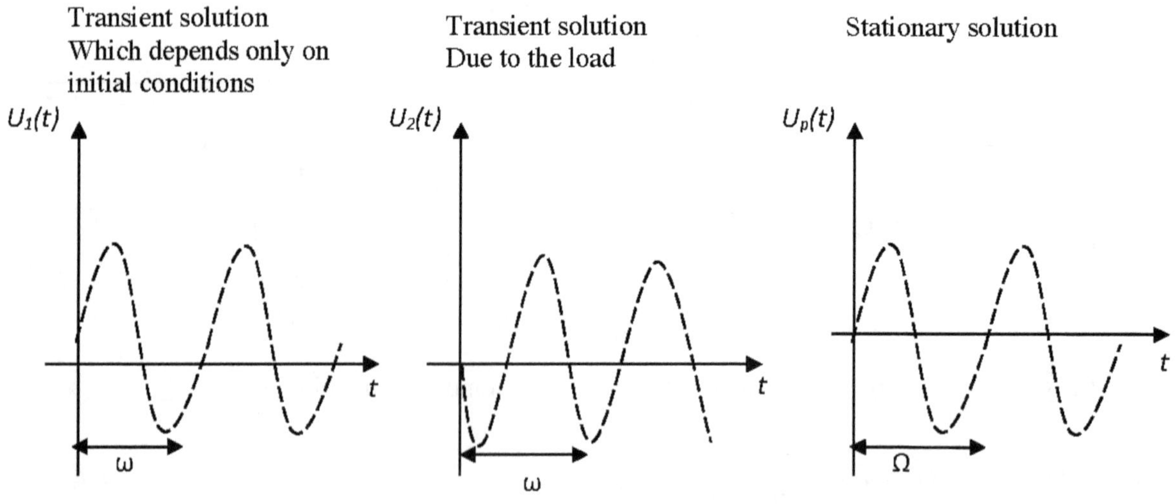

With $\frac{P_0}{K} = U_{st}$

Represente static displacement

$$F.A.D = \frac{1}{1-\beta^2}$$

Is the amplification factor that represents the effect of the dynamic amplification of the applied harmonic load.

$$\beta = \frac{\Omega}{\omega}$$

Is the frequency ration

$$\frac{U(t)}{U_{st}} = R(t)$$

Is the dynamic influence factor.

Assuming the initial conditions are equal to zero; $U_0 = \dot{U} = 0$

$$U(t) = \underbrace{0\,Cons\omega t + 0\ \mathrm{Sin}\,\omega t}_{\text{the transient solution will be equal to zero}} - \underbrace{\frac{P_0\beta}{K(1-\beta^2)}\sin\omega t}_{\text{Transient solution due to applied load}} + \underbrace{\frac{P_0}{K(1-\beta^2)}\sin\Omega t}_{\textit{Trasient solution}}$$

$$U(t) = \underbrace{\frac{-P_0\beta}{K(1-\beta^2)}\sin\omega t}_{\text{Transient solution due to applied load}} + \underbrace{\frac{P_0}{K(1-\beta^2)}\sin\Omega t}_{\text{Transient solution}}$$

$$U(t) = \frac{P_0}{K(1-\beta^2)}(\sin\Omega t - \beta\sin\omega t)$$

$$R(t) = \frac{U(t)}{Ust} = \frac{1}{(1-\beta^2)}(\sin\Omega t - \beta\sin\omega t) \qquad (5)$$

DYNAMIC INFLUENCE FACTOR STUDY R(T)

Very low frequency solicitation compared to its own pulsation $\beta \ll 1$. If

$$\frac{\Omega}{\omega} << 1 \Rightarrow \quad \beta << 1 \ so \quad \Omega << \omega$$

In this case

$R(t) \approx 1(\sin\Omega t - \beta\sin\omega t)$

Figure 4. Graphical Representation of the Influence Factor for the Case β<<1

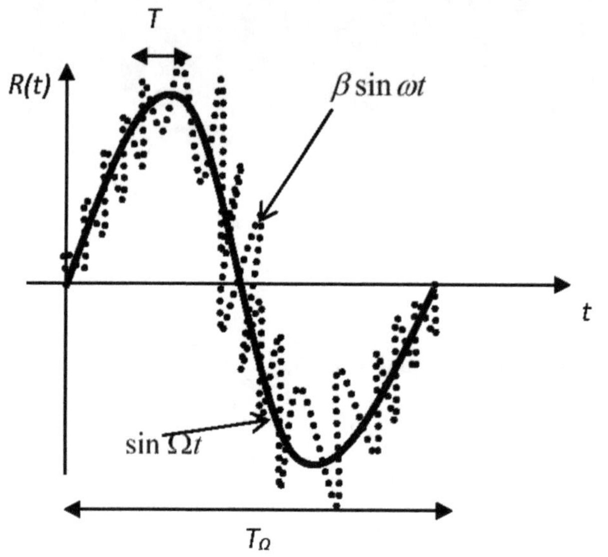

Very High Solicitation Frequency with Respect to its own Pulsation $\beta \gg 1$ If

$$\frac{\Omega}{\omega} >> 1 \Rightarrow \quad \beta >> 1 \, so \quad \Omega >> \omega$$

In this case $R(t) \approx 0$ because

$$\frac{1}{(1-\beta^2)} \approx 0$$

Solicitation Frequency Equal to the own Reasoning Heartbeat $\beta=1$

If $\Omega=\omega$ there is resonance phenomenon which causes the plastification of the materials before reaching extreme displacements. In this case,

$R(t) \rightarrow \infty$ et $u(t) \rightarrow \infty$ $\beta = 1 \quad R(t) = \frac{0}{0};$

This is the case of indeterminacy ; to remove this indetermination, we apply the Hopital rule which states:

$$\lim_{x\to\alpha}\frac{f(x)}{g(x)}=\lim_{x\to\alpha}\frac{f'(x)}{g'(x)}$$

Knowing that $\Omega=\beta\omega$ the equation becomes as follows

$$\lim_{\beta\to 1}R(t)=\lim_{\beta\to 1}\frac{\omega t\cos\beta wt-\sin\omega t}{-2\beta}=\frac{1}{2}\left(\sin\omega t-\omega t\cos\omega t\right) \tag{6}$$

The term $\frac{1}{2}\sin\omega t$ is small and can be neglected, in this case, the dynamic influence factor R(t) amplitude increases linearly see figure 5.

$$U_p(t)=\frac{-P_0}{2K}\omega t\cos\omega t$$

Figure 5. Resonance case of an and amped forced system

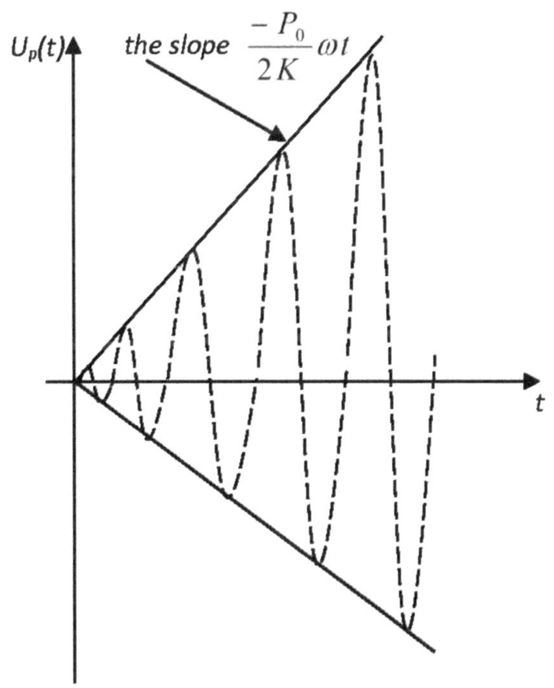

Solicitation Frequency Tends Towards the proper Pulsation Beat Case $\beta\approx 1$, $\Omega\to\omega$ Beat case Figure bellow allow as to have an idea of beat case:

Figure 6. Beat case of a free forced system (Jerome 2012)

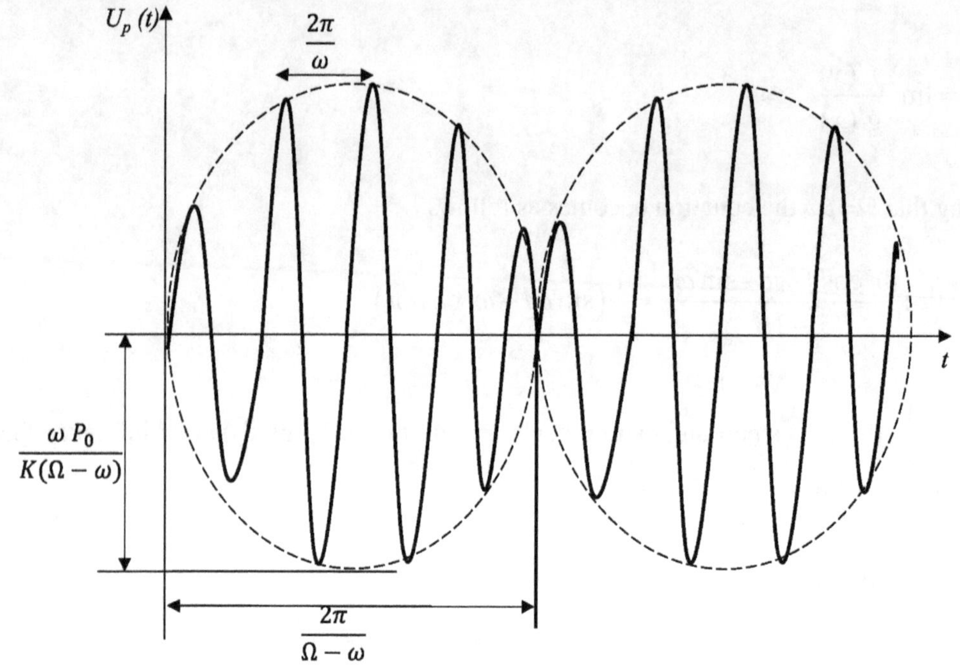

DYNAMIC AMPLIFICATION FACTOR PROPERTY VIEWED IN A DIFFERENT WAY

We can reformulate equation (4) as follows

$$U(t) = U_0 \cos \omega t + \frac{\dot{U}_0}{\omega} \sin \omega t - \frac{P_0}{\frac{K}{\omega^2}\left(\omega^2 - \Omega^2\right)} (\frac{\Omega}{\omega}) \sin \omega t + \frac{P_0}{\frac{K}{\omega^2}\left(\omega^2 - \Omega^2\right)} \text{Sin}\, \Omega t$$

$$U(t) = U_0 \cos \omega t + \left[\frac{\dot{U}_0}{\omega} - \frac{\frac{P_0}{M}}{\omega^2 - \Omega^2} (\frac{\Omega}{\omega}) \right] \sin \omega t + \frac{\frac{P_0}{M}}{\omega^2 - \Omega^2} \text{Sin}\, \Omega t$$

$$U(t) = U_0 \cos \omega t + \frac{\dot{U}_0}{\omega} \sin \omega t + \frac{\frac{P_0}{M}}{\omega^2 - \Omega^2} \left[\text{Sin}\, \Omega t - \frac{\Omega}{\omega} \sin \omega t \right]$$

If the initial conditions are zero:

$$U(t) = \frac{\frac{P_0}{M}}{\omega^2 - \Omega^2}\left[\operatorname{Sin}\Omega t - \frac{\Omega}{\omega}\operatorname{Cos}\omega t\right] \quad (5)$$

Note that the function is composed of two harmonic functions but the resultant is not necessarily harmonic, this is due to the difference of the frequencies ($\omega \neq \Omega$).

We assume the $\beta = \frac{\Omega}{\omega}$ from where rewriting equation (5), gives the following equation

$$U(t) = \frac{P_0}{K\left(1-\beta^2\right)}\left[\operatorname{Sin}\Omega t - \beta \operatorname{Cos}\omega t\right] \quad (6)$$

Assuming also that

$$U_{static} = \frac{P_0}{K}$$

and dynamic amplification factor

$$D.A.F = \frac{1}{(1-\beta^2)}$$

Three cases are to be studied:

First case $\beta \to 0$ implies that dynamic amplification factor (D.A.F) =1 that is to say

The most disadvantageous case to design a structure is when u (t) is maximum, so it is the case if $\mathrm{Sin}\Omega t$ is maximum it means equal to 1.

In that case $U(t) = U_{static}$ the static response is the same as the dynamic response.

Second case $\beta \to \infty$ implies that D.A.F =0, U(t) =0, that means the mass is stationary (it does not move).

Third case $\beta \to 1$ ($\Omega = \omega$) implies that D.A.F $=\infty$, U(t)= ∞, that means resonance state (no damping to dissipate energy).

The shape of the D.A.F curve according to Ω / ω is as follows:

The sign of β depends on the denominator if $\Omega/\omega < 1$ this implies $1-(\Omega/\omega)^2 > 0$ two cases may arise from taking into account the sign

First case $\Omega < \omega$: The angular frequency of the exciter force is lower than the angular frequency of the system. In this case, the period of the exciter force is greater than the system period. That means that the system cycles several times while the force makes a single cycle; the force and the response are in the same phase because $1-(\Omega/\omega)^2 > 0$ (see figure 8) .

Figure 7. Curve representing D.A.F according to β

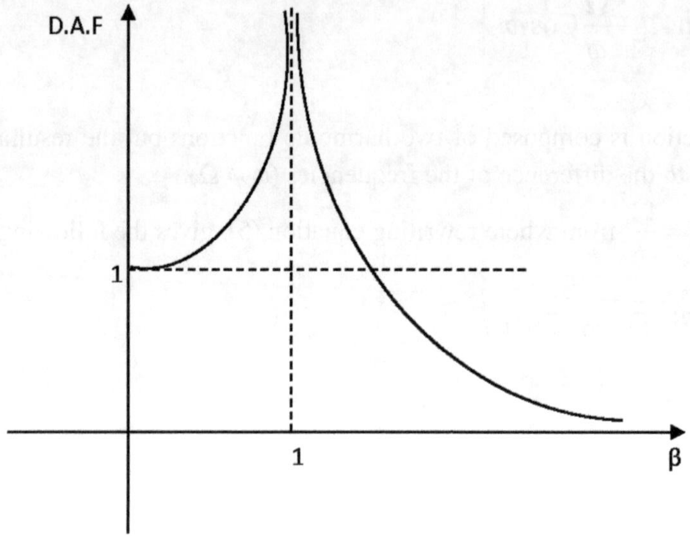

Figure 8. Graphical comparison between the load and the response of the structure for Ω <ω

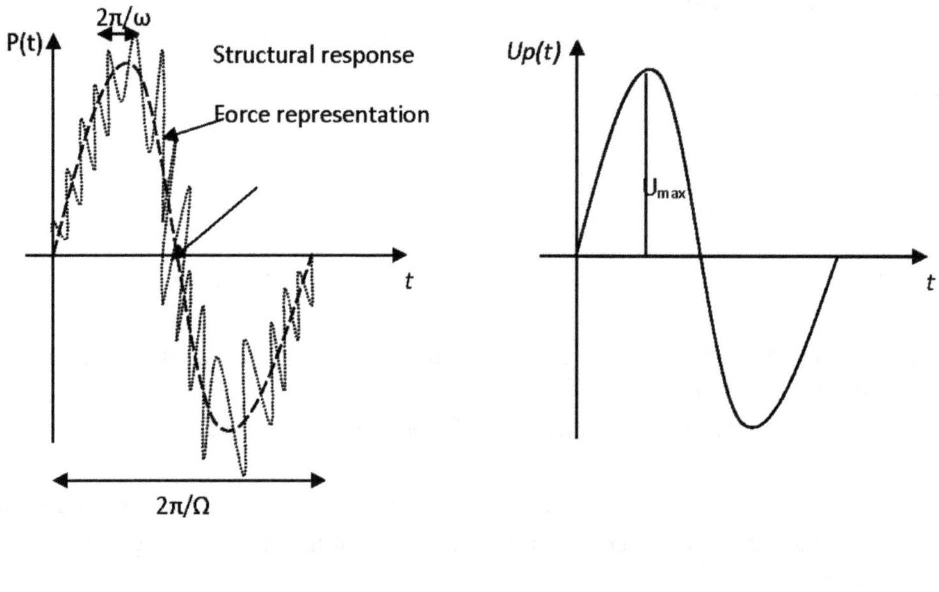

Second case Ω/ ω > 1 from where Ω> ω implies that 1- $(\Omega/\omega)^2<0$

The angular frequency of the exciter force is greater than the angular frequency of the system.

Therefore, the period of the exciter force is inferior to the period of the C.A.D system that while the system is doing a single cycle the exciter force makes several cycles. As long as 1- $(\Omega / \omega)^2 <0$ the force and the response are out of phase (see figure 9).

Figure 9. Graphical comparison between load and structure response for Ω> ω

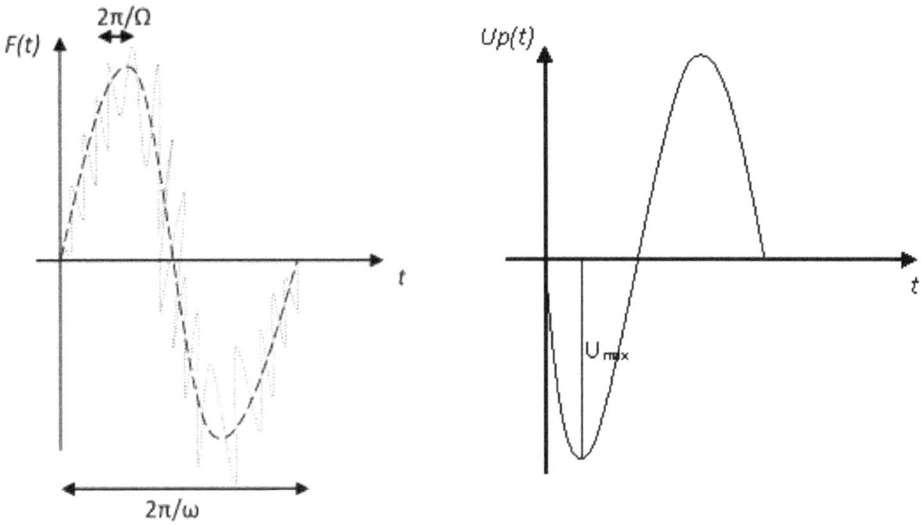

SINGLE DEGREE OF FREEDOM DAMPED SYSTEM UNDER HARMONIC LOAD

Figure bellow represents one degree of freedom damped system:

Figure 10. Model of damped system with a single degree of freedom

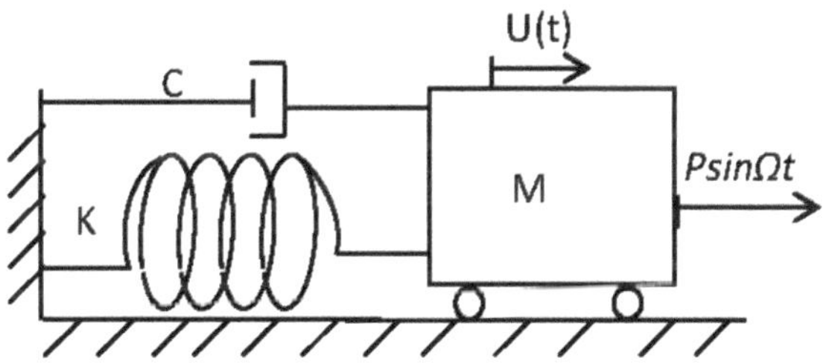

To write the equation of motion we will isolate the mass as follows:

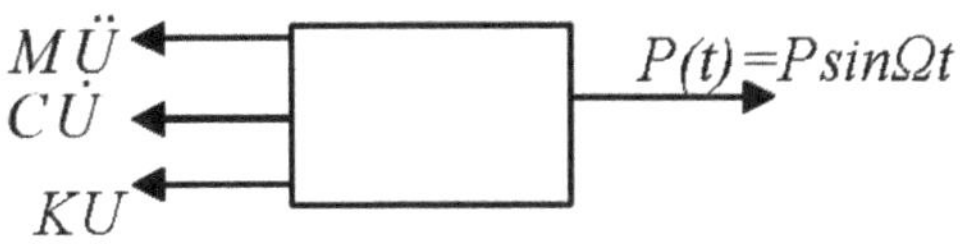

The equation of equilibrium is as follows:

$$M\ddot{U} + C\dot{U} + KU = P\,\mathrm{Sin}\,\Omega t$$

$$\ddot{U} + \frac{C}{M}\dot{U} + \frac{K}{M}U = \frac{P}{M}\text{Sin}\,\Omega t \tag{11}$$

Equation (11) is the second order differential equation of damped second member systems Knowing that

$$2\eta = 2\xi\omega = \frac{C}{M} \text{ and } \omega^2 = \frac{K}{M}$$

The total solution is:

$U(t)=U_h(t)+ U_p(t)$

The homogeneous solution in the case where $C \neq 0$ is:

$$U_h(t) = e^{-nt}\left[C_1 \text{Sin}\,\omega_d t + C_2 \cos\omega_d t\right] \tag{12}$$

The expression of the particular solution may be the same as the excitatory force, or it may take the general part, which is the sum of function sinus and cosinus as follows for further reading see (Mohannad 2013).

$$U_p(t) = C_3 \sin\Omega t + C_4 \cos\Omega t \tag{13}$$

$$\dot{U}_p(t) = +\Omega C_3 \cos\Omega t - \Omega C_4 \text{Sin}\,\Omega t$$

$$\ddot{U}_p(t) = -\Omega^2 C_3 \text{Sin}\,\Omega t - \Omega^2 C_4 \text{Cos}\,\Omega t$$

We replace in equation (11) the expressions of $U_p(t)$, $\dot{U}_p(t), \ddot{U}_p(t)$ we find:

$$-C_3\Omega^2 \text{Sin}\,\Omega t - C_4\Omega^2 \text{Cos}\,\Omega t + 2\eta C_3\Omega \text{Cos}\,\Omega t$$
$$-2\eta C_4\Omega \text{Sin}\,\Omega t + \omega^2 C_3 \text{Sin}\,\Omega t + \omega^2 C_4 \text{Cos}\,\Omega t = \frac{P}{M}\text{Sin}\,\Omega t$$

$$\text{Sin}\,\Omega t\left[-C_3\Omega^2 - 2\eta\Omega C_4 + \omega^2 C_3\right] + \left[-C_4\Omega^2 + 2\eta\Omega C_3 + \omega^2 C_4\right]\text{Cos}\,\Omega t = \frac{P}{M}\text{Sin}\,\Omega t$$

$$\text{Sin}\,\Omega t\left[-C_3\Omega^2 - 2\eta\Omega C_4 + \omega^2 C_3 - \frac{P}{M}\right] + \text{Cos}\,\Omega t\left[-C_4\Omega^2 + 2\eta\Omega C_3 + \omega^2 C_4\right] = 0$$

To have a solution it is necessary that the terms in brackets are equal to zero from where:

$$\begin{cases} -C_3\Omega^2 - 2\eta\Omega C_4 + \omega^2 C_3 - \dfrac{P}{M} = 0 \\ -C_4\Omega^2 + 2\eta\Omega C_3 + \omega^2 C_4 = 0 \end{cases}$$

The solution of this equation system gives us:

$$C_3 = \frac{(\omega^2 - \Omega^2)\dfrac{P}{M}}{(\omega^2-\Omega^2)^2 + 4\eta^2\Omega^2} \text{ and } C_4 = \frac{-2\eta\Omega\dfrac{P}{M}}{(\omega^2-\Omega^2)^2 + 4\eta^2\Omega^2}$$

$$C_3 = \frac{(\omega^2 - \Omega^2)\dfrac{P_0}{M}}{\omega^4\left(1-\dfrac{\Omega^2}{\omega^2}\right)^2 + 4\xi^2\omega^2\Omega^2} = \frac{\dfrac{(\omega^2 - \Omega^2)\dfrac{P_0}{M}}{\omega^4}}{\left(1-\dfrac{\Omega^2}{\omega^2}\right)^2 + \left[2\xi\dfrac{\Omega}{\omega}\right]^2}$$

$$= \frac{(1-\dfrac{\Omega^2}{\omega^2})\dfrac{P_0}{\omega^2 M}}{\left(1-\dfrac{\Omega^2}{\omega^2}\right)^2 + \left[2\xi\dfrac{\Omega}{\omega}\right]^2} = \frac{(1-\dfrac{\Omega^2}{\omega^2})\dfrac{P_0}{K}}{\left(1-\dfrac{\Omega^2}{\omega^2}\right)^2 + \left[2\xi\dfrac{\Omega}{\omega}\right]^2}$$

$$C_4 = \frac{-2\eta\Omega\dfrac{P_0}{M}}{(\omega^2-\Omega^2)^2 + 4\eta^2\Omega^2} = \frac{-2\xi\omega\Omega\dfrac{P_0}{M}}{\omega^4\left(1-\dfrac{\Omega^2}{\omega^2}\right)^2 + 4\xi^2\omega^2\Omega^2}$$

$$= \frac{-2\xi\omega\Omega\dfrac{P_0}{\omega^4 M}}{\left(1-\dfrac{\Omega^2}{\omega^2}\right)^2 + 4\xi^2\dfrac{\Omega^2}{\omega^2}} = \frac{-2\xi\dfrac{\Omega}{\omega}\dfrac{P_0}{K}}{\left(1-\dfrac{\Omega^2}{\omega^2}\right)^2 + \left(2\xi\dfrac{\Omega}{\omega}\right)^2}$$

The general solution is:

$$U(t) = \underbrace{e^{-\eta t}\left[C_1 \operatorname{Sin}\omega_d t + C_2 \operatorname{Cos}\omega_d t\right]}_{\text{free vibration that will disappear with time (transient response)}} \underbrace{+ C_3 \operatorname{Sin}\Omega t + C_4 \operatorname{Cos}\Omega t}_{\text{stationary response}}$$

Determination of the constants C_1 and C_2 through the initial conditions.

U(t)= 0 at t=0 this implies:

$$C_2 = U(0) - C_4$$

$$C_2 = U(0) + \frac{P_0}{K}\frac{2\xi\frac{\Omega}{\omega}}{\left(1-\frac{\Omega^2}{\omega^2}\right)^2+\left(2\xi\frac{\Omega}{\omega}\right)^2}$$

$\dot{U}(t) = 0$ at t=0 this implies

$$\dot{U}(t) = -\eta e^{-\eta t}\left[C_1 \operatorname{Sin}\omega_d t + C_2 \operatorname{Cos}\omega_d t\right]$$
$$+e^{-\eta t}\left[C_1\omega_d \cos\omega_d t - C_2\omega_d \operatorname{Sin}\omega_d t\right] + C_3\Omega\cos\Omega t - C_4\Omega\sin\Omega t$$

$$C_1 = \frac{\dot{U}(0) + \eta C_2 - C_3\Omega}{\omega_d}$$

$$U(t) = \underbrace{e^{-\xi\omega t}(C_1\cos\omega_d t + C_2\sin\omega_d t)}_{\text{free vibration that will disappear with time (transient response)}}$$

$$\underbrace{+\frac{P_0}{K}\frac{(1-\frac{\Omega^2}{\omega^2})}{\left(1-\frac{\Omega^2}{\omega^2}\right)^2+\left[2\xi\frac{\Omega}{\omega}\right]^2}\sin\Omega t + \frac{P_0}{K}\frac{-2\xi\frac{\Omega}{\omega}}{\left(1-\frac{\Omega^2}{\omega^2}\right)^2+\left(2\xi\frac{\Omega}{\omega}\right)^2}\cos\Omega t}_{\text{stationary response}} \quad (14)$$

The response U (t) contains two distinct components of vibration: forced vibration or stationary vibration and transient vibration.

U_p (t) is considered as a stationary solution.

$$U_p(t) = U_{st}\frac{1}{\sqrt{\left(1-(\beta)^2\right)^2+(2\xi\beta)^2}}\frac{(1-\beta^2)\sin\Omega t - 2\xi\beta\cos\Omega t}{\sqrt{\left(1-(\beta)^2\right)^2+(2\xi\beta)^2}} \quad (15)$$

with

$$\beta = \frac{\Omega}{\omega}, U_{st} = \frac{P_0}{K} \text{ and } D = \frac{1}{\sqrt{\left(1-(\beta)^2\right)^2+(2\xi\beta)^2}}$$

$$U_p(t) = U_{st} D \frac{(1-\beta^2)\,\mathrm{Cos}\,\Omega t - 2\xi\beta\,\mathrm{Sin}\,\Omega t}{\sqrt{\left(1-\beta^2\right)^2 + \left(2\xi\beta\right)^2}}$$

$$\cos\theta = \frac{1-\beta^2}{\sqrt{\left(1-(\beta)^2\right)^2 + \left(2\xi\beta\right)^2}} \quad \sin\theta = \frac{2\xi\beta}{\sqrt{\left(1-(\beta)^2\right)^2 + \left(2\xi\beta\right)^2}} \quad \theta = \tan^{-1}\left[\frac{2\xi\beta}{1-\beta^2}\right] \tag{16}$$

From where

$$U_p(t) = U_{st}\ D\ \left[\cos\theta \sin\Omega t - \sin\theta\cos\Omega t\right] = U_{st}\ D\sin(\Omega t - \theta)$$

$$U_p(t) = U_{st}\ D\sin(\Omega t - \theta) \tag{17}$$

We assume that U_{st} $D=U_0$. From where

$$\frac{U_{st}}{U_0} = D \tag{18}$$

With θ is the phase angle and D is the dynamic amplification factor

Figure 11. Variation of dynamic amplification factor as a function of damping and frequency

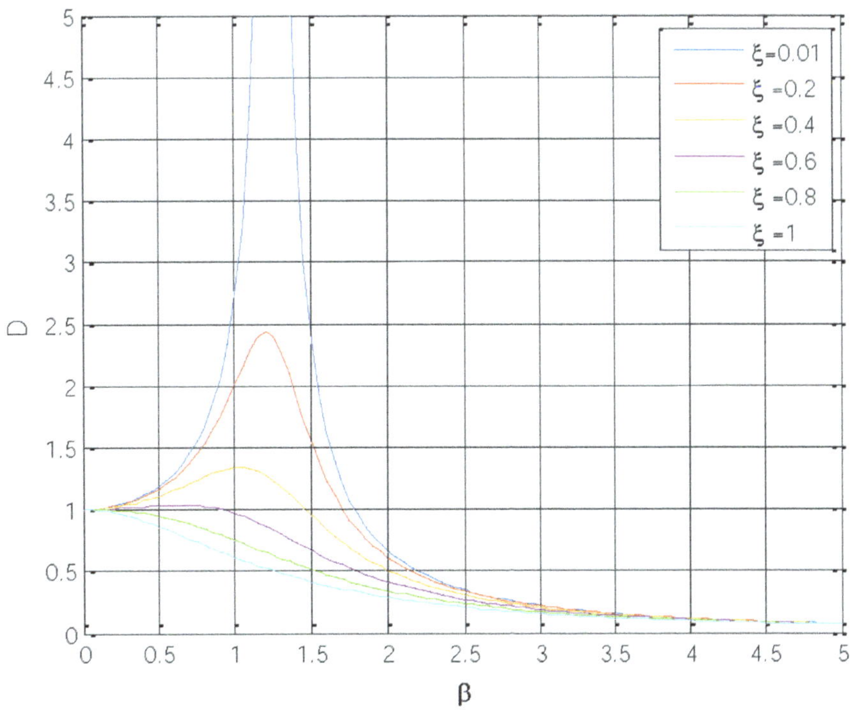

For high damping values, the dynamic displacement is always less than the static displacement ($D < 1$); and whatever the value of β.

Figure 12. Variation of phase angle with damping and frequency (Clough & Penzien, 1995)

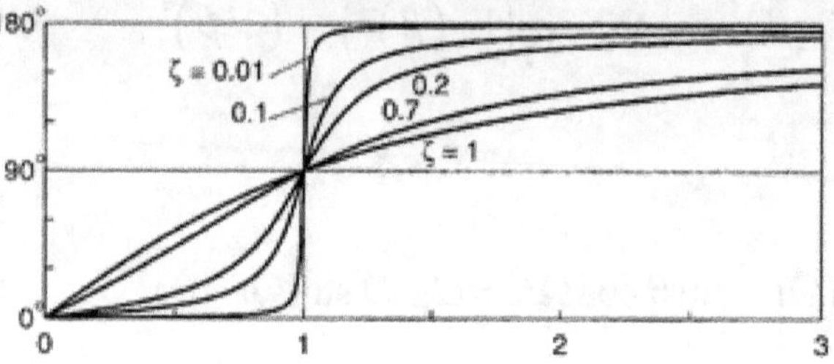

SOME INTERESTING CASES BETWEEN THE PARTICULAR SOLUTION AND THE STATIC SOLUTION

Very low Solicitation Frequency Compared to the own frequency with an a Phase Angle that Tends to Zero $\beta \ll 1$ $\theta \to 0$

If

$$\beta \prec\prec 1 \quad D \approx 1 \quad \& \quad \theta \to 0$$

In this case

$$U_p(t) = U_{st} D \sin(\Omega t - \theta) \approx U_{st} \sin(\Omega t)$$

Figure 13. Solution controlled by stiffness

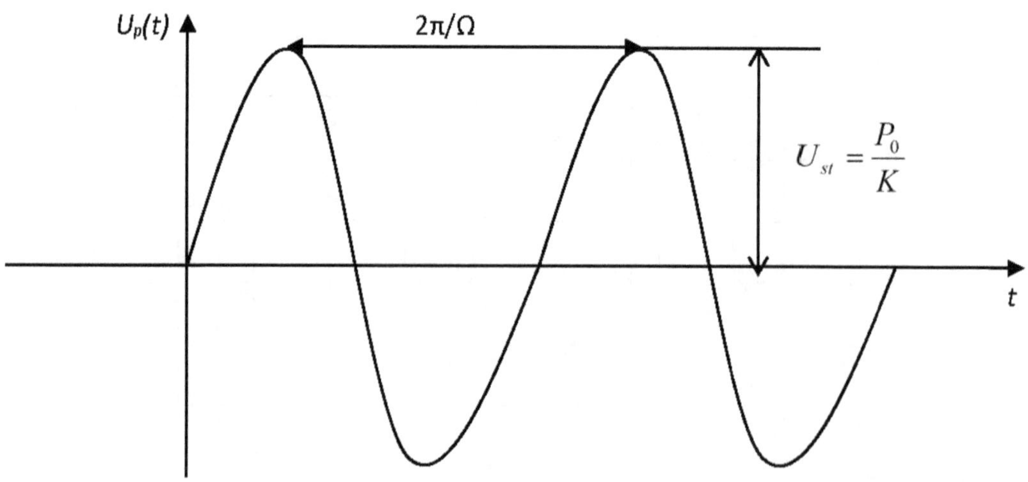

Note that the amplitude depends on rigidity. This case is controlled by rigidity. In other words, in this case, ($\beta \prec\prec 1 \quad D \approx 1 \quad \& \quad \theta \to 0$), donc Ω/ω→0 0 implies that ω>> (ω is very big). Therefore, the period of the system is small, but the force period is very large. In this case the system is rigid and $D \approx 1$, the force of inertia $M\ddot{U} = 0$ (there is no motion), as long as θ = 0; the excitatory force and the structure response have the same phase.

Very High Solicitation Frequency with Respect to the own Pulsation with a Phase Angle Which Tends Towards 180⁰

If

$$\beta \succ\succ 1 \quad D \to 0 \quad \& \quad \theta \to 180^0$$

This is the case of out of phase vibration:

$$U_p(t) = U_{st} \frac{1}{\sqrt{\beta^4 + \beta^2}} \sin(\Omega t - \pi)$$

$$U_p(t) = U_{st} \frac{1}{\sqrt{\beta^4 (1 + \frac{1}{\beta^2})}} \sin(\Omega t - \pi) = U_{st} \frac{1}{\beta^2} \sin(\Omega t - \pi)$$

$$U_p(t) = \frac{P_0}{K\,\beta^2} \sin(\Omega t - \pi) = \frac{P_0}{K \frac{\Omega^2}{\omega^2}} \sin(\Omega t - \pi)$$

$$U_p(t) = \frac{P_0}{M\,\Omega^2} \sin(\Omega t - \pi) \tag{19}$$

Note that the amplitude depends on the mass, this case is controlled by mass.

In other words, in this case ($\beta \succ\succ 1 \quad D \to 0 \quad \& \quad \theta \to 180^0$) therefore Ω / ω → ∞ implies ω → 0, hence the period of the system that tends to ∞, and the period of the exciter force tends to 0.

In this case the system is flexible; the elastic force (Fe = KU → 0) is negligible because there is no force that allows the system to return to equilibrium.

For this same case Ω / ω >> 1 so Ω / ω → ∞ implies that Tanθ = ∞ / ∞ is the case of indeterminacy; to remove this indetermination we derive the numerator and the denominator with respect to (Ω / ω).

$$\underbrace{\lim}_{\frac{\Omega}{\omega} \to n} (Tg\theta) = \frac{2\xi}{-2\frac{\Omega}{\omega}} = \frac{\xi}{-\frac{\Omega}{\omega}}$$

When $\Omega/\omega \rightarrow \infty$ $Tg \approx \theta^{-}$ (because of sige - Ω/ω)

Therefore $Tg\theta \approx 0$ implies that $\theta=\pi$, hence the following conclusion: the system response and the exciter force are in out of phase.

Figure 14. Solution controlled by mass

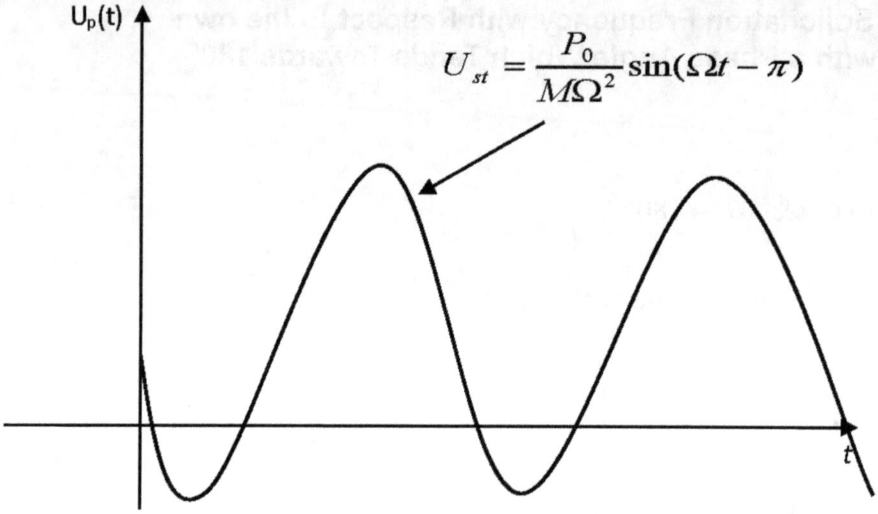

SOLICITATION FREQUENCY EQUAL TO THE OWN PULSATION (RESONANCE)

$\Omega/\omega \rightarrow 1$ implies that $D=1/(2\xi)$

In this case, we note that the damping ξ has an influence on the response of the structure.

Figure 15. Response of a damped system in resonance case (Chopra 2001)

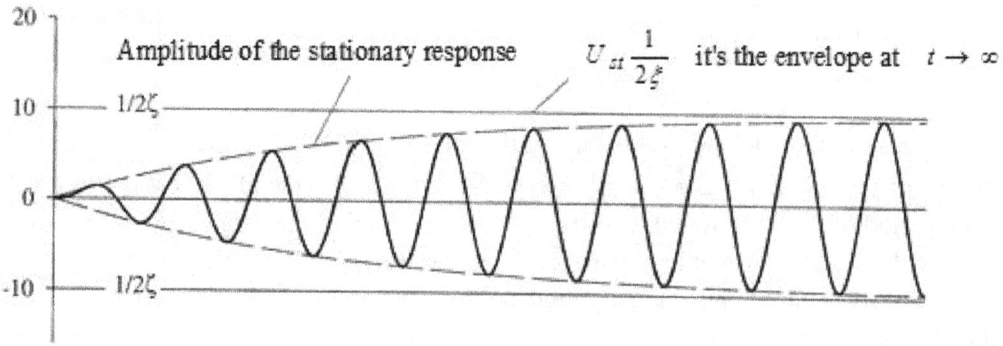

HARMONIC MOTION AT THE BASE (EXAMPLE OF AN EARTHQUAKE)

We can have an excitatory force, but that is applied indirectly ; this is the case of the displacement of the soil. Let Ug = d SinΩt (didactic case) be a harmonic ground displacement. Instead of applying a force, the ground will undergo a supposedly harmonic motion, see Figure 16.

Figure 16. Model of single degree of freedom damped system under the effect of the ground displacement

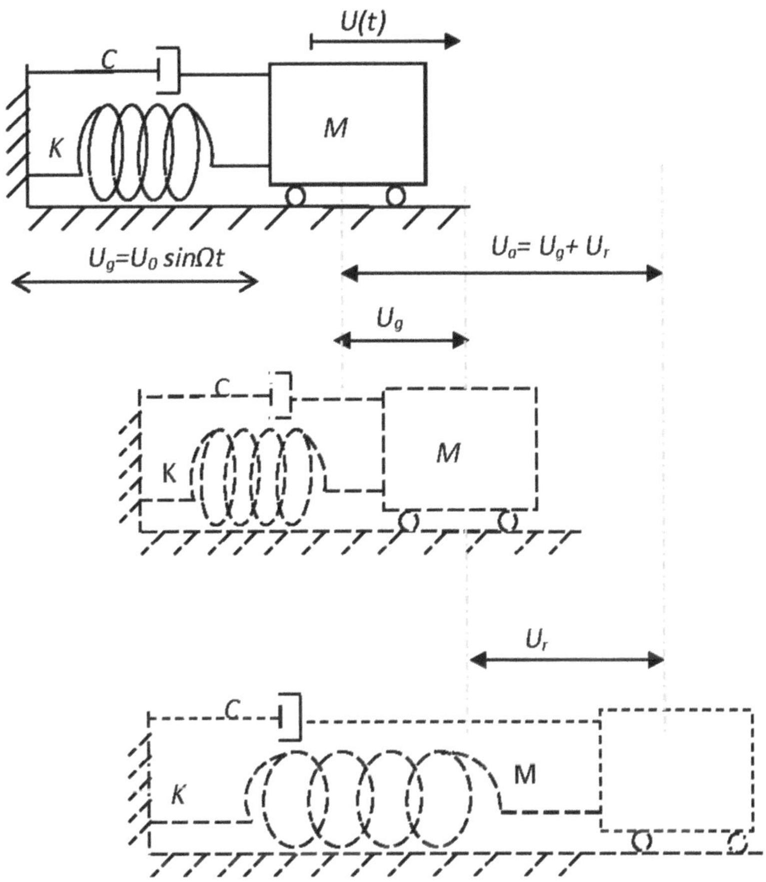

The absolute displacement of the mass is the total displacement and it has like expression:

U_a=U_g+U_r from where U_r= U-U_g

With U_a: Total displacement (absolute)

U_g: Ground displacement

U_r: Relative displacement of the mass with respect to the ground displacement.

The balance of forces gives:

$$M\ddot{U}_a + CU_r + KU_r = 0$$

$$M\ddot{U}_a + C(\dot{U}_a - \dot{U}_g) + K(U_a - U_g) = 0$$

$$M\ddot{U}_a + C\dot{U}_a + KU_a = C\Omega U_0 \cos\Omega t + KU_0 \sin\Omega t \tag{20}$$

Knowing that

$$a\ \cos(x) + b\sin(x) = \sqrt{a^2+b^2}\ \sin(x+\phi)$$

with $tag\phi = \frac{a}{b}$

Which is a trigonometric formula; which allows us to reduce the linear combination of a sinus and a cosinus to a sinus; we can write the equation of motion as function of absolute (total) displacement

$$M\ddot{U}_a + C\dot{U}_a + KU_a = U_0\sqrt{K^2+(C\Omega)^2}\ \sin(\Omega t + \bar{\theta})$$

With $\tan\bar{\theta} = \frac{C\Omega}{K} = 2\xi\beta$

$$M\ddot{U} + C\dot{U} + KU = P_0 \sin(\Omega t + \bar{\theta})$$

With $P_0 = U_0\sqrt{K^2+(C\Omega)^2}$

Therefore, we can write the particular solution as follows:

$$U_p = U_{st}\ D\ \sin(\Omega t + \bar{\theta} + \theta)$$

$$U_p = \frac{P_0}{K}\frac{1}{\sqrt{(1-\beta^2)^2+(2\xi\beta)^2}}\ \sin(\Omega t + \bar{\theta} + \theta) \tag{21}$$

$$U_p = \frac{U_0}{\sqrt{K^2}}\frac{\sqrt{K^2+(C\Omega)^2}}{\sqrt{(1-\beta^2)^2+(2\xi\beta)^2}}\ \sin(\Omega t + \bar{\theta} + \theta)$$

$$U_p = \frac{U_0\sqrt{1+(2\xi\beta)^2}}{\sqrt{(1-\beta^2)^2+(2\xi\beta)^2}}\ \sin(\Omega t + \bar{\theta} + \theta)$$

$$U_p = U_0 T_r\ \sin(\Omega t + \bar{\theta} + \theta) \tag{22}$$

With

$$T_r = Transmi\tan ce = \frac{\sqrt{1+(2\xi\beta)^2}}{\sqrt{(1-\beta^2)^2+(2\xi\beta)^2}} = D\sqrt{1+(2\xi\beta)^2}$$

The transmittance, denoted by T_r, which physically represents the fraction of the applied force that is transmitted to the support through the system, it is very useful for the isolation of vibratory systems.

Figure 17. Transmittance of harmonic excitation

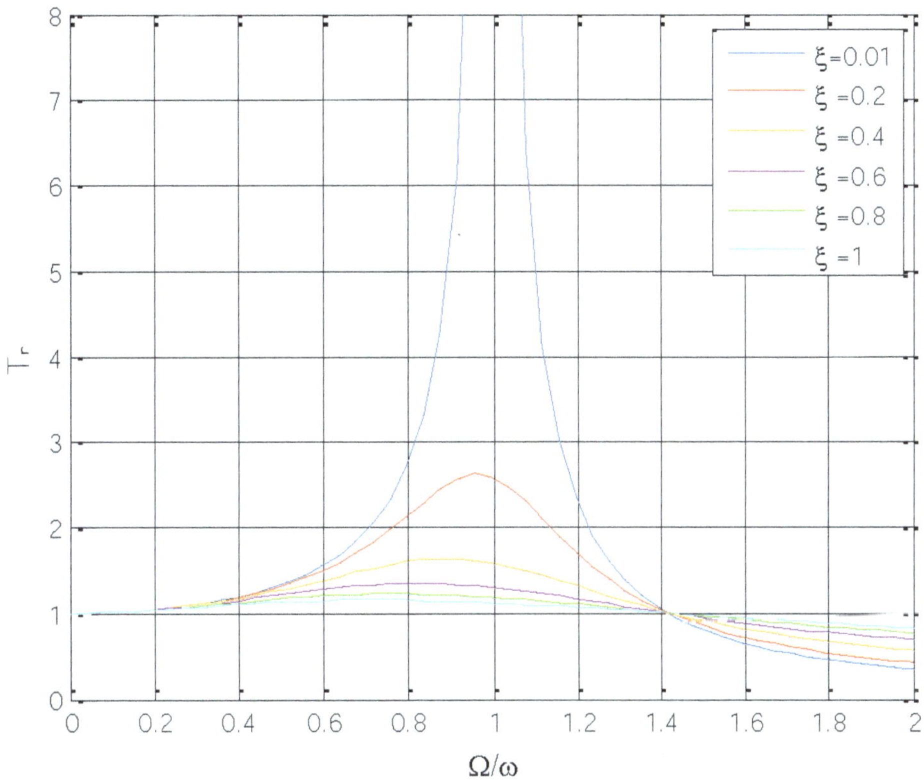

The ordinate axis is put in a logarithmic scale, we note that for only one frequency ratio, $\frac{\Omega}{\omega} \prec \sqrt{2}$, damping reduces the transmitted force.

Table 1. Summary of solicitations and structure responses

Solicitation	Structure	Response
Free Vibration	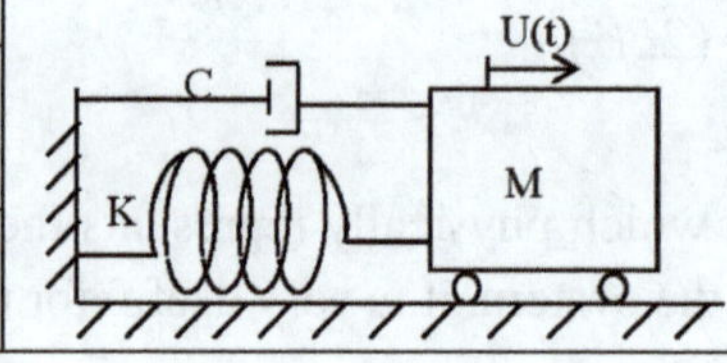	Transient
Harmonic load		Transient + Stationary
Periodic load		Stationary

PERIODIC LOAD REPRESENTATION IN FOURIER SERIES

Each periodic function can be written as a Fourier series

Figure 18. Damped forced system under periodic loading

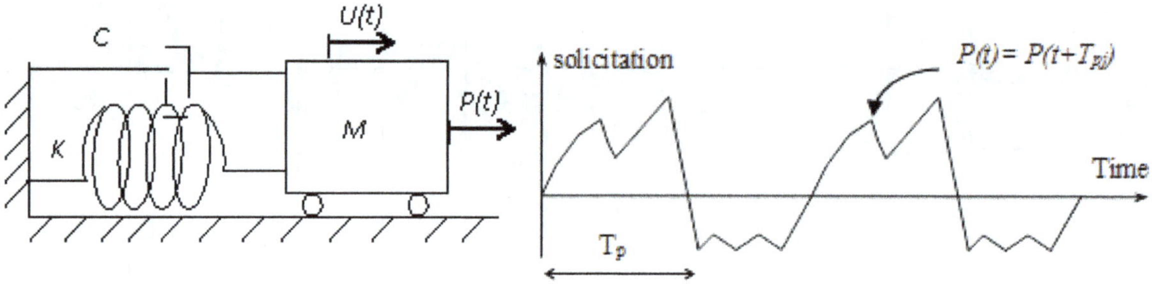

The mathematical expression of the Fourier series is as follows:

$$p(t) = a_0 + \sum_{j=1}^{\infty} a_j \cos\left(j\omega_0 t\right) + \sum_{j=1}^{\infty} b_j \sin\left(j\omega_0 t\right) \qquad avec\ \omega_0 = \frac{2\pi}{T_0} \tag{23}$$

$$a_0 = \frac{1}{T_p}\int_0^{T_p} p(t)dt \quad ; \quad a_j = \frac{2}{T_p}\int_0^{T_p} p(t)\cos n\Omega t\ dt \quad ; \quad b_n = \frac{2}{T_p}\int_0^{T_p} p(t)\sin n\Omega t\ dt$$

Take as an example the following periodic load

$$p(t) = \frac{p_0}{T_p}t$$

Figure 19. Periodic triangular loading case

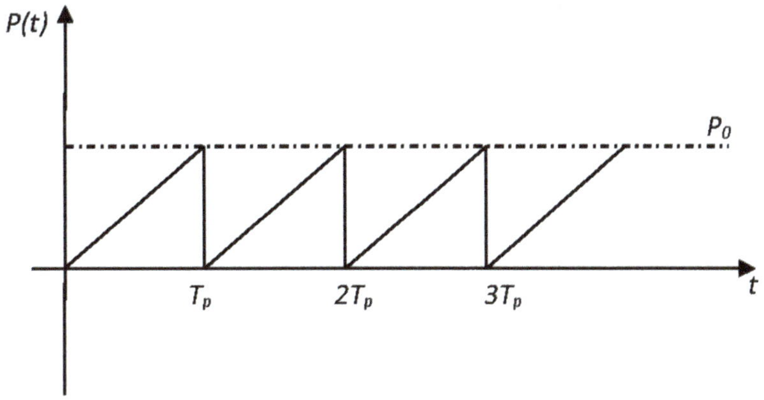

Figure 20. Representation of the periodic triangular load in Fourier series (Jerome 2012)

$$p(t) = \frac{p_0}{2} + \sum_{j=1}^{\infty} -\frac{p_0}{j\pi}\sin(\omega_0 jt) \tag{24}$$

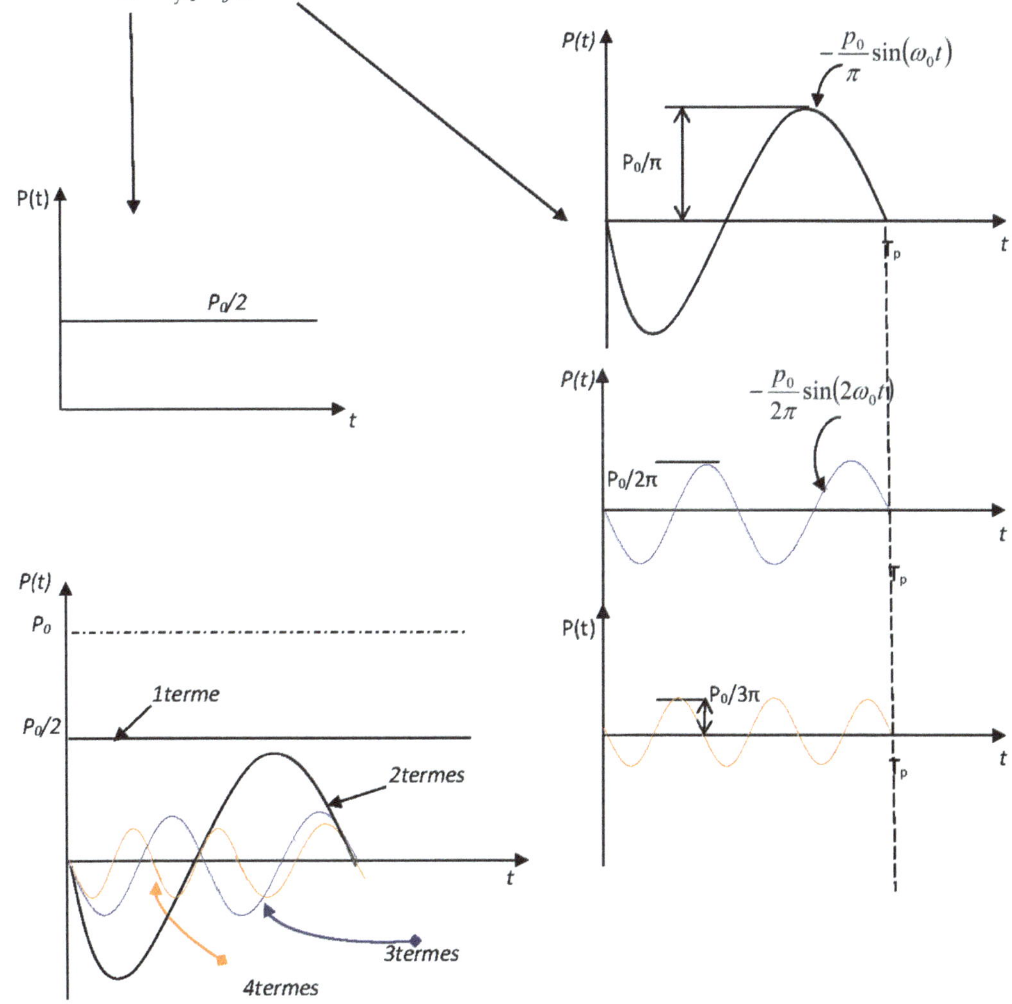

$$a_0 = \frac{1}{T_p}\int_0^{T_p} p(t)dt = \frac{P_0}{2} \quad a_j = \frac{2}{T_p}\int_0^{T_p} p(t)\cos n\Omega t\ dt = -\frac{p_0}{j\pi}$$

The load in terms of Fourier series will have the following expression:

$$p(t) = \frac{p_0}{2} + \sum_{j=1}^{\infty} -\frac{p_0}{j\pi}\sin\left(\omega_0 jt\right) \tag{24}$$

Response to Periodic Load

$$M\ddot{U}_r(t) + C\dot{U}_r(t) + KU_r(t) = a_0 + \sum_{j=1}^{\infty} a_j \cos\left(j\omega_0 t\right) + \sum_{j=1}^{\infty} b_j \sin\left(j\omega_0 t\right) \tag{25}$$

Remarks

If the periodic charge exists for a very long time, we will not consider the transient solution.

When we have a linear system, we can consider the superposition. It means if we consider the response R1 under the load 1, and the response R_2 under the load 2, at that moment, under the effect of the load 1 + the load 2, the response will be the response R_1 + response R_2. Let us go back to our case where there are three loads, which are:

$$p(t) = \underbrace{a_0}_{\text{first load a constant}} + \underbrace{\sum_{j=1}^{\infty} a_j \cos\left(j\omega_0 t\right)}_{\text{Second load}} + \underbrace{\sum_{j=1}^{\infty} b_j \sin\left(j\omega_0 t\right)}_{\textit{Third load}}$$

In this case, following the supposition rule mentioned earlier as a remark, we can write:

$$U(t) = U_p(t) = U_{p_0} + \sum_{j=1}^{\infty} U'_J + \sum_{j=1}^{\infty} U''_J \tag{26}$$

Load 1: a_0 (a constant, static problem) $U_p = \frac{a_0}{K}$

Load 2: (a_j cos ($j\omega_0 t$))

$$M\ddot{U}(t) + C\dot{U}(t) + KU(t) = a_j \cos\left(j\omega_0 t\right)$$

With $P_0 = a_j$ *et* $\Omega = j\omega 0$

$$U'_j = \frac{a_j}{K} \cdot \frac{-2\xi j\beta \sin j\omega_0 t + \left(1-(j\beta)^2 \cos j\omega_0 t\right)}{\left[1-(j\beta)^2\right]^2 + \left[2\xi j\beta\right]^2} \tag{27}$$

Load 3: (a_j sin($j\omega_0$t))

$$M\ddot{U}(t) + C\dot{U}(t) + KU(t) = a_j \cos\left(j\omega_0 t\right)$$

With $P_0 = a_j$ *et* $\Omega = j\omega 0$

$$U''_j = \frac{b_j}{K} \cdot \frac{\left(1-(j\beta)^2 \sin j\omega_0 t\right) - 2\xi j\beta \cos j\omega_0 t}{\left[1-(j\beta)^2\right]^2 + \left[2\xi j\beta\right]^2} \tag{28}$$

The total response of the system is:

$$U_p(t) = \frac{a_0}{K} + \sum_{j=1}^{\infty} U'_j + \sum_{j=1}^{\infty} U''_j \tag{29}$$

Figure 21. Representation of the impulsive load explosion case

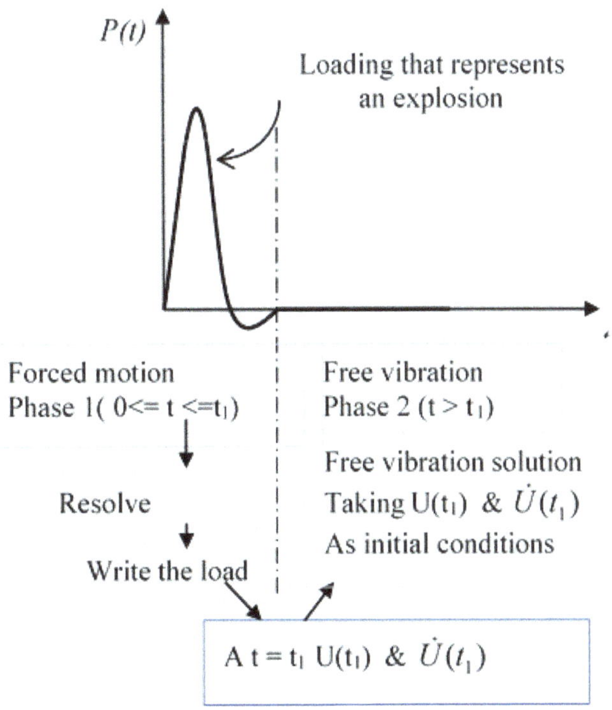

IMPULSIVE LOAD

The impulsive charge appears for a very short duration, for example an explosion.

DUHAMEL'S INTEGRAL

In forced vibrations, we have studied the case of harmonic forces, but in reality, this is not the case, because the excitatory force is often arbitrary.

Figure 22. Arbitrary load

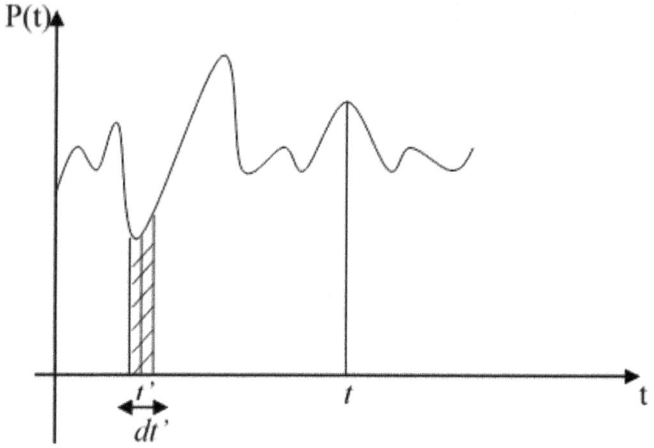

We suppose we are at time t' and we want to determine the response at time t, to do this it is necessary to first determine the force at time t'

$$M\ddot{U}_r(t) + C\dot{U}_r(t) + KU_r(t) = P(t')$$

We call the force P(t') impulsive force, which will give the mass a certain velocity

$P\,dt' = M\,d\dot{U}$ ($d\dot{U}$ is a velocity increment).

Note: we look for the response at time t due to a force P (t ') applied to the time t'. The system response is of the form:

$$U_r(t) = e^{-\xi\omega t}\left[U_0 \operatorname{Cos}\omega_d t + \frac{\dot{U}_0 + \xi\omega U_0}{\omega_d}\operatorname{Sin}\omega_d t\right]$$

At time t'=0 we can equalize $d\dot{U}_0$ to $\dot{U}_0$ with U_0=0 so the equation above becomes as follows:

$$dU(t) = e^{-\xi\omega(t-t')}\frac{P}{M\omega_d}\operatorname{Sin}\omega_d(t-t')\,dt' \tag{32}$$

We can do the same work for each impulse, and have the total response by integrating.

$$\int_0^{U(t)} dU(t) = \frac{P}{M\omega_d}\int_0^t e^{-\xi\omega(t-t')}\operatorname{Sin}\omega_d(t-t')\,dt' \tag{33}$$

Equation (33) is called the Duhamel integral.
For the particular case where $\xi = 0$ (undamped system) the equation is written

$$U_r(t) = \frac{1}{\omega}\int_0^t \frac{P}{M}\operatorname{Sin}\omega(t-t')dt' \tag{34}$$

Note: $U(t)=U_r(t)$ is the relative displacement

RECTANGULAR LOAD EXAMPLE

$$P(t) = \begin{cases} P_0 & 0 \le t \le t_d \\ 0 & t > t_d \end{cases}$$

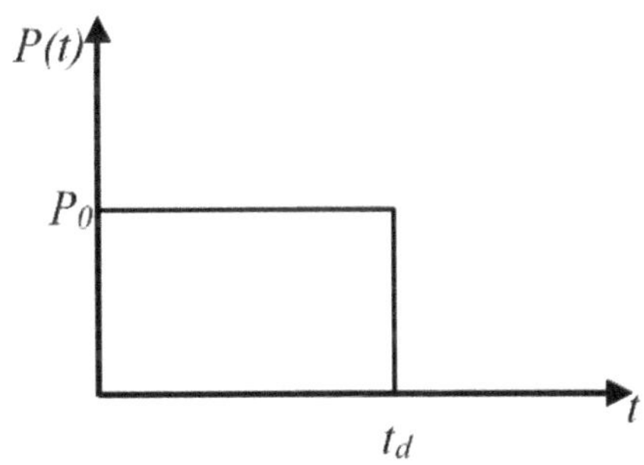

First phase response to forced vibration ($0 \le t \le t_d$)

$$M\ddot{U} + KU = P_0$$

$$U_h(t) = A\cos\omega t + B\sin\omega t$$

$$U_p = \frac{P_0}{K}$$

$$U(t) = U_h(t) + U_p(t) = A\cos\omega t + B\sin\omega t + \frac{P_0}{K}$$

From the initial conditions

$$U(0) = A + \frac{P_0}{K} = 0 \quad \Rightarrow A = -\frac{P_0}{K}$$

$$\dot{U}(0) = 0 \Rightarrow B = 0$$

$$U(t) = \frac{P_0}{K} - \frac{P_0}{K}\cos\omega t$$

$$U(t) = \frac{P_0}{K}\left(1 - \cos\omega t\right)$$

$$\frac{U(t)}{U_{st}} = 1 - \cos\omega t = 1 - \cos\frac{2\pi}{T}t \; t \le t_d \tag{30}$$

Figure 23. Response of the first phase (Chopra 2001)

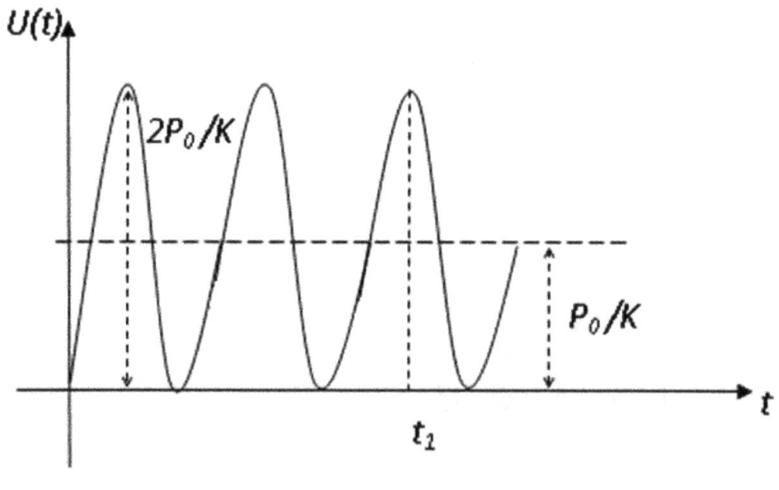

At=t=t_d $U(t_d) = \frac{P_0}{K}\left(1 - \cos\omega t_d\right) \quad \dot{U}(t_d) = \frac{P_0}{K}\omega\sin\omega t_d$

Second phase response to forced vibration ($t>t_d$)

$$M\ddot{U} + KU = 0$$

Let's Define a new time variable $t' = t - t_d$

$U(t'=0) = U(t_d)$ and $\dot{U}(t'=0) = \dot{U}(t_d)$

$$U(t) = \frac{P_0}{K}\left(1-\cos\omega t_d\right)\cos\omega(t-t_d) + \frac{P_0}{K}\sin\omega t_d \sin\omega t'$$

$$\frac{U(t)}{U_{st}} = \left(1-\cos\omega t_d\right)\cos\omega(t-t_d) + \sin\omega t_d \sin\omega(t-t_d)\, t \geq t_d \quad (31)$$

Which can be simplified using trigonometric transformations

$$\frac{U(t)}{U_{st}} = \cos\omega(t-t_d) - \cos\omega t \;\; t \geq t_d$$

By posing $\omega = \frac{2\pi}{T}$, using trigonometric transformations we will be able to write the equation (31) as follows:

$$\frac{U(t)}{U_{st}} = \left(2\sin\frac{\pi\, t_d}{T}\right)\sin\left[\left(2\pi\left(\frac{1}{T} - \frac{1}{2}\frac{t_d}{T}\right)\right)\right] t \geq t_d \quad (32)$$

APPLICATIONS

Example 1 (Commend 2001)

We have a horizontal oscillator with damping $\xi = 0.05$, subjected to a sinusoidal acceleration

$$\overline{\ddot{u}_s} = A\sin\left(\overline{\omega}t - \varphi\right)$$

with A = 1 and $\varphi = 0$. Knowing that m = 50 kg, k = 200 N/m

1. Determine the numerical response for Ω = 0.5, 1.0, 1.5, 2.0, 2.5, 3.0. As well as maximum displacement max u_{max}
2. Draw u_{max} according to Ω/ω.
3. Complete the figure of response spectrum of load, which have a signal type as follows:

Figure 24. Dynamic response of an undamped system under a rectangular impulsive force the static solution is shown by broken lines (Chopra 2001)

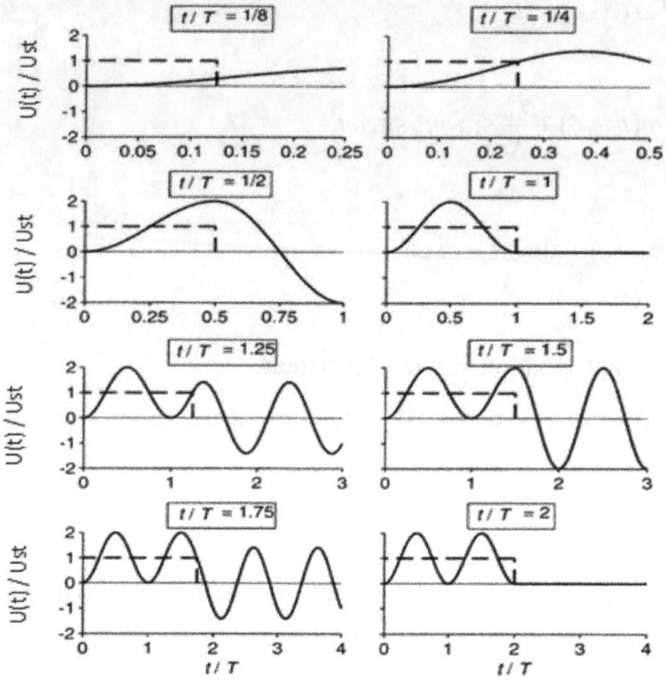

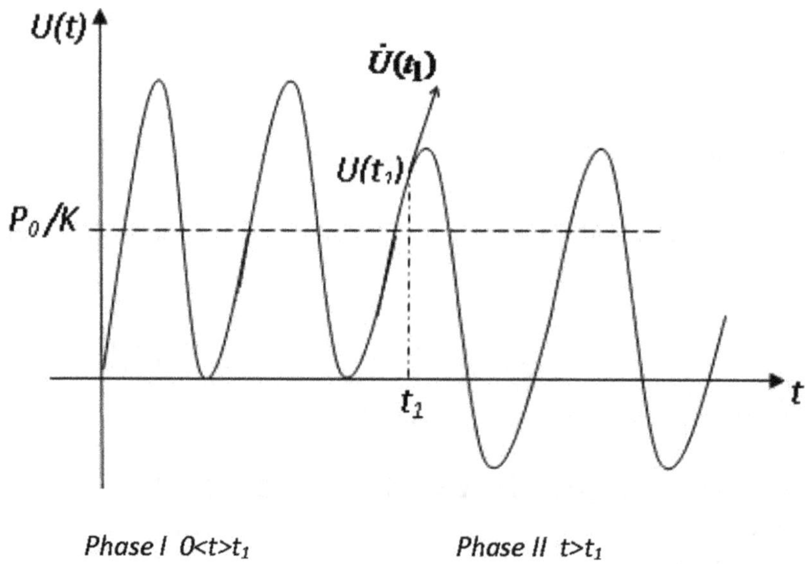

Phase I $0<t>t_1$ *Phase II* $t>t_1$

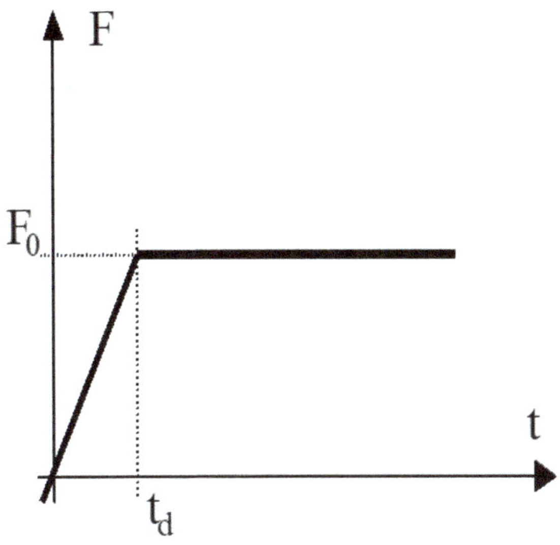

Solution

$$\ddot{u}_s(t) = A\sin(\Omega t - \varphi)$$

With A=1, Ω=0.5;1;1.5;2;2.5;3, φ=0
Motion equation:

$$m\ddot{u}_r + c\dot{u}_r + ku_r = -m\ddot{u}_s(t)$$

Numerical response with

m = 50 kg, k = 200 N/m et Ω=0.5:

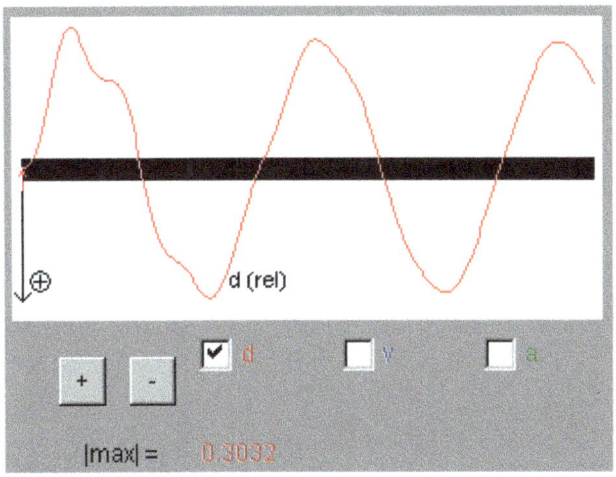

Representation of the maximum relative displacement $u_{r,\max}$ According to the ratio

$$\frac{\Omega}{\omega}(\text{with}\ \omega = \sqrt{\frac{k}{m}} = 2):$$

Box 1.

$\frac{\Omega}{\omega}$	0	0.25	0.5	1.0	1.5	2.0	2.5	3.0
$u_{rel,\max}$	0.25	0.3032	0.4044	2.4861	0.4031	0.1966	0.1474	0.1123

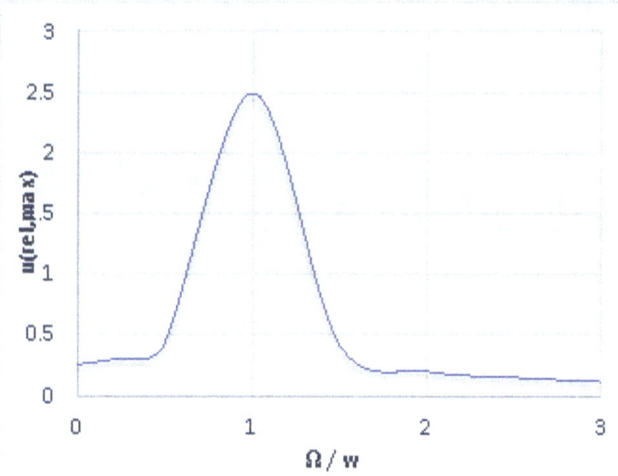

Observations:

1) The maximum response corresponds to $\frac{\Omega}{\omega} = 1$

2) for $\frac{\Omega}{\omega} = 0, u_{\max} = u_{stat} = \frac{mA}{k} = 0.25$

3) for $\frac{\Omega}{\omega} >>$, $u_{\max}$ tends to zero

Establishment a response spectrum of load, which have a signal type:

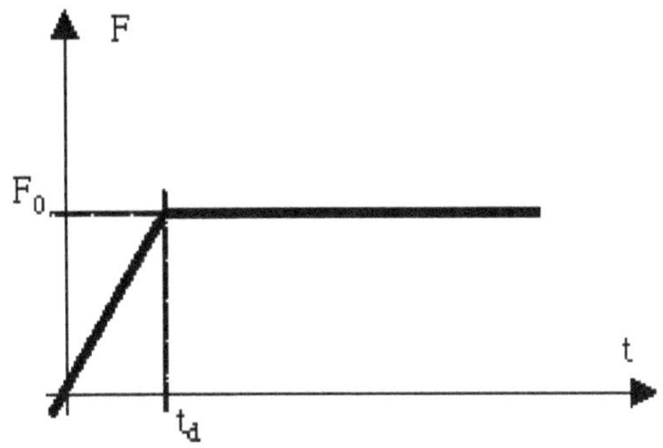

We want to represent the dynamic amplification factor (D.A.F) according to the ratio $\frac{t_d}{T}$, of an oscillator with:

$$T = \frac{2\pi}{\omega}, FAD = \frac{u_{\max}}{u_{stat}}, \text{k} = 1\text{N/m, m} = 1\text{kg, c} = 0 \text{ and } \text{F}_0 = 1\text{N,}$$

we obtain:

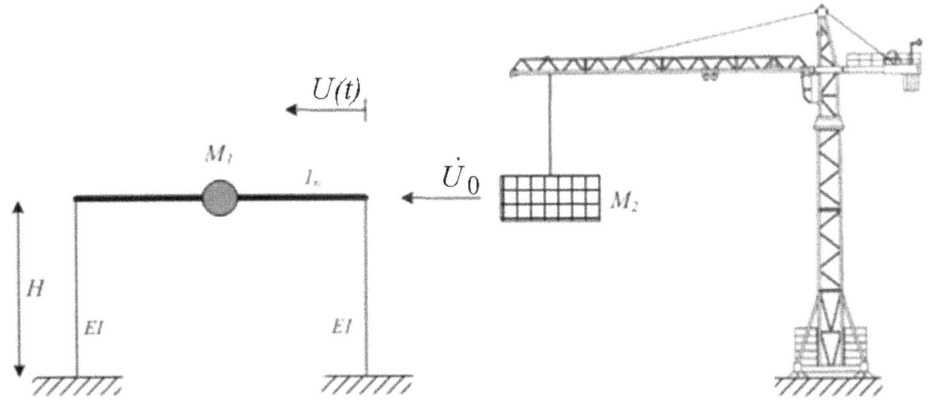

Example 2 (Smith 2010)

Due to handling error of a crane, a mass load M2 hits the top of beam column structure system under construction with a speed V, see figure below. The load remains in contact with the structure after impact. The oscillations generated are only in the x direction.

We will neglect the mass of the columns compared to that of the beam.

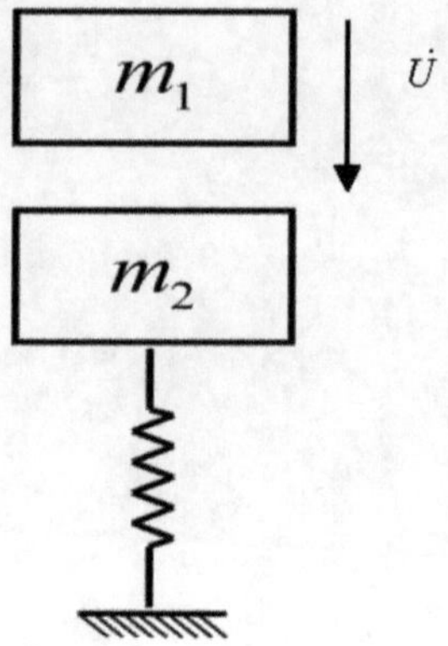

$\dot{U}_0$ = 15 m/s
E = 200 000 MPa
I = 300 10^6 mm^4
H = 4 m
M_1 = 2500 kg
M_2 = 5500 kg

Write the motion equation of the beam column system (response expression), considering a damping of ζ = 0,1.

Solution

K_{equ} the equivalent stiffness of the structure:

$$K_{equ} = 2.\frac{12EI}{H^3} = 22500\ kN\ /\ m$$

The proper frequency of the system after the shock:

$$\omega = \sqrt{\frac{K_{equ}}{M_1 + M_2}} = \sqrt{\frac{22500}{2.5+5.5}} = 53.03 ras / s$$

The response of the structure is:

$$U(t) = e^{-\xi\omega t}\left[A \mathrm{Cos}\,\omega_d t + B \mathrm{Sin}\,\omega_d t\right]$$

The velocity is obtained by deriving U(t) with respect to time,

$$\dot{U}(t) = \left(B\omega_d - A\xi\right)e^{-\xi\omega t}\cos\omega_d t - \left(B\xi - A\omega_d\right)e^{-\xi\omega t}\sin\omega_d t$$

Reminder: the linear momentun is preserved

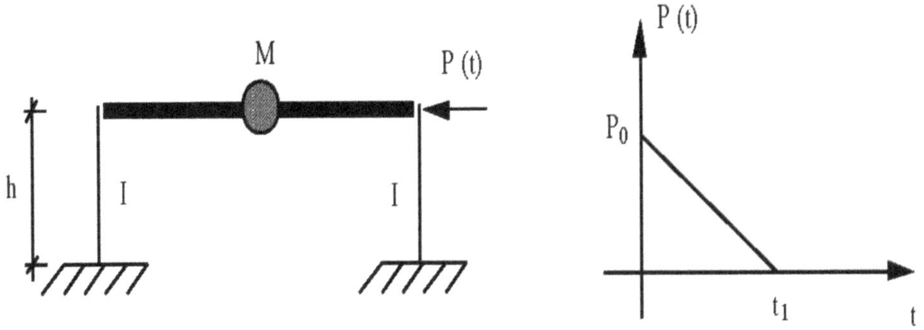

$$m_1\dot{U}_1 = (m_1 + m_2)\dot{U}_2$$

$$\dot{U}_2 = \dot{U}_0 = \frac{m_1}{(m_1 + m_2)}\dot{U}_1$$

The coefficients A and B are determined using the initial conditions:

$U(0)=0=A$

$$\dot{U}(0) = \frac{M_1}{M_2 + M_2}\dot{U}_0 = B\omega_d \Rightarrow B = \frac{M_1}{[M_2 + M_2]\omega_d}\dot{U}_0$$

$$\omega_d = \omega\sqrt{1-\xi^2} = 52.764 rad/s$$

B=0.1954cm

The equation of motion (damped oscillations) thus becomes:

$$U(t) = 0.1954\ e^{-5.303\,t}\ \text{Sin}\ (52.764\,t)$$

Example 3 (Smith 2010)

Determine analytically the displacement at the column head as a function of time when the structure is subjected to an explosive load.

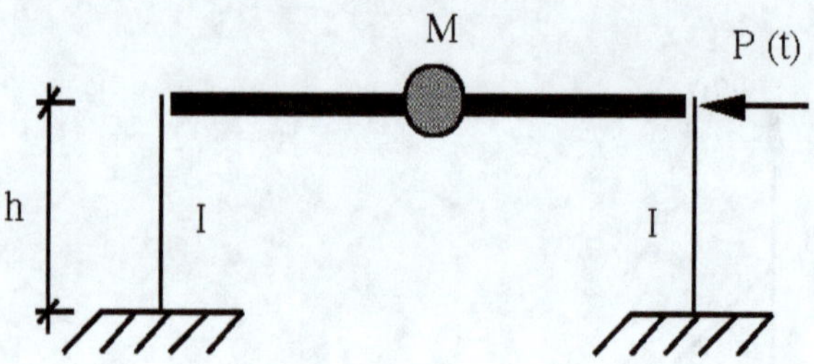

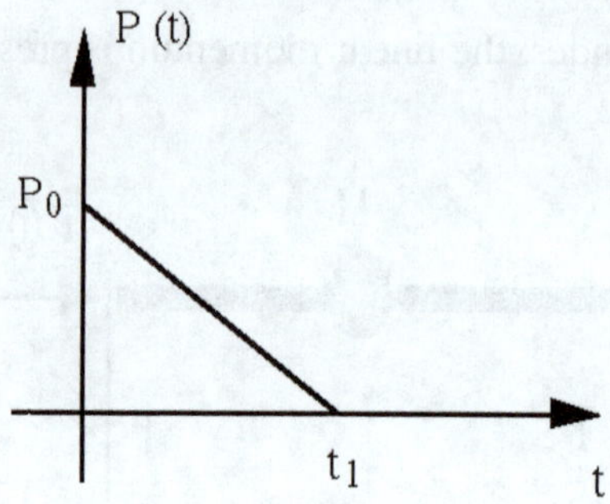

$$P(t)=\begin{cases}P_0\left(1-\dfrac{t}{t_1}\right) & \text{for } t\le t_1\\ 0 & \text{for } t>t_1\end{cases}$$

$P_0 = 100$ kN $I = 5\ 10^7$ mm^4 $E = 210\ 10^3$ N/mm^2

$M = 5000$ kg $h = 5$ m

Solution

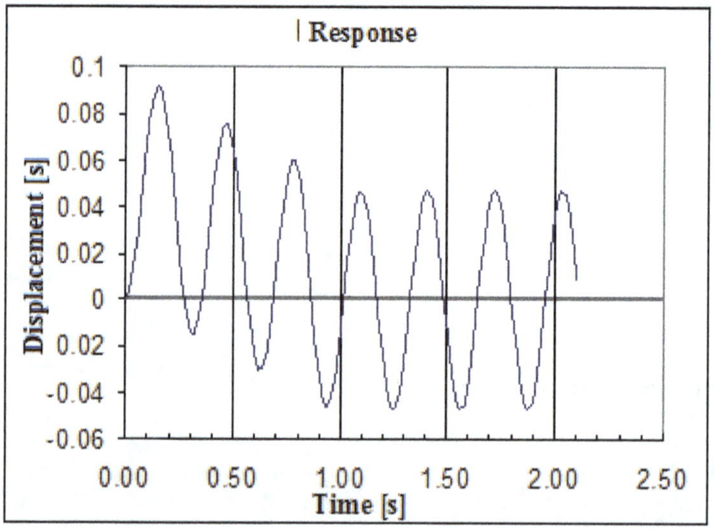

Stiffness :

$$K = 2\cdot\frac{12\cdot EI}{h^3} = 2.016\cdot 10^6\ \frac{N}{m}$$

$$\omega = \sqrt{\frac{K}{M}} \cong 20.08\ rad/s$$

The integral of Duhamel is written (no damping):

$$U(t) = \frac{1}{M\omega}\int_0^t P(t')\,\mathrm{Sin}\,\omega(t-t')dt'$$

With:

$$P(t') = P_0\left[1 - \frac{t'}{t_1}\right] \quad 0 \le t' \le t_1$$

$$P(t') = 0 \quad t' \ge t_1$$

First phase $0 \le t' \le t_1$:

$$U(t) = \frac{P_0}{M\omega}\int_0^t \left[1 - \frac{t'}{t_1}\right]\mathrm{Sin}\,\omega_d(t-t')dt'$$

$$U(t) = \frac{P_0}{M\omega}\int_0^t \mathrm{Sin}\,\omega(t-t')dt' - \frac{P_0}{M\omega t_1}\int_0^t t'\,\mathrm{Sin}\,\omega\,(t-t')dt'$$

Reminder: integration by part

$$\int u(t')v'(t')dt' = u(t')v(t') - \int u'(t')v(t')dt'$$

With

$$u(t') = t' \quad v'(t') = \sin(\omega(t-t'))$$
$$u'(t') = 1 \quad v(t') = \frac{\cos(\omega(t-t'))}{\omega}$$

$$U(t) = \frac{P_0}{M\omega^2}\left[\cos\omega(t-t')\right]_{t'=0}^{t'=1} - \frac{P_0}{M\omega^2 t_1}\left\{\left[t'\cos\omega(t-t')\right]_{t'=0}^{t'=1} - \int_0^t \cos\omega(t-t')dt'\right\}$$

$$U(t) = \frac{P_0}{M\omega^2}(1-\cos\omega t) - \frac{P_0}{M\omega^2 t_1}\left\{t + \left[\frac{1}{\omega}\sin\omega(t-t')\right]_{t'=0}^{t'=1}\right\}$$

$$U(t) = \frac{P_0}{M\omega^2}(1-\cos\omega t) - \frac{P_0}{M\omega^2}\left(\frac{t}{t_1} - \frac{1}{\omega t_1}\sin\omega t\right)$$

$$U(t) = \frac{P_0}{K}\left\{1 - \cos\omega t - \frac{t}{t_1} + \frac{1}{\omega t_1}\sin\omega t\right\}$$

$$\dot{U}(t) = \frac{P_0}{K}\left\{\omega \sin\omega t - \frac{1}{t_1} + \frac{1}{t_1}\cos\omega t\right\}$$

Second phase $t' \geq t_1$: The system is a harmonic oscillator whose initial conditions are:

$$U_0 = U(t_1) = \frac{P_0}{K}\left\{-\cos\omega t_1 + \frac{1}{\omega t_1}\sin\omega t_1\right\}$$

$$\dot{U}_0 = \dot{U}(t_1) = \frac{P_0}{K}\left\{\omega \sin\omega t_1 - \frac{1}{t_1} + \frac{1}{t_1}\cos\omega t_1\right\}$$

Reminder: Harmonic oscillator

$U(t) = A\sin(\omega t) + B\cos(\omega t)$

So we find:

$$U(\bar{t}) = +U_0\cos(\omega\bar{t}) + \frac{\dot{U}_0}{\omega}\sin(\omega\bar{t})$$

with $\bar{t} = t - t_1$

The static displacement

$$\Delta_{sta} = U_{sta} = \frac{P_0}{K} \approx 4.96cm$$

En $t = t_1$:

$x(t_1) \cong -0.287\Delta_{stat}$ $U(t_1) \approx -0.287\Delta_{sta}$

$\dot{U}(t_1) \approx 0.906 m/s^2$

Maximum displacement:

$\dot{U}(t) = 0 \Rightarrow t \approx 0.15 s$

$U_{\max} \approx 1.85\Delta_{sta} = 9.176cm$

REFERENCES

Chopra, A. K. (2001). *Dynamics of Structures: Theory and Applications to Earthquake Engineering* (2nd ed.). Prentice Hall.

Clough, R. W. & Penzien, J. (1995). *Dynamic of structures.* Compute and Structure, Inc.

Commend, S., & Zimmermann, Th. (2001). *LSCDYN 1DOF and nDOF: two applications for computer-assisted teaching of structural dynamics. LSC Internal Report.* Swiss Federal Institute of Technology.

Jerome, P. L. (2012). *Structure dynamics course note.* Civil and Environmental Engineering University of Michigan.

Mohammed, I. Y. (2013). *MEMS Linear and nonlinear Static and Dynamics.* Springer-verlag.

Smith, I. (2010). *Structural dynamics, Internal course.* Swiss Federal Institute of Technology.

KEY TERMS AND DEFINITIONS

Damped Forced Oscillator: Is an oscillator which is subjected to an external load in which its oscillation decreases with time.

Dynamic Influence Factor: Is the ratio of dynamic displacement to static displacement.

Impulsive Load: Is a charge of very short duration compared to the period of vibration of structures.

Periodic Load: Is a charge that repeat at regular time intervals.

Undamped Forced Oscillator: Oscillator, without any friction.

Section 2

Multi-Degree of Freedom System Study and Hydrodynamic Principle

Chapter 3
Multi-Degrees of Freedom System and Hydrodynamic Principle

ABSTRACT

A system with one degree of freedom is far from reality, because we do not take into account all the degrees of freedom. In order to be close to the reality, it is necessary to use a system with several degrees of freedom. Efforts in this chapter are concentrated to the study of multi-degrees of freedom system, whether for a free undamped and forced damped system, by detailing the modal superposition method as well as a coupled coordinates. We finish the chapter with hydrodynamic study using Hozner method as well as some applications.

INTRODUCTION

A system with one degree of freedom is far from reality, because we do not take into account all the degree of freedom. In more complicated structures, it is necessary to use a system with several degrees of freedom, to obtain a satisfactory dynamic model. A real system, generally comprises several masses, connected together by elements of the spring and damper type, in general, there is six degree of freedom for each node three displacements and three rotations. This chapter is devoted solely to the development of all equations concerning construction of an analytical as well as mathematical model of free, forced damped and undamped multi-degrees of freedom vibration linear system linear at the end the hydrodynamic study is detailed using Hozner method .

DOI: 10.4018/978-1-7998-4399-3.ch003

Copyright © 2021, IGI Global. Copying or distributing in print or electronic forms without written permission of IGI Global is prohibited.

Figure 1. Six space degree of freedom

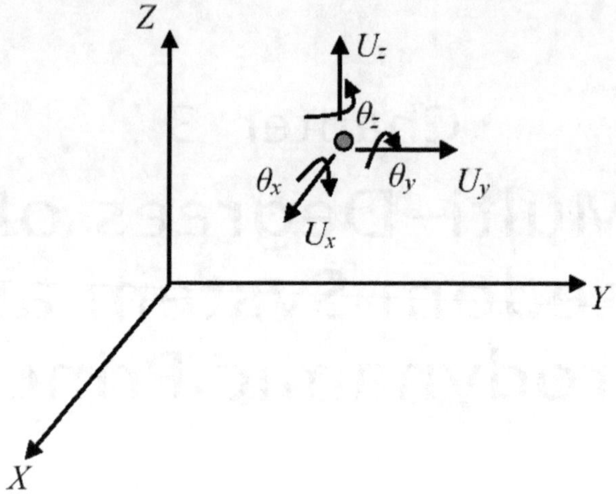

In the case of an undamped free vibration the equation of motion of a single degree of freedom system (SDOF) is: $M\ddot{U} + KU = 0$ is it supposed.

$U(t) = Ge^{st} = A \cos \omega t + \beta \sin \omega t$

We found

$$U(t) = \sqrt{U(0)^2 + \left(\frac{\dot{U}(0)}{\omega}\right)^2} \cos(\omega t - \theta)$$

With

$$\theta = \tan^{-1}\left(\frac{\frac{\dot{U}(0)}{\omega}}{U(0)}\right)$$

On the other hand, for a system with several degrees of freedom we have to manipulate matrices and vectors because the equation of motion becomes.

$$[M]\{\ddot{U}\} + [K]\{U\} = 0$$

UNDAMPED FREE VIBRATION

For each degree of freedom, the conditions of equilibrium, including the effects due to the acceleration and those due to the other external actions of the system (Newton), will be established at considered point. Figure 2a and 2b show an example of a system with several degrees of freedom, in which the beam is infinitely rigid.

Figure 2. (a) Two-degree free system; (b) Model of an infinitely rigid beam

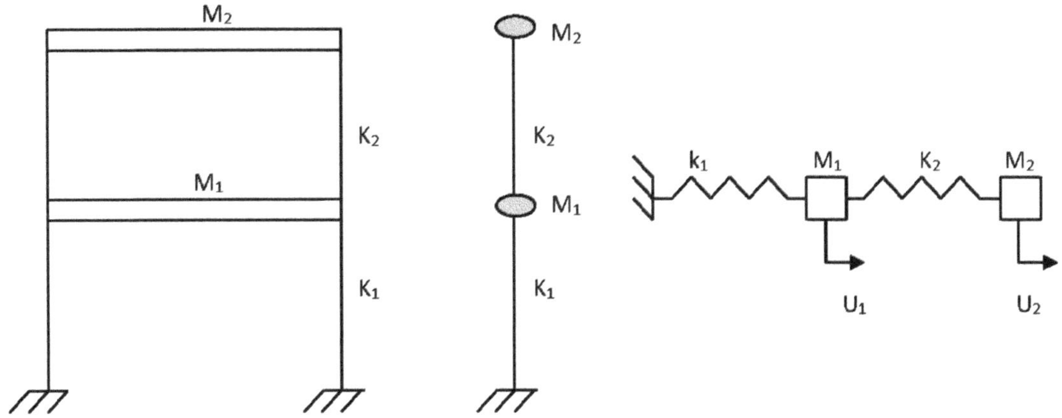

To write the motion equations, we use the principle of DALAMBERT (Newton's second law) in other words, for each degree of freedom, Newton's law is applied.

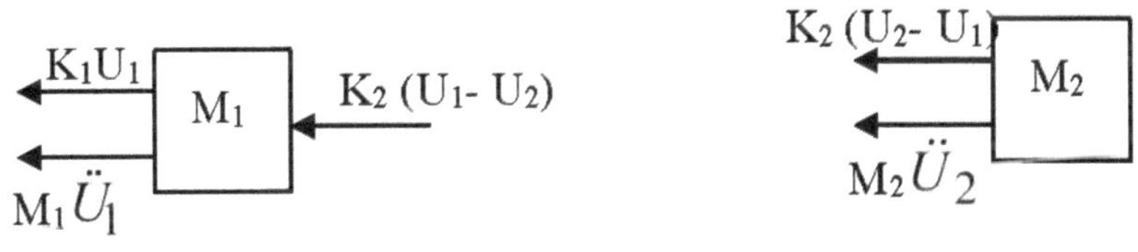

$$\Sigma F/M_1=0 \Rightarrow M_1\ddot{U}_1 + K_1U_1 + K_2(U_1 - U_2) = 0$$

$$\Rightarrow M_1\ddot{U}_1 + (K_1 + K_2)U_1 - K_2U_2 = 0$$

$$\Sigma F/M_2=0 \Rightarrow M_2\ddot{U}_2 + K_2(U_2 - U_1) = 0$$

$$\Rightarrow M_2\ddot{U}_2 - K_2U_1 + K_2U_2 = 0$$

$$\begin{cases} M_1\ddot{U}_1 + (K_1 + K_2)U_1 - K_2U_2 = 0 \\ M_2\ddot{U}_2 - K_2U_1 + K_2U_2 = 0 \end{cases} \tag{1}$$

The equation (1) is coupled equation system, which means that each equation contains several unknowns in this case there are two unknowns U_1 and U_2.

Equation (1) can be written in matrix form as follows:

$$\begin{bmatrix} M_1 & 0 \\ 0 & M_2 \end{bmatrix}\begin{Bmatrix} \ddot{U}_1 \\ \ddot{U}_2 \end{Bmatrix} + \begin{bmatrix} K_1+K_2 & -K_2 \\ -K_2 & K_2 \end{bmatrix}\begin{Bmatrix} U_1 \\ U_2 \end{Bmatrix} = \begin{Bmatrix} 0 \\ 0 \end{Bmatrix} \tag{2}$$

$$[M]\{\ddot{U}\} + [K]\{U\} = \{0\} \tag{3}$$

EIGENVALUES AND EIGENVECTORS

The principle of calculating eigenvalues and eigenvectors will be detailed in the next chapters.

To solve equation (1), assume the following solution (the method of separating variables):

$$\{U(t)\} = \{\varphi\} \sin(\omega t + \theta) \tag{4}$$

With {U(t)} Displacement vector ;

{φ} Is the vibration amplitude (time independent).

ω: is the frequency vibration ;

θ is the phase angle.

As long as $\{U(t)\} = \{\varphi\} \sin(\omega t + \theta)$

From where

$$\{\dot{U}(t)\} = \omega\{\varphi\}\cos(\omega t + \theta)$$

And

$$\{\ddot{U}(t)\} = -\omega^2\{\varphi\}\sin(\omega t + \theta) = -\omega^2\{U(t)\} \tag{5}$$

By replacing equations (4) and (5) in equation (3) we found:

$$[M](-\omega 2^{)}\{U\} + [K]\{U\} = \{0\}$$

$$[M](-\omega 2^{)}\{\varphi\} \sin(\omega t+\theta) + [K]\{\varphi\} \sin(\omega t+\theta) = \{0\}$$

$$([K] - \omega 2[M])\{\varphi^{\}} \sin(\omega t + \theta) = \{0\} \tag{6}$$

Equation (6) can be satisfied in two ways

First way:

$\sin(\omega t + \theta)=0$

From where {U(t)}=0 what physically means that there is no motion (trivial solution), not interesting solution.

Second way:

$$([K] - \omega 2[M])\{\varphi\} = 0 \tag{7}$$

In this case, the natural frequencies ωi and the modes φi must satisfy the following algebraic equation

Equation (7) can be interpreted as N homogeneous algebraic system equations, this system equation has a trivial solution which is {φ} = 0; this is not usable because it implies that there is no motion, but on the other hand the system equations has non-trivial solutions only if the determinant of [K] -ω2 [M] is null.

$$\det([K] - \omega 2[M]) = \{0\} \tag{8}$$

Equation (8) is known as the characteristic equation, and has N real and positive roots (ω2); because [K] and [M] (stiffness and mass matrix), are symmetrical and positive definite; the positive definite stiffness and mass matrix property is assumed for all structures, in order to prevent rigid body motion. This Kind of motion must be avoided for civil engineering structures. The property of positive definite leads to a stable system and real or positive roots (frequencies) of characteristic equation.

In another way, a matrix of size n x n is positive definite, when these sub-matrices of size 2x2, 3x3 ... up to n x n have a positive determinant. If one of the sub-matrix has a determinant equal to zero, at that moment it is called a positive semi-definite matrix. The stiffness matrix is real and positive definite, however if the structure undergoes a rigid body motion, in this case the matrix is semi-definite positive.

The roots of the characteristic equation are also known as eigenvalues; when the natural frequencies ω are known, the equation (7) can be solved for the corresponding vector {φ}.

The problem with the eigenvalues is that we do not fix the absolute amplitude of the vector {φ} only his shape. In summary, a vibratory system with N dof has N natural vibration frequencies ω_i, arranged from the smallest to the largest value ($\omega_1 < \omega_2 < \omega_3 .... < \omega_n$); corresponding to natural periods T_n, and natural modes φ_n. The term natural is used to describe all these properties of vibrations, and to insist that they are natural properties; structures under free vibration; and that they depend only on the properties of their masses and rigidities.

The development of equation (8) is as follows

$$\left| \begin{bmatrix} K_1 + K_2 & -K_2 \\ -K_2 & K_2 \end{bmatrix} - \omega_i^2 \begin{bmatrix} M_1 & 0 \\ 0 & M_2 \end{bmatrix} \right| = 0$$

$$\left|\begin{bmatrix} K_1+K_2-\omega_i^2 M_1 & -K_2 \\ -K_2 & K_2-\omega_i^2 M_2 \end{bmatrix}\right| = 0$$

The determinant is a 4th degree equation as a function of ω with ω_1^2 and ω_2^2 are the eigenvalues of the system; for each ω we determine φ; through equation (7). From where:

$$\begin{cases} (K_1+K_2)\varphi_1 - K_2\varphi_2 = M_1\varphi_1\omega_1^{\ 2} \\ -K_2\varphi_1 + K_2\varphi_2 = M_2\varphi_2\omega_2^{\ 2} \end{cases}$$

In matrix form

$$\begin{bmatrix} K_1+K_2 & -K_2 \\ -K_2 & K_2 \end{bmatrix}\begin{Bmatrix} \varphi_1 \\ \varphi_2 \end{Bmatrix} - \omega^2 \begin{bmatrix} M_1 & 0 \\ 0 & M_2 \end{bmatrix}\begin{Bmatrix} \varphi_1 \\ \varphi_2 \end{Bmatrix} = 0$$

↓ stiffness matrix ↘ amplitude vector ↘ mass matrix

We obtain

$$\omega_1(1^{et}\ \text{mod}\,e) \rightarrow \varphi_1 \begin{Bmatrix} \varphi_{11} \\ \varphi_{21} \end{Bmatrix}$$

Eigen vector of the mode (I)

$$\omega_2(2^{nd}\ \text{mod}\,e) \rightarrow \varphi_2 \begin{Bmatrix} \varphi_{12} \\ \varphi_{22} \end{Bmatrix}$$

Eigen vector of the mode (II)
The final solution is as follows

$$\{U(t)\} = \sum_{i=1}^{n} \{\varphi_i\} \sin\left(\omega_i t + \theta_i\right) \tag{9}$$

MULTI-DEGREE OF FREEDOM SYSTEM

After dealing, with the case of a system with two degrees of freedom, now we will proceed to a more general treatment of the problems. In the previous chapters, it has been revealed that a system with two degrees of freedom is more involved, in the design than the system with one degree of freedom; but when the number of mass to be considered increases in the structure, the mathematical formulation becomes more complicated.

The theory of the multi-degree of freedom system is of great importance; because it is applicable not only to systems classified as discrete; but also in the case of numerical solutions of continuous systems converted into a discrete one; by concentrating the mass in significant points. A real representation of such a case can only be done with a system with several degrees of freedom.

Many structures, such as frame, have essentially a concentrated mass, since the mass of the columns is often negligible compared to the floors, the system can be treated as a system with several degrees of freedom.

Assuming that a frame structure with a rigid floor can be reduced to an equivalent mass-spring-damping system see figure bellow.

Figure 3. (a) Multi-degree of freedom system; (b) Frame model with infinitely rigid beam

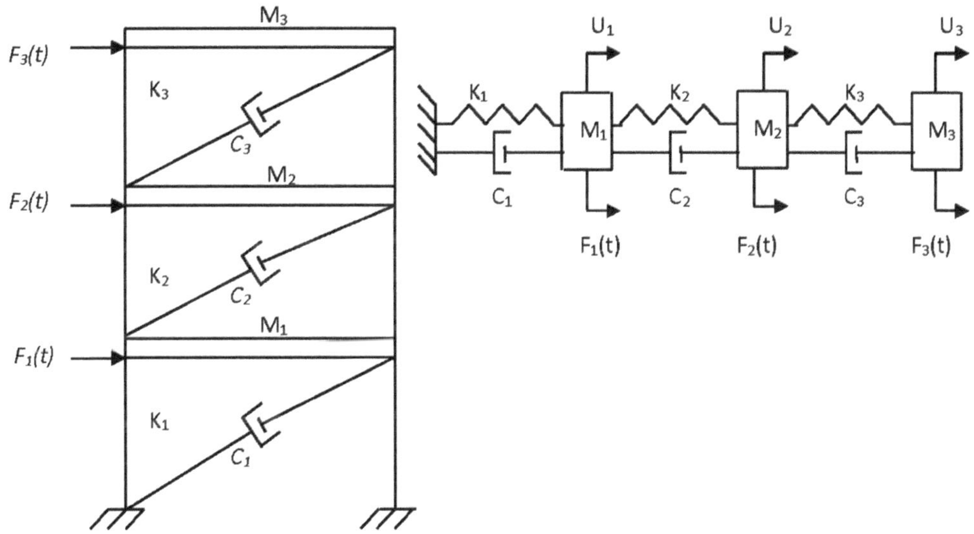

In order to determine motion equations system, one of the principles already explained in chapter 1 is chosen; which is the principle of DALAMBERT. We begins by isolating each mass and replacing the removed part system by equivalent forces to ensure balance

$$\begin{bmatrix} K_1+K_2 & -K_2 \\ -K_2 & K_2 \end{bmatrix} \begin{Bmatrix} \phi_1 \\ \phi_2 \end{Bmatrix} - \omega^2 \begin{bmatrix} M_1 & 0 \\ 0 & M_2 \end{bmatrix} \begin{Bmatrix} \phi_1 \\ \phi_2 \end{Bmatrix} = 0$$

Stiffness matrix   Amplitude vector   Mass matrix

$$\begin{cases} M_1 \ddot{U}_1 + C_1\dot{U}_1 + C_2 (\dot{U}_1 - \dot{U}_2) + K_1 U_1 + K_2(U_1 - U_2) = F_1(t) \\ M_2 \ddot{U}_2 + C_2 (\dot{U}_2 - \dot{U}_1) + C_3 (\dot{U}_2 - \dot{U}_3) + K_2(U_2 - U_1) + K_3(U_2 - U_3) = F_2(t) \\ M_3 \ddot{U}_3 + C_3(\dot{U}_3 - \dot{U}_2) + K_3(U_3 - U_2) = F_3(t) \end{cases}$$

$$\begin{cases} M_1 \ddot{U}_1 + (C_1 + C_2) \dot{U}_1 - C_2\dot{U}_2 + (K_1 + K_2)U_1 - k_2 U_2 = F_1(t) \\ M_2 \ddot{U}_2 - C_2 \dot{U}_1 + (C_2 + C_3)\dot{U}_2 - C_3\dot{U}_3 - K_2 U_1 + (K_2 + K_3)U_2 - K_3 U_3 = F_2(t) \\ M_3 \ddot{U}_3 - C_3\dot{U}_2 + C_3\dot{U}_3 - K_3 U_2 + K_3 U_3 = F_3(t) \end{cases} \quad (10)$$

The equation system (10) can be written in matrix form as follows.

$$\begin{bmatrix} m_1 & 0 & 0 \\ 0 & m_2 & 0 \\ 0 & 0 & m_3 \end{bmatrix} \begin{Bmatrix} \ddot{U}_1 \\ \ddot{U}_2 \\ \ddot{U}_3 \end{Bmatrix} + \begin{bmatrix} C_1 + C_2 & -C_2 & 0 \\ -C_2 & C_2 + C_3 & -C_3 \\ 0 & -C_3 & C_3 \end{bmatrix} \begin{Bmatrix} \dot{U}_1 \\ \dot{U}_2 \\ \dot{U}_3 \end{Bmatrix} + \begin{bmatrix} K_1 + K_2 & -K_2 & 0 \\ -K_2 & K_2 + K_3 & -K_3 \\ 0 & -K_3 & K_3 \end{bmatrix} \begin{Bmatrix} U_1 \\ U_2 \\ U_3 \end{Bmatrix} = \begin{Bmatrix} F_1(t) \\ F_2(t) \\ F_3(t) \end{Bmatrix}$$

The same equation can be written in a compact form, as follows:

$$[M]\{\ddot{U}\} + [C]\{\dot{U}\} + [K]\{U\} = \{F(t)\}$$

MANUAL EIGENVALUE CALCULATIONS EXAMPLE

The following example illustrates the use of Cramer's law for calculating eigenvalues of frame structure shown in figure 2a.

$$M=\begin{bmatrix}4&0&0\\0&4&0\\0&0&2\end{bmatrix}\quad K=\begin{bmatrix}14&-6&0\\-6&10&-4\\0&-4&4\end{bmatrix}$$

The calculation of eigenvalues using of Cramer's law is as follows

$$K-\omega^2M=\begin{bmatrix}(14-4\omega^2)&-6&0\\-6&(10-4\omega^2)&-4\\0&-4&(14-2\omega^2)\end{bmatrix}$$

The determinant of a 3x3 matrix is given:

$$\left|K-\omega^2M\right|=(-1)^{(1+1)}(14-4\omega^2)\begin{bmatrix}(10-4\omega^2)&-4\\-4&(10-2\omega^2)\end{bmatrix}$$
$$+(-1)^{(1+2)}(-6)\begin{bmatrix}-6&-4\\0&(10-2\omega^2)\end{bmatrix}+(-1)^{(1+3)}(0)\begin{bmatrix}-6&(10-4\omega^2)\\0&-4\end{bmatrix}$$

$$\left|K-\omega^2M\right|=(14-4\omega^2)\left[(10-4\omega^2)(10-2\omega^2)-(-4)^2\right]-(-6)\left[(-6)(10-2\omega^2)\right]$$

$$\left|K-\omega^2M\right|=(14-4\omega^2)\left[8\omega^4-36\omega^2+24\right]-36\left[4-2\omega^2\right]$$

$$\left|K-\omega^2M\right|=-32\omega^6+256\omega^4-528\omega^2+192$$

$$\omega_1^{\ 2}=0.46056\quad,\omega_2^{\ 2}=2.681993\quad\omega_3^{\ 2}=4.857448$$

Another more complicated example is represented, which considers a four-storey beam column system, the mass and stiffness matrix of so-called system are:

$$M=\begin{bmatrix}4&0&0&0\\0&4&0&0\\0&0&2&0\\0&0&0&2\end{bmatrix}\quad K=\begin{bmatrix}14&-6&0&0\\-6&10&-4&0\\0&-4&8&-4\\0&0&-4&4\end{bmatrix}$$

$$K-\omega^2 M=\begin{bmatrix}(14-4\omega^2) & -6 & 0 & 0\\ -6 & (10-4\omega^2) & -4 & 0\\ 0 & -4 & (8-2\omega^2) & -4\\ 0 & 0 & -4 & (4-2\omega^2)\end{bmatrix}$$

$$\left|K-\omega^2 M\right|=(-1)^{(1+1)}(14-4\omega^2)\begin{bmatrix}(10-4\omega^2) & -4 & 0\\ -4 & (8-2\omega^2) & -4\\ 0 & -4 & (4-2\omega^2)\end{bmatrix}$$

$$+(-1)^{(1+2)}(-6)\begin{bmatrix}-6 & -4 & 0\\ -4 & (8-2\omega^2) & -4\\ 0 & -4 & (4-2\omega^2)\end{bmatrix}$$

$$\left|K-\omega^2 M\right|=(14-4\omega^2)\left[(10-4\omega^2)\left\{(8-2\omega^2)(4-2\omega^2)-16\right\}\right]$$
$$+(-1)(-4)\left\{-4(4-2\omega^2\right\}+(-1)(-6)\left[-6\left\{(8-2\omega^2)(4-2\omega^2)-16\right\}\right]$$

$$\left|K-\omega^2 M\right|=64\omega^8-768\omega^6+284\omega^4-3328\omega^2+768$$

$$\omega_1^{\ 2}=0.303146\quad ,\omega_2^{\ 2}=1.591674\quad \omega_3^{\ 2}=4.241005\quad \omega_4^{\ 2}=5.864176$$

It is clear, that the manual calculation becomes very difficult as soon as one goes beyond three levels. From where, the need to resort to the numerical methods for the calculation of the eigenvectors and eigenvalues which will be developed in chapter 7.

Figure 4. 3D Dynamic motion

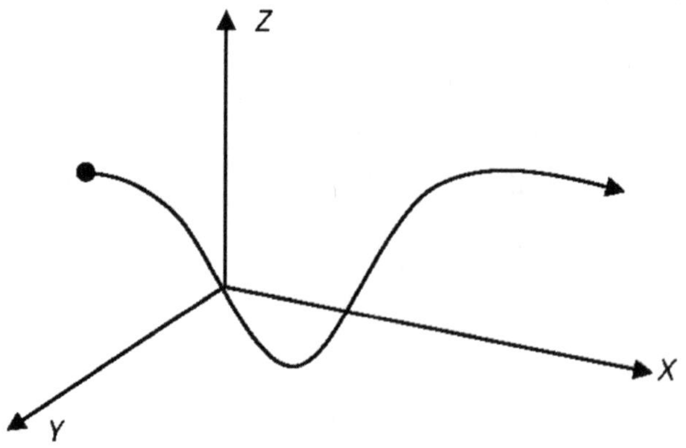

EIGEN MODES ORTHOGONALITY

Figure 4. Shows a dynamic motion described in a 3D space through the X, Y, Z coordinate system The coordinate system is defined by three independent vectors

$$\begin{Bmatrix}1\\0\\0\end{Bmatrix},\ \begin{Bmatrix}0\\1\\0\end{Bmatrix},\ \begin{Bmatrix}0\\0\\1\end{Bmatrix},$$

which are three orthogonal basic functions. Let's do an orthogonality test from the X, Y plane of figure 2.

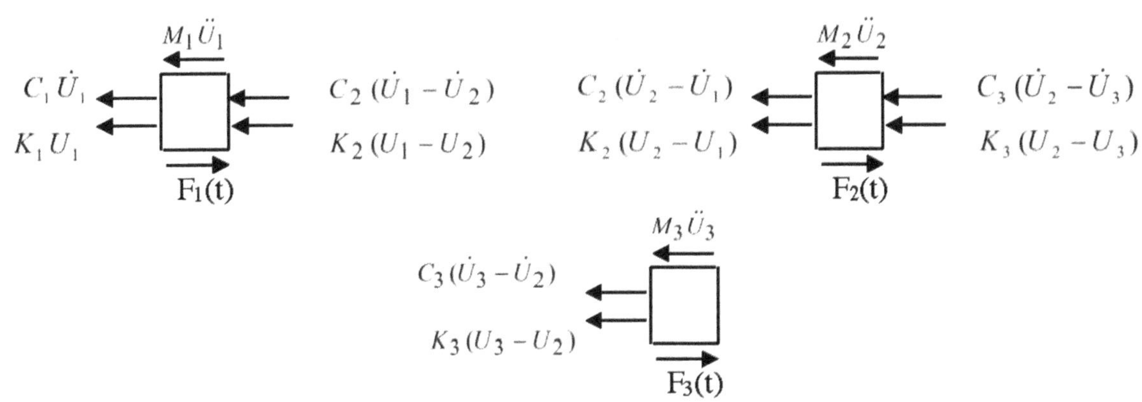

$$\{X\}^T\{Y\}=\{1\quad 0\quad 0\}\begin{Bmatrix}0\\1\\0\end{Bmatrix}=0$$

It is shown, considering the properties of the mass and stiffness matrix (M and K), that the eigen modes (the modal vectors) are orthogonal, and constitute a complete base. In an other words, the response of the structure can be decomposed into a sum of the modal responses (Chopra 2001).

3D COORDINATE SYSTEM ANALOGY

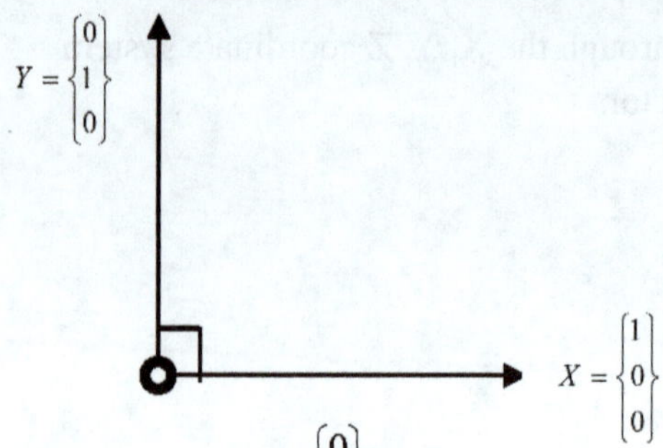

$$\{X\}^T\{Y\} = \{1 \quad 0 \quad 0\}\begin{Bmatrix} 0 \\ 1 \\ 0 \end{Bmatrix} = 0$$

$$\begin{Bmatrix} x \\ y \\ z \end{Bmatrix} = X(t)\begin{Bmatrix} 1 \\ 0 \\ 0 \end{Bmatrix} + Y(t)\begin{Bmatrix} 0 \\ 1 \\ 0 \end{Bmatrix} + Z(t)\begin{Bmatrix} 0 \\ 0 \\ 1 \end{Bmatrix}$$

From where

$$\begin{Bmatrix} 1 \\ 0 \\ 0 \end{Bmatrix}, \begin{Bmatrix} 0 \\ 1 \\ 0 \end{Bmatrix}, \begin{Bmatrix} 0 \\ 0 \\ 1 \end{Bmatrix}$$

is the orthonormal basis

Figure 5. Displacement vector decomposition in an orthonormal data-base

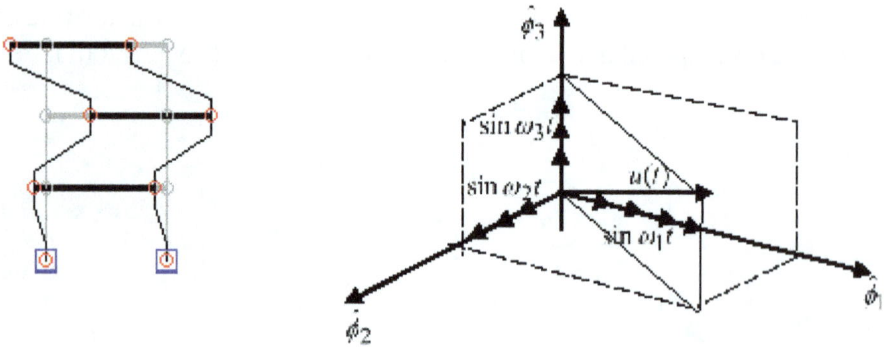

$$\begin{Bmatrix} u_1 \\ u_2 \\ u_3 \end{Bmatrix} = \sin\omega_1 t \quad \left\{\hat{\varphi}_1\right\} +, \sin\omega_2 t \quad \left\{\hat{\varphi}_2\right\} + \sin\omega_3 t \quad \left\{\hat{\varphi}_3\right\}$$

From where $\left\{\hat{\varphi}_1\right\}, \quad \left\{\hat{\varphi}_2\right\}, \quad \left\{\hat{\varphi}_3\right\}$ is the orthonormal basis

Note $\left\{\hat{\varphi}\right\}$ Represents the normalized normal vector; the standardization of the eigenvectors will be treated later.

Take the following characteristic equation:

$$\left\{[K] - \omega_i^2 [M]\right\}\{\varphi\} = 0$$

$$\downarrow \qquad \searrow$$

eigenvalues eignevectors

We have an equation system coupled, it means that equation depends on several unknowns; this coupling comes from the stiffness matrix, either of the mass matrix, or of both sets.

Consistent mass

$$[K] = \begin{bmatrix} k_{11} & k_{12} & \dots & k_{1n} \\ k_{21} & \cdot & & \vdots \\ \vdots & & \cdot & \vdots \\ k_{n1} & & & k_{nn} \end{bmatrix} \qquad [M] = \begin{bmatrix} m_{11} & m_{12} & \dots & m_{1n} \\ m_{21} & \cdot & & \vdots \\ \vdots & & \cdot & \vdots \\ m_{n1} & & & m_{nn} \end{bmatrix}$$

Concentrated mass system

$$[M] = \begin{bmatrix} m_{11} & 0 & \dots & 0 \\ 0 & m_{22} & 0 & 0 \\ \vdots & .. & .. & \vdots \\ 0 & 0 & .. & m_{nn} \end{bmatrix}$$

For mode i we have

$$\left\{[K] - \omega_i^2 [M]\right\}\{\varphi_i\} = 0$$

Thus

$$[K]\{\varphi_i\}-\omega_i^2[M]\{\varphi_i\}=0 \tag{11}$$

For mode j we have

$$\{[K]-\omega_j^2[M]\}\{\varphi_j\}=0$$

From where

$$[K]\{\varphi_j\}-\omega_j^2[M]\{\varphi_j\}=0 \tag{12}$$

By pre-multiplying equation (11) by $\{\varphi j_j t$ from either side

$$\{\varphi_j\}^t[K]\{\varphi_i\}=\omega_i^2\{\varphi_j\}^t[M]\{\varphi_i\} \tag{13}$$

By pre multiplying the equation (12) $\{\varphi i_j t$ from either side

$$\{\varphi_i\}^t[K]\{\varphi_j\}=\omega_j^2\{\varphi_i\}^t[M]\{\varphi_j\} \tag{14}$$

If we take the transpose of equation (14)

$$\left[\{\varphi_i\}^t[K]\{\varphi_j\}\right]^t=\{\varphi_j\}^t[K]^t\{\varphi_i\}=\{\varphi_j\}^t[K]\{\varphi_i\}$$

$[K]^t=[K]$ Because the stiffness matrix is symmetrical and

$$\left[\omega_j^2\{\varphi_i\}^t[M]\{\varphi_j\}\right]^t=\omega_j^2\{\varphi_j\}^t[M]^t\{\varphi_i\}=\omega_j^2\{\varphi_j\}^t[M]\{\varphi_i\}$$

$\{M\}^t=[M]$ Because the mass matrix is symmetrical
We will have

$$\{\varphi_j\}^t[K]\{\varphi_i\}=\omega_j^2\{\varphi_j\}^t[M]\{\varphi_i\} \tag{15}$$

We will subtract (12) from (15)

$$0=(\omega_j^2-\omega_i^2)\{\varphi_j\}^t[M]\{\varphi_i\}$$

We know that

$\omega j \neq \omega i \Rightarrow \{\varphi j\}t[M^{/}\{\varphi i\} = 0$

So

$$\{\varphi j\}t[M^{/}\{\varphi i\} = \{\varphi j\}t[K]\{\varphi i\} = 0 \tag{16}$$

Equation (16) represents the orthogonality relation that allows decoupling equations.
If I write the matrix Φ as follows:

$[\varphi] = [\varphi 1\ \varphi 2 \ldots \varphi n]$

With $[\varphi]$ is the matrix of the eigenmodes (modal matrix)

$$[\varphi]=\begin{bmatrix}\varphi_{11} & \varphi_{12} & \cdots\cdots & \varphi_{1n}\\ \varphi_{21} & \varphi_{22} & \cdots\cdots & \varphi_{2n}\\ \vdots & \vdots & \cdots\cdots & \vdots\\ \varphi_{n1} & \varphi_{n2} & \cdots\cdots & \varphi_{nn}\end{bmatrix}$$

The orthogonality relations are written as follows

$$[\varphi]^{t}[M][\varphi]=\begin{bmatrix}\backslash & 0 & 0 & 0\\ 0 & \backslash & 0 & 0\\ : & .. & \backslash & ..\\ 0 & 0 & 0 & \backslash\end{bmatrix}$$

Diagonal matrix (all terms off diagonal are equal to zero)

$$[\varphi]^{t}[K][\varphi]=\begin{bmatrix}\backslash & 0 & 0 & 0\\ 0 & \backslash & 0 & 0\\ : & .. & \backslash & ..\\ 0 & 0 & 0 & \backslash\end{bmatrix}$$

Diagonal matrix (all terms off diagonal are equal to zero)

FREE UNDAMPED VIBRATIONS RESPONSE

The equation that governs this system is as follows:

$$[M]\{\ddot{U}\}+[K]\{U\}=0 \tag{17}$$

We assume that the initial conditions are:

$$U(0)=\begin{Bmatrix} U_1(0) \\ U_2(0) \\ : \\ U_n(0) \end{Bmatrix};\ \dot{U}(0)=\begin{Bmatrix} \dot{U}_1(0) \\ \dot{U}_2(0) \\ : \\ \dot{U}_n(0) \end{Bmatrix}$$

The differential equation (17) is a coupled equation, which is due to the mass, stiffness and damping matrix in the damped case. This coupling comes from to the fact that the matrices are not diagonal; if for example m_{ij} exists for i 1 j, that means that the system has a dynamic coupling, in the other hand if the stiffness matrix is non diagonal then we speak of static coupling.

In order to remove this coupling a modal analysis is envisaged and which has the principle of introducing a linear transformation of the variables by using the modal matrix.

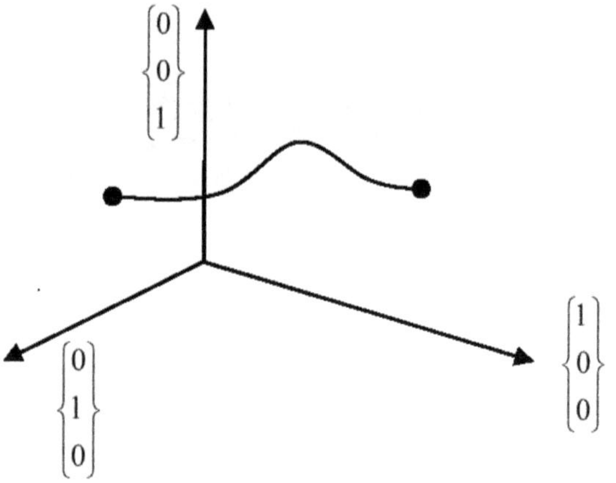

By replacing the expression of U in equation (17)

$$[M][\varphi]\{\ddot{y}\}+[K][\varphi]\{y\}=0 \tag{18}$$

By multiplying the equation (18) by $[\varphi]t$ we have

$$[\varphi]^t[M][\varphi]\{\ddot{y}\}+[\varphi]^t[K][\varphi]\{y\}=0$$

$$\underbrace{[\varphi]^t[M][\varphi]}_{\text{diagonal matrix}}\{\ddot{y}\}+\underbrace{[\varphi]^t[K][\varphi]}_{\text{diagonal matrix}}\{y\}=0$$

$$\begin{bmatrix} \backslash & 0 & 0 & 0 \\ 0 & \backslash & 0 & 0 \\ : & .. & \backslash & .. \\ 0 & 0 & 0 & \backslash \end{bmatrix} \{\ddot{y}\} + \begin{bmatrix} \backslash & 0 & 0 & 0 \\ 0 & \backslash & 0 & 0 \\ : & .. & \backslash & .. \\ 0 & 0 & 0 & \backslash \end{bmatrix} \{y\} = \{0\}$$

Thus we obtain a system of equations which contains only elements of the diagonal

$$\begin{cases} \{\varphi_1\}^t [M]\{\varphi_1\}\ddot{y}_1 + \{\varphi_1\}^t [K]\{\varphi_1\} y_1 = 0 \\ \{\varphi_2\}^t [M]\{\varphi_2\}\ddot{y}_2 + \{\varphi_2\}^t [K]\{\varphi_2\} y_2 = 0 \\ : \\ \{\varphi_n\}^t [M]\{\varphi_n\}\ddot{y}_n + \{\varphi_n\}^t [K]\{\varphi_n\} y_n = 0 \end{cases}$$

We have just decomposed a system with several degrees of freedom to a system with several equations with one degree of freedom each

$$\begin{cases} \ddot{y}_1 + \dfrac{\{\varphi_1\}^t [K]\{\varphi_1\}}{\{\varphi_1\}^t [M]\{\varphi_1\}} y_1 = 0 \\ \ddot{y}_2 + \dfrac{\{\varphi_n\}^t [K][\varphi_2]}{\{\varphi_n\}^t [M]\{\varphi_2\}} y_2 = 0 \\ : \\ \ddot{y}_n + \dfrac{\{\varphi_n\}^t [K][\varphi_n]}{\{\varphi_n\}^t [M]\{\varphi_n\}} y_n = 0 \end{cases}$$

With $\dfrac{\{\varphi_1\}^t [K]\{\varphi_1\}}{\{\varphi_1\}^t [M]\{\varphi_1\}} = \omega_1^2$ Rayleigh ratio

$$\begin{cases} \ddot{y}_1 + \omega_1^2 y_1 = 0 \\ \ddot{y}_2 + \omega_2^2 y_2 = 0 \\ : \\ \ddot{y}_n + \omega_n^2 y_n = 0 \end{cases} \underset{\text{the solution is}}{\Rightarrow} \begin{cases} y_1 = y_1(0) cons\omega_1 t + \dfrac{\dot{y}_1(0)}{\omega_1} \sin \omega_1 t \\ y_2 = y_2(0) cons\omega_2 t + \dfrac{\dot{y}_2(0)}{\omega_2} \sin \omega_2 t \\ : \\ y_n = y_n(0) cons\omega_n t + \dfrac{\dot{y}_n(0)}{\omega_n} \sin \omega_n t \end{cases} \quad (18)$$

$y_1(0), y_2(0),\ldots, y_n(0)$ and $\dot{y}_1(0), \dot{y}_2(0), \dot{y}_3(0)$, are constants that can be determined by the initial conditions

If eigenvectors are normalized to mass

$$\{\hat{\varphi}_i\}^t [M]\{\hat{\varphi}_i\} = 1$$

$$\{\hat{\varphi}_i\} = \frac{1}{c_i}\{\varphi_i\}$$

With c_1 is an arbitrary constant

The determination of c_1 is as follows

If $\{\hat{\varphi}_i\}^t [M]\{\hat{\varphi}_i\} = 1$

From the Rayleigh coefficient, see equation (18)

$$\{\hat{\varphi}_i\}^t [K]\{\hat{\varphi}_i\} = \omega_i^2$$

$$\frac{1}{C_i}\{\varphi_i\}^t [M]\{\varphi_i\}\frac{1}{C_i} = 1$$

$$\{\varphi_i\}^t [M]\{\varphi_i\} = C_i^2$$

$$C_i = \pm\sqrt{\{\varphi_i\}^t [M]\{\varphi_i\}}$$

$[\varphi]t^t M][\varphi] = [I]$

Matrix identity the elements of the diagonal are equal to 1

DISPLACEMENT MODAL EXPANSION

Any n-independent vector series can be used for the representation of any N-order vector. In this section, the eigenmodes are used as a basis for a modal expansion of any displacement vector U has the following form

$$\{U\}=\sum_{r-1}^{N}\varphi_r y_r=[\varphi]\{y\} \tag{19}$$

Where y_r is a scalar multiplier called modal or normal coordinate with $y=(y_1\, y_2 \dots y_n)^t$ when the eigen-vectors φt are known for a given displacement U it is possible to evaluate yr by multiplying the two sides of the equation (19) by $\{\varphi n\}_t[^M]$ using the orthogonality property we find:

$$\{\varphi_n\}^t[M]\{U\}=\sum_{r=1}^{N}\varphi_n{}^t M\varphi_r y_r=\{\varphi_n\}^t[M][\varphi]\{y\}$$

Applying the orthogonality property

$$\{\varphi_n\}^t[M]\{U\}=\{\varphi_n\}^t[M]\{\varphi_n\}\{y_n\}$$

$$\begin{cases}\{y_n(0)\}=\dfrac{\{\varphi_n\}^t[M]\{U(0)\}}{\{\varphi_n\}^t[M]\varphi_n}=\dfrac{\{\varphi_n\}^t[M]\{U(0)\}}{M_n^*}\\ \{\dot{y}_n(0)\}=\dfrac{\{\varphi_n\}^t[M]\{\dot{U}(0)\}}{\{\varphi_n\}^t[M]\{\varphi_n\}}=\dfrac{\varphi_n{}^t M\{\dot{U}(0)\}}{M_n^*}\end{cases} \tag{20}$$

Thanks to the equation (20), we can have the initial conditions of the normal (modal) coordinates as follows:

$$\begin{cases}y_n(0)=\dfrac{\{\varphi_n\}^t[M]\{U(0)\}}{\{\varphi_n\}^t[M]\varphi_n}=\dfrac{\{\varphi_n\}^t[M]\{U(0)\}}{M_n^*}\\ \dot{y}_n(0)=\dfrac{\{\varphi_n\}^t[M]\{\dot{U}(0)\}}{\{\varphi_n\}^t[M]\{\varphi_n\}}=\dfrac{\{\varphi_n{}^t\}[M]\{\dot{U}(0)\}}{M_n^*}\end{cases}$$

as $U(0)$ and $\dot{U}(0)$ are known ⇒ we can calculate $y_1(0), y_2(0),\dots, y_n(0)$ and, $\dot{y}_1(0), \dot{y}_2(0),\dots, \dot{y}_n(0)$ hence the possibility of determining $y_1\, y_2 \dots y_n$. So we can finally calculate the real variable U $\{U\}=[\varphi]\{y\}$

UNDAMPED FORCED VIBRATIONS RESPONSE

The equation of motion of undamped forced system is:

$$[M]\{\ddot{U}\}+[K]\{U\}=\{f\} \tag{21}$$

$[M]_{(nxn)}$, $\{\ddot{U}\}_{(n)}$, $[K]_{(nxn)}$, $\{U\}_{(n)}$, $\{f\}_{(n)}$

We assume that $\{U\}=[\varphi]\{y\}$

We replace the expression of $\{U\}$ in the equation (21) and we pre multiplied by $[\varphi]$t we have

$$\underbrace{[\varphi]^t[M][\varphi]}_{\text{diagonal matrix}}\{\ddot{y}\}+\underbrace{[\varphi]^t[K][\varphi]}_{\text{diagonal matrix}}\{y\}=[\varphi]^t\{F\}$$

$$\begin{bmatrix} \backslash & 0 & 0 & 0 \\ 0 & \backslash & 0 & 0 \\ \vdots & .. & \backslash & .. \\ 0 & 0 & 0 & \backslash \end{bmatrix}\{\ddot{y}\}+\begin{bmatrix} \backslash & 0 & 0 & 0 \\ 0 & \backslash & 0 & 0 \\ \vdots & .. & \backslash & .. \\ 0 & 0 & 0 & \backslash \end{bmatrix}\{y\}=[\varphi]^t\{F\}$$

Thus, we obtain an equation system, which contains only elements of the diagonal

$$\begin{cases} \{\varphi_1\}^t[M]\{\varphi_1\}\ddot{y}_1+\{\varphi_1\}^t[K]\{\varphi_1\}y_1=\{\varphi_1\}^t\{F\} \\ \{\varphi_2\}^t[M]\{\varphi_2\}\ddot{y}_2+\{\varphi_2\}^t[K]\{\varphi_2\}y_2=\{\varphi_2\}^t\{F\} \\ \vdots \\ \{\varphi_n\}^t[M]\{\varphi_n\}\ddot{y}_n+\{\varphi_n\}^t[K]\{\varphi_n\}y_n=\{\varphi_n\}^t\{F\} \end{cases}$$

We have just decomposed the multi-degree system to a multi-equation system with a single degree of freedom

$$\begin{cases} \ddot{y}_1+\dfrac{\{\varphi_1\}^t[K]\{\varphi_1\}}{\{\varphi_1\}^t[M]\{\varphi_1\}}y_1=\dfrac{\{\varphi_1\}^t}{\{\varphi_1\}^t[M]\{\varphi_1\}}\{F\} \\ \ddot{y}_2+\dfrac{\{\varphi_2\}^t[K]\{\varphi_2\}}{\{\varphi_2\}^t[M]\{\varphi_2\}}y_2=\dfrac{\{\varphi_2\}^t}{\{\varphi_2\}^t[M]\{\varphi_2\}}\{F\} \\ \ddot{y}_n+\dfrac{\{\varphi_n\}^t[K]\{\varphi_n\}}{\{\varphi_n\}^t[M]\{\varphi_n\}}y_n=\dfrac{\{\varphi_n\}^t}{\{\varphi_n\}^t[M]\{\varphi_n\}}\{F\} \end{cases}$$

We assume, $M_n^*=\{\varphi_n\}^t[M]\{\varphi_n\}$ which is a generalized mass

And $F_n^*(t)=\{\varphi_n\}^t F(t)$ which is the generalized force

$$\begin{cases} \ddot{y}_1+\omega_1^2 y_1=\dfrac{\{\varphi_1\}^t}{\{\varphi_1\}^t[M]\{\varphi_1\}}\{F\} \\ \ddot{y}_2+\omega_2^2 y_2=\dfrac{\{\varphi_1\}^t}{\{\varphi_2\}^t[M]\{\varphi_2\}}\{F\} \\ \ddot{y}_n+\omega_n^2 y_n=\dfrac{\{\varphi_n\}^t}{\{\varphi_n\}^t[M]\{\varphi_n\}}\{F\} \end{cases} \tag{22}$$

The solution of the equation system (22) is as follows

$$\begin{cases} y_1(t)=y_1(0)\cos\omega_1 t+\dfrac{\dot{y}_1(0)}{\omega_1}\sin\omega_1 t+\dfrac{1}{M_1^*\omega_1}\displaystyle\int_0^t F_1^*(t')\sin\omega_1(t-t')dt' \\ y_2(t)=y_2(0)\cos\omega_2 t+\dfrac{\dot{y}_2(0)}{\omega_2}\sin\omega_2 t+\dfrac{1}{M_2^*\omega_2}\displaystyle\int_0^t F_2^*(t')\sin\omega_2(t-t')dt' \\ \vdots \\ y_n(t)=y_n(0)\cos\omega_n t+\dfrac{\dot{y}_n(0)}{\omega_n}\sin\omega_n t+\dfrac{1}{M_n^*\omega_n}\displaystyle\int_0^t F_n^*(t')\sin\omega_n(t-t')dt' \end{cases} \tag{23}$$

To define the modal vectors, it is possible to choose one of the components of the vector to an arbitrary value. It should be noted that it is necessary to normalize at 1 the largest component. Generally, most of the design programs retain the following normalization (Clough & Penzien, 1995):

$[\varphi n]t[M][\varphi n] = [I]$

FORCED AND DAMPED VIBRATION SYSTEM RESPONSE

The equation of motion that governs this system is as follows:

$$[M]_{(nxn)}\{\ddot{U}\}_{(n)}+[C]_{(nxn)}\{\dot{U}\}_{(n)}+[K]_{(nxn)}\{U\}_{(n)}=\{f\}_{(n)} \tag{24}$$

It is a system with several degrees of freedom to solve it we have two methods:

- Modal superposition method valid for elastic linear structures
- Numerical method valid for linear and nonlinear structures, which will be the detailed in chapter 8

MODAL SUPERPOSITION METHOD

The basic general assumption is: the deformed structure can be expressed as a linear combination of all modes.

$$U_i(t) = \varphi_{i1}y_1 + \varphi_{i2}y_2 + \varphi_{i3}y_3 + \ldots + \varphi_{in}y_n = \sum_{j=1}^{n} \varphi_i^j y_j(t)$$

$\varphi 1$: Represents the eigen-mode of mode1
$y_1, y_2, y_3 \ldots y_n$ represent the modal amplitude that varies with time

$\{U\} = [\phi]\{y\}$ ⟶ Vector of modal amplitudes

↓

Modal matrix

$$\{y\} = \begin{Bmatrix} y_1 \\ y_2 \\ y_3 \\ \vdots \\ y_4 \end{Bmatrix}$$

So-called generalized system coordinates

While U coordinates are called natural coordinates of the system.

Modal analysis is a decomposition process of the dynamic equilibrium equation using generalized coordinates.

We replace U by its value in the dynamic motion equation (24) we find:

$$[M][\varphi]\{\ddot{y}\} + [C][\varphi]\{\dot{y}\} + [K][\varphi]\{y\} = \{F(t)\}$$

$$[\varphi]^t[M][\varphi]\{\ddot{y}\} + [\varphi]^t[C][\varphi]\{\dot{y}\} + [\varphi]^t[K][\varphi]\{y\} = [\varphi]^t\{F(t)\}$$

$$\begin{bmatrix} \backslash & 0 & 0 & 0 \\ 0 & \backslash & 0 & 0 \\ \vdots & .. & \backslash & .. \\ 0 & 0 & 0 & \backslash \end{bmatrix}\{\ddot{y}\} + \begin{bmatrix} \backslash & 0 & 0 & 0 \\ 0 & \backslash & 0 & 0 \\ \vdots & .. & \backslash & .. \\ 0 & 0 & 0 & \backslash \end{bmatrix}\{\dot{y}\} + \begin{bmatrix} \backslash & 0 & 0 & 0 \\ 0 & \backslash & 0 & 0 \\ \vdots & .. & \backslash & .. \\ 0 & 0 & 0 & \backslash \end{bmatrix}\{y\} = [\varphi]^t\{F(t)\}$$

$$\begin{cases} \{\varphi_1\}^t[M]\{\varphi_1\}\ddot{y}_1+\{\varphi_1\}^t[C]\{\varphi_1\}\dot{y}_1+\{\varphi_1\}^t[K]\{\varphi_1\}y_1=\{\varphi_1\}^t F(t) \\ \{\varphi_2\}^t[M]\{\varphi_2\}\ddot{y}_2+\{\varphi_2\}^t[C]\{\varphi_2\}\dot{y}_2+[\varphi_2]^t[K][\varphi_2]y_2=\{\varphi_2\}^t F(t) \\ \vdots \\ \{\varphi_n\}^t[M]\{\varphi_n\}\ddot{y}_n+\{\varphi_n\}^t[C]\{\varphi_n\}\dot{y}_n+\{\varphi_n\}^t[K]\{\varphi_n\}y_n=\{\varphi_n\}^t F(t) \end{cases}$$

The only problem that arises is for the damping matrix [*C*] (because there is no orthogonality). For that we will assume that [*C*] is proportional to mass matrix [*M*] or stiffness matrix [*K*] or a combination of both [*M*] and [*K*]

$$\begin{cases} [C]=\alpha[M] \\ [C]=\beta[K] \\ [C]=\alpha[M]+\beta[K] \end{cases} \Rightarrow \text{based on experience}$$

Knowing that:

$M_n^*=\{\varphi_n\}^t[M]\{\varphi_n\}$ Generalized mass of mode n

$K_n^*=\{\varphi_n\}^t[K]\{\varphi_n\}=\omega_n^2M_n$ Generalized stiffness of mode n

$C_n^*=\{\varphi_n\}^t C\{\varphi_n\}=2\xi_n\omega_nM_n$ Generalized damping of mode n

$F_n^*(t)=\{\varphi_n\}^t F(t)$ Generalized force of mode n

We can write the equation system as follows

$$\begin{cases} M_1^*\ddot{y}_1+C_1^*\dot{y}_1+K_1^*y_1=\{\varphi_1\}^t F_1^*(t) \\ M_1^*\ddot{y}_2+C_2^*\dot{y}_2+K_2^*y_2=\{\varphi_2\}^t F_2^*(t) \\ \vdots \\ M_n^*\ddot{y}_n+C_n^*\dot{y}_n+K_n^*y_n=\{\varphi_n\}^t F_n^*(t) \end{cases} \tag{25}$$

This equation system represents at the structure level the following:

Figure 6. Decomposition of a multi-degree of freedom system into multiple systems with a single degree of freedom

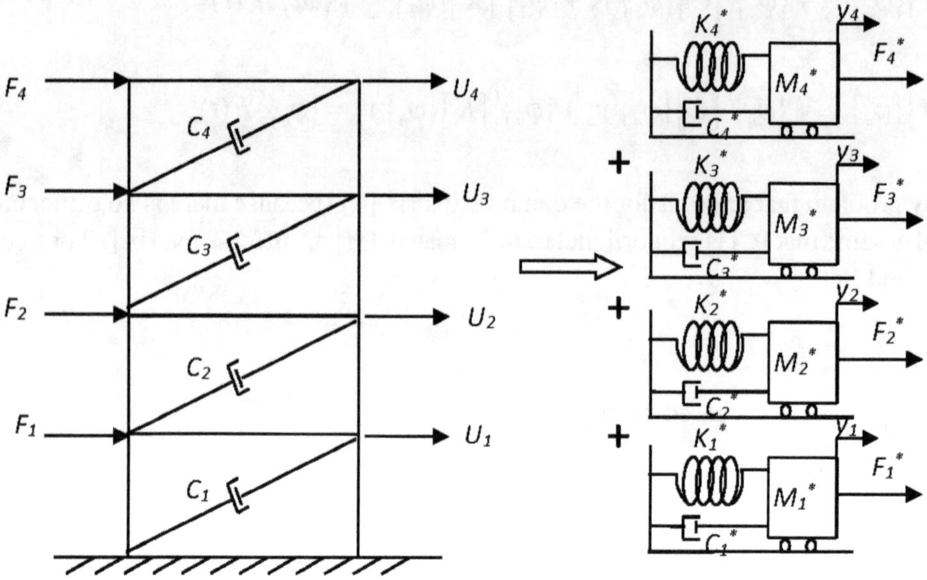

We can write the same system of equation as follows

$$
\begin{cases}
\ddot{y}_1 + \dfrac{\{\varphi_1\}^t[C]\{\varphi_1\}}{\{\varphi_1\}^t[M]\{\varphi_1\}}\dot{y}_2 + \dfrac{\{\varphi_1\}^t[K]\{\varphi_1\}}{\{\varphi_1\}^t[M]\{\varphi_1\}}y_1 = \dfrac{\{\varphi_1\}^t}{\{\varphi_1\}^t[M]\{\varphi_1\}}F(t) \\
\ddot{y}_2 + \dfrac{\{\varphi_2\}^t[C]\{\varphi_2\}}{\{\varphi_2\}^t[M]\{\varphi_2\}}\dot{y}_2 + \dfrac{[\varphi_2]^t[K]\{\varphi_2\}}{\{\varphi_2\}^t[M]\{\varphi_2\}}y_2 = \dfrac{\{\varphi_2\}^t}{\{\varphi_2\}^t[M]\{\varphi_2\}}F(t) \\
\vdots \\
\ddot{y}_n + \dfrac{\{\varphi_n\}^t[C]\{\varphi_n\}}{\{\varphi_n\}^t[M]\{\varphi_n\}}\dot{y}_2 + \dfrac{\{\varphi_n\}^t[K]\{\varphi_n\}}{\{\varphi_n\}^t[M]\{\varphi_n\}}y_n = \dfrac{\{\varphi_n\}^t}{\{\varphi_n\}^t[M]\{\varphi_n\}}F(t)
\end{cases}
$$

$$
\begin{cases}
\ddot{y}_1 + 2\xi_1\omega_1\dot{y}_1 + \omega_1^{\,2}y_1 = \dfrac{F_1^*(t)}{M_1^*} \\
\ddot{y}_2 + 2\xi_2\omega_2\dot{y}_2 + \omega_2^{\,2}y_2 = \dfrac{F_2^*(t)}{M_2^*} \\
\vdots \\
\ddot{y}_n + 2\xi_n\omega_n\dot{y}_n + \omega_n^{\,2}y_n = \dfrac{F_n^*(t)}{M_n^*}
\end{cases}
$$

The solution of this equation system is as follows:

$$\begin{cases} y_1(t) = e^{\xi_1\omega_1 t}\left(y_1(0)\cos\omega_{d1}t + \dfrac{\xi_1\omega y_1(0) + \dot{y}_1(0)}{\omega_{d1}}\sin\omega_{d1}t \right) + \dfrac{1}{M_1^*\omega_{d1}}\int_0^t e^{\xi_1\omega_1(t-t')}F_1^*(t')\sin\omega_{d1}(t-t')dt' \\ y_2(t) = e^{\xi_2\omega_2 t}\left(y_2(0)\cos\omega_{d2}t + \dfrac{\xi_2\omega_2 y_2(0) + \dot{y}_2(0)}{\omega_{d2}}\sin\omega_{d2}t \right) + \dfrac{1}{M_2^*\omega_{d1}}\int_0^t e^{\xi_2\omega_2(t-t')}F_2^*(t')\sin\omega_{d2}(t-t')dt' \\ \vdots \\ y_n(t) = e^{\xi_n\omega_n t}\left(y_n(0)\cos\omega_{dn}t + \dfrac{\xi_n\omega_n y_n(0) + \dot{y}_n(0)}{\omega_{dn}}\sin\omega_{dn}t \right) + \dfrac{1}{M_n^*\omega_{dn}}\int_0^t e^{\xi_n\omega_n(t-t')}F_n^*(t')\sin\omega_{dn}(t-t')dt' \end{cases} \tag{26}$$

We have just decomposed a system with several degrees of freedom to a system with several equations with only one degree of freedom for each equation.

MODAL ANALYSIS SUMMARY

Modal analysis represents a powerful tool in the analysis of any linear structure with concentrated mass subjected to any dynamic loading.

The steps to follow during a modal analysis are:

Step 1 Choice of (degree of freedom) Dof and calculation of mass and stiffness matrix:

a. Choose the most important Dof;
b. Built mass and stiffness matrix;
c. Assign a critical damping fraction for each mode.

Step 2 Calculation of natural frequencies and vibration modes:

The problem of eigenvalues and eigenvectors is obtained by:

$$\left[[K] - \omega_i^2[M]\right] = 0$$

Step 3 Calculation of mass and generalized dynamic loads:

The generalized mass is calculated, M_i, and the generalized dynamic load, $P_i(t)$ for each mode i:

$$M_n^* = \{\varphi_i\}^t[M]\{\varphi_i\} = 0......P_i^* = \{\varphi_i\}^t\{F(t)\} = 0$$

Step 4 Calculation of the initial conditions (if necessary):

If the structure is not previously resting, the initial conditions are calculated:

$$\{\varphi_i\}^t[M]\{u(t)\} = \{\varphi_i\}^t[M][\varphi]\{y(t)\}$$

$$\{y(0)\} = \frac{\{\varphi_i\}^t [M]\{u(0)\}}{\{\varphi_i\}^t [M]\{\varphi_i\}} = \frac{\{\varphi_i\}^t [M]\{u(0)\}}{M_i^*}$$

$$\{\dot{y}(0)\} = \frac{\{\varphi_i\}^t [M]\{\dot{u}(0)\}}{\{\varphi_i\}^t [M]\{\varphi_i\}} = \frac{\{\varphi_i\}^t [M]\{\dot{u}(0)\}}{M_i^*}$$

$$y_{i0} = \frac{\{\varphi_i\}^t [M]\{u(0)\}}{\{\varphi_i\}^t [M]\{\varphi_i\}} = \frac{\{\varphi_i\}^t [M]\{u(0)\}}{M_i^*}$$

$$\dot{y}_{i0} = \frac{\{\varphi_i\}^t [M]\{\dot{u}(0)\}}{\{\varphi_i\}^t [M]\{\varphi_i\}} = \frac{\{\varphi_i\}^t [M]\{\dot{u}(0)\}}{M_i^*}$$

Step 5 Calculation of undamped modal responses case:

The answer in each mode is as follows

$$y_i(t) = \underbrace{y_{i0} \operatorname{Cos} \omega_i t + \frac{\dot{y}_{i0}}{\omega_i} \operatorname{Sin} \omega_i t}_{\text{homogeneous solution (free case)}} + \underbrace{\frac{1}{M_i^* \omega_i} \int_0^t P_i(t) \operatorname{Sin} \omega_i (t - t')dt'}_{\text{particular solution}}$$

In a damped case

$$y_i(t) = \underbrace{e^{\xi_1 \omega_1 t} \left(y_i(0) \cos \omega_{di} t + \frac{\xi_1 \omega y_i(0) + \dot{y}_i(0)}{\omega_{di}} \sin \omega_{di} t \right)}_{\text{homogeneous solution (free case)}} + \underbrace{\frac{1}{M_i^* \omega_{di}} \int_0^t e^{\xi_i \omega_i (t-t')} F_i^*(t') \sin \omega_{di} (t - t')dt'}_{\text{particular solution}}$$

Step 6: Calculation of the response in the real coordinates

We superpose the modal responses to obtain the answer of each degree of freedom U_i (t).

$$\{U(t)\} = [\varphi]\{y(t)\} = 0$$

$$U_i(t) = \sum_{j=1}^{N_m} \varphi_i^{(j)} y_j(t)$$

COUPLED COORDINATES

We have an idealized system having two masses m and 2m respectively connected to two rigid bars with negligible mass, two stiffness springs K and 2K respectively connected to the end of these said bars.

The system thus proposed possesses two degrees of freedom; a vertical translation y and a rotation θ, as shown in figure (7a), the forces acting on the system free vibration are shown in figure (7b).

Figure 7a. Coupled coordinate system

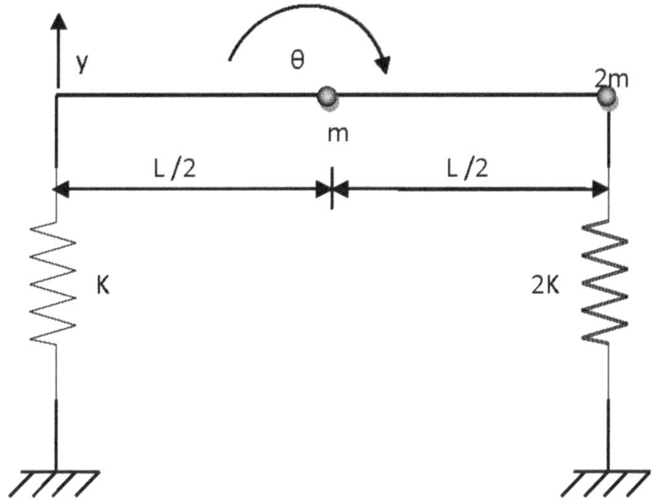

Figure 7b. Acting force on the system

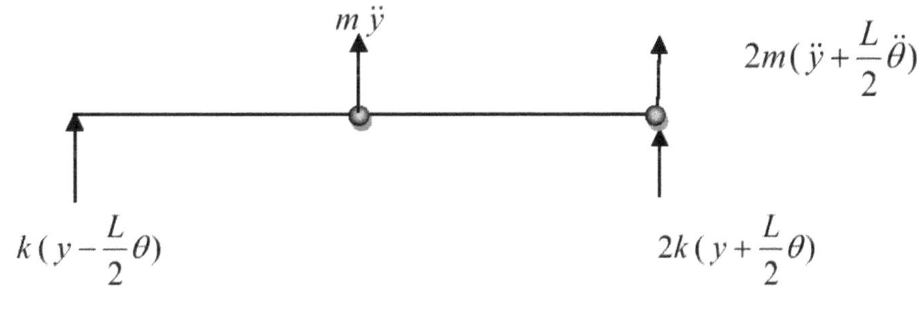

From figure 7b. We can deduce the motion equation system from the vertical component

$$m\ddot{y}+2m(\ddot{y}+\frac{L}{2}\ddot{\theta})+k(y-\frac{L}{2}\theta)+2k(y+\frac{L}{2}\theta)=0$$

From where

$$3m\,\ddot{y} + mL\ddot{\theta} + 3k\,y + \frac{kL}{2}\theta = 0 \tag{27}$$

Applying the moment of the forces with respect to the medium (the point of application of the mass m) we find:

$$-2m(\ddot{y} + \frac{L}{2}\ddot{\theta})\frac{L}{2} + k(y - \frac{L}{2}\theta)\frac{L}{2} - 2k(y + \frac{L}{2}\theta)\frac{L}{2} = 0$$

We have

$$mL\ddot{y} + \frac{1}{2}mL^2\ddot{\theta} + \frac{kL}{2}y + \frac{3}{4}kL^2\theta = 0 \tag{28}$$

Equation (27) and (28) are grouped together in a system of the following equation

$$\begin{cases} 3m\,\ddot{y} + mL\ddot{\theta} + 3k\,y + \dfrac{kL}{2}\theta = 0 \\ mL\ddot{y} + \dfrac{1}{2}mL^2\ddot{\theta} + \dfrac{kL}{2}y + \dfrac{3}{4}kL^2\theta = 0 \end{cases} \tag{29}$$

In matrix form, the system of equation (29) can be written as follows:

$$\begin{bmatrix} 3m & mL \\ mL & \frac{1}{2}mL^2 \end{bmatrix} \begin{Bmatrix} \ddot{y} \\ \ddot{\theta} \end{Bmatrix} + \begin{bmatrix} 3k & \frac{1}{2}kL \\ \frac{1}{2}kL & \frac{3}{2}kL^2 \end{bmatrix} \begin{Bmatrix} y \\ \theta \end{Bmatrix} = \begin{Bmatrix} 0 \\ 0 \end{Bmatrix} \tag{30}$$

Equation (30) shows the existence of a coupling between the coordinates; if the mass matrix is non-diagonal, we can speak about dynamic coupling, and if the stiffness matrix isnon-diagonal then we speak of static coupling.

Equation (30) reveals the existence of both dynamic and static coupling; the decoupling technique has been demonstrated in previous sections.

DISTRIBUTED MASS AND STIFFNESS SYSTEM

Theoretically, distributed parameter systems have an infinite number of degrees of freedom, modeled through partial differential equations.

In practice, multi-degree of freedom systems represent idealized continuous systems.

Figure 8. Beam with distributed parameters

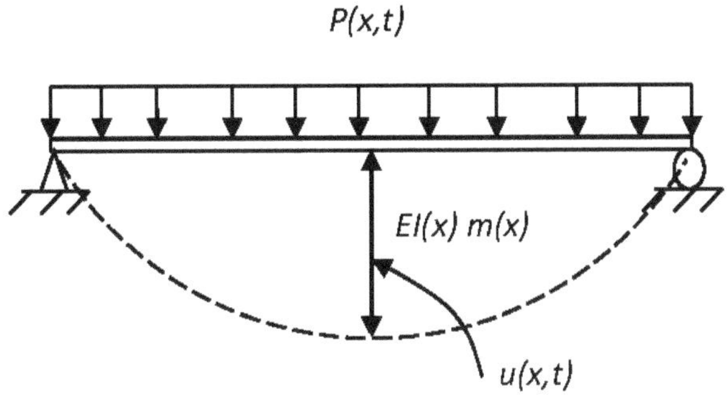

The equation of motion is presented as a superposition of the shape function. This latter is chosen to represent displacements; it must be unique, and satisfy the boundary conditions.

To do this we must define a series of generalized coordinates U_i that will serve as unknowns

$$U(x,t)=\sum_{i=1}^{n}U_i(t)\phi_i(x)=U_1(t)\phi_1(x)t)+U_2(t)\phi_2(x)+......U_n(t)\phi_n(x)$$

$\phi(x)$ Represents the shape function

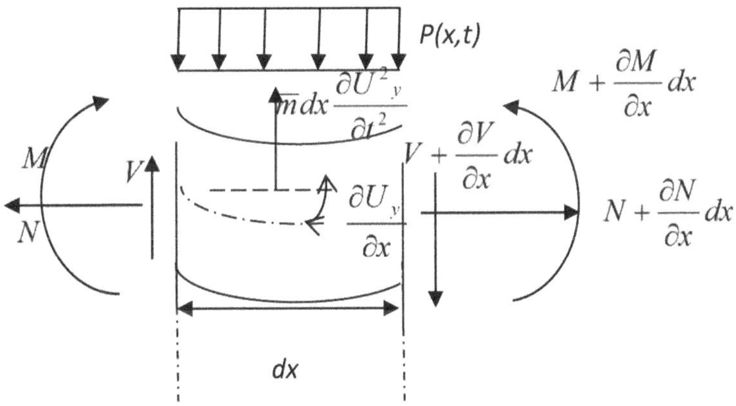

Equilibrium equation

$$\sum f_x : V-(V+\frac{\partial V}{\partial x}dx)+\bar{m}dx\frac{\partial^2 U_y}{\partial t^2}=P_y dx$$

$$\frac{\partial V}{\partial x}=-P_y+\bar{m}\frac{\partial^2 U_y}{\partial t^2} \tag{31}$$

$$\sum M: M-(M+\frac{\partial M}{\partial x}dx)+(V+\frac{\partial V}{\partial x}dx)dx-(N+\frac{\partial N}{\partial x}dx)\frac{\partial U_y}{\partial x}dx+P_y dx\frac{dx}{2}-\bar{m}dx\frac{\partial^2 U_y}{\partial t^2}\frac{dx}{2}$$

Ignoring higher order terms (dx^2) and (dx^3)

$$-\frac{\partial M}{\partial x}+V-N\frac{\partial U_y}{\partial x}=0 \quad (32)$$

With $M=-EI\frac{\partial^2 U_y}{\partial x^2}$

$$\frac{\partial}{\partial x}\left[\frac{\partial}{\partial x}(EI\frac{\partial^2 U_y}{\partial x^2})+V-N\frac{\partial U_y}{\partial x}=0\right]$$

$$\frac{\partial^2}{\partial x^2}(EI\frac{\partial^2 U_y}{\partial x^2})+\frac{\partial V}{\partial x}-\frac{\partial}{\partial x}(N\frac{\partial U_y}{\partial x})=0 \quad (33)$$

Replacing the expression (31) in (33), we find.

$$\bar{m}\frac{\partial^2 U_y}{\partial t^2}+\frac{\partial^2}{\partial x^2}(EI\frac{\partial^2 U_y}{\partial x^2})-\frac{\partial}{\partial x}(N\frac{\partial U_y}{\partial x})=P_y \quad (34)$$

Equation (34) is the strong form of the equation of equilibrium

VIRTUAL DISPLACEMENTS PRINCIPLE

We assume $U=U_y$ and $P=P_y$

$$\int_0^L \delta U\left(\bar{m}\frac{\partial^2 U}{\partial t^2}+\frac{\partial^2}{\partial x^2}(EI\frac{\partial^2 U}{\partial x^2})-\frac{\partial}{\partial x}(N\frac{\partial U}{\partial x})-P(x,t)\right)dx$$

$$\int_0^L \delta U\left(\bar{m}\frac{\partial^2 U}{\partial t^2}-P(x,t)\right)dx+\int_0^L\frac{\partial^2(\delta U)}{\partial x^2}(EI\frac{\partial^2 U}{\partial x^2})dx-\int_0^L\frac{\partial(\delta U)}{\partial x}(N\frac{\partial U}{\partial x})dx+$$
$$+\left[\delta U\frac{\partial}{\partial x}(EI\frac{\partial^2 U}{\partial x^2})-\frac{\partial(\delta U)}{\partial x}(EI\frac{\partial^2 U}{\partial x^2})-\delta U(N\frac{\partial U}{\partial x}\right]_0^L=0$$

Assuming that the boundary conditions (the limit terms) are equal to zero:

$$\left[\delta U \frac{\partial}{\partial x}(EI\frac{\partial^2 U}{\partial x^2}) - \frac{\partial(\delta U)}{\partial x}(EI\frac{\partial^2 U}{\partial x^2}) - \delta U(N\frac{\partial U}{\partial x}\right]_0^L = 0$$

$$\int_0^L \delta U\left(\bar{m}\frac{\partial^2 U}{\partial t^2} - P(x,t)\right)dx + \int_0^L \frac{\partial^2(\delta U)}{\partial x^2}(EI\frac{\partial^2 U}{\partial x^2})dx + \int_0^L \frac{\partial(\delta U)}{\partial x}(N\frac{\partial U}{\partial x})dx = 0 \quad (35)$$

Equation (35) is the weak form of the equation of equilibrium. Assuming:

$$\begin{cases} U(x,t) = \sum U_i(t)\phi_i(x) \\ \delta U(x,t) = \delta U_i(t)\phi_i(x) \end{cases}$$

By replacing in equation (35)

$$\sum \delta U_i \left[\sum \ddot{U}_j(t)\int_0^L \Psi_i(x)\bar{m}\Psi_j(x)dx + \sum U_j(t)\int_0^L \Psi_i''(x)EI\Psi_j''(x)dx \right.$$
$$\left. + \sum U_j(t)\int_0^L \Psi_i'(x)N\Psi_j'(x)dx - \int_0^L \Psi_i P(x,t)dx \right] = 0$$

Assuming that δUi = 1 and all the others δUj = 0 with i ¹ j

$$\sum \ddot{U}_j(t)\underbrace{\int_0^L \Psi_i(x)\bar{m}\Psi_j(x)dx}_{m_{ij}} + \sum U_j(t)\underbrace{\int_0^L \Psi_i''(x)EI\Psi_j''(x)dx}_{K_{ij}}$$
$$+ \sum U_j(t)\underbrace{\int_0^L \Psi_i'(x)N\Psi_j'(x)dx}_{K_{ij}^G} = \underbrace{\int_0^L \Psi_i P(x,t)dx}_{P_i}$$

$$\int_0^L \Psi_i(x)\bar{m}\Psi_j(x)dx = m_{ij} \quad (36)$$

$$\int_0^L \Psi_i''(x)EI\Psi_j''(x)dx = K_{ij} \quad (37)$$

$$\int_0^L \Psi'_i(x) N \Psi'_j(x) dx = K^G{}_{ij} \tag{38}$$

$$\int_0^L \Psi_i P(x,t) dx = P_i \tag{39}$$

$$\begin{Bmatrix} P_1 \\ P_2 \\ \vdots \\ P_n \end{Bmatrix} = \begin{bmatrix} m_{11} & m_{12} & \cdots & m_{1n} \\ m_{11} & m_{22} & \cdots & m_{2n} \\ \vdots & \vdots & \vdots & \vdots \\ m_{n1} & m_{n2} & \cdots & m_{nn} \end{bmatrix} \begin{Bmatrix} \ddot{U}_1 \\ \ddot{U}_2 \\ \vdots \\ \ddot{U}_n \end{Bmatrix} + \begin{bmatrix} k_{11} & k_{12} & \cdots & k_{1n} \\ k_{11} & k_{22} & \cdots & k_{2n} \\ \vdots & \vdots & \vdots & \vdots \\ k_{n1} & k_{n2} & \cdots & k_{nn} \end{bmatrix} \begin{Bmatrix} U_1 \\ U_2 \\ \vdots \\ U_n \end{Bmatrix} + \begin{bmatrix} k^G{}_{11} & k^G{}_{12} & \cdots & k^G{}_{1n} \\ k^G{}_{11} & k^G{}_{22} & \cdots & k^G{}_{2n} \\ \vdots & \vdots & \vdots & \vdots \\ k^G{}_{n1} & k^G{}_{n2} & \cdots & k^G{}_{nn} \end{bmatrix} \begin{Bmatrix} U_1 \\ U_2 \\ \vdots \\ U_n \end{Bmatrix}$$

Where n is the number of degrees of freedom

$$\{P(t)\} = [M]\{\ddot{U}(t)\} + ([K] + [K^G])\{U(t)\} = \{P(t)\} \tag{40}$$

Note

- The equations above did not take into account the dumping because it is not a realistic way to evaluate it; approximate methods assume that the dumping is proportional to the rate of deformation.
- The normal force N is positive (traction), if it is negative (compression), in this case the stiffness of system will be reduced.

HYDRODYNAMIC STUDY

Introduction

When a covered tank is completely full, there is naturally no relative movement of the fluid with respect to the tank because of excitation. From a dynamic point of view, everything happens if the fluid-reservoir assembly constituted a single mass. On the other hand, in partially filled tanks, the excitation puts part of the fluid in motion, which leads to the formation of waves on the surface.

In fact, Newmark has shown that a filling defect of 2% in height is sufficient for closed tanks to behave like free surface tanks from the point of view of wave formation. This observation will justify taking into account only tanks with free surfaces.

There are methods, which aim at ensuring an acceptable protection of tank against the effects of seismic actions by an appropriate design and dimensioning.

Among these methods, we will use the Houzner method, given its simplicity and the fact that it applies regardless of the filling rate of the tank.

APPROXIMATE MODEL CALCULATION ACCORDING TO HOUZNER

We know that in general, it is when the tank is partially filled that the dynamic problem is important. In this case, the excitation puts part of the fluid in motion, creating waves on the surface. The liquid in the tank will be considered incompressible, and the tank will be assumed to be rigidly linked to the ground, which gives it the same acceleration of the ground.

In Houzner's method, we break down the action of the liquid into:

- Passive action causing impulse effects
- An active action causing oscillation efforts

The impulse force comes from the fact that part of the fluid mass (called passive mass) reacts by inertia to the translation of walls tank. Its equivalent mechanical system is obtained by considering a mass (M_i) rigidly linked to the tank at a height hi such that it exerts on the walls the same horizontal forces as the equivalent mass of water.

The oscillation forces come from a part of the fluid mass (called active) which starts to oscillate under the action of the earthquake.

Its equivalent mechanical system is a mass (M_0) retained by stiffness springs K_0 at a level h_0 whose horizontal oscillations exert the same vibratory forces as the active mass of water.

We will therefore have:

Figure 9. Equivalent mechanical system of an active and passive mass

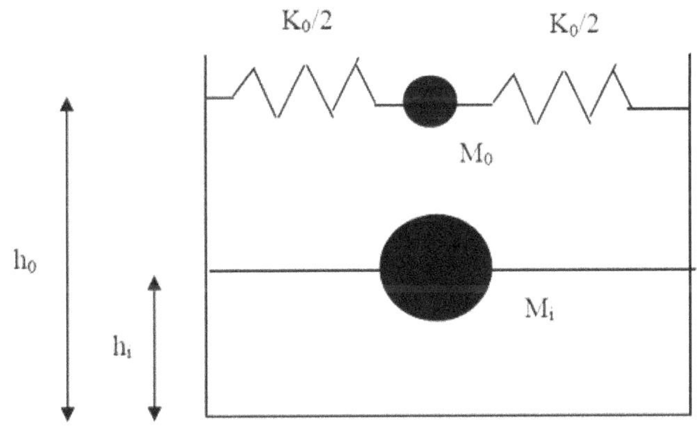

In a general case, the equivalent mechanical system is obtained by considering n masses M_{0n} retained by stiffness springs K_n at levels des at levels h_{on} or h^*_{on}, whose horizontal oscillations exert the same vibratory forces as the active mass of fluid.

Figure 10. Cylindrical tank physical system and equivalent mechanical system
a) Equivalent mechanical system of impulse pressures b) Equivalent mechanical system of the oscillation pressures: action on the walls, c) Equivalent mechanical system of oscillation pressures: action on the walls and on the base

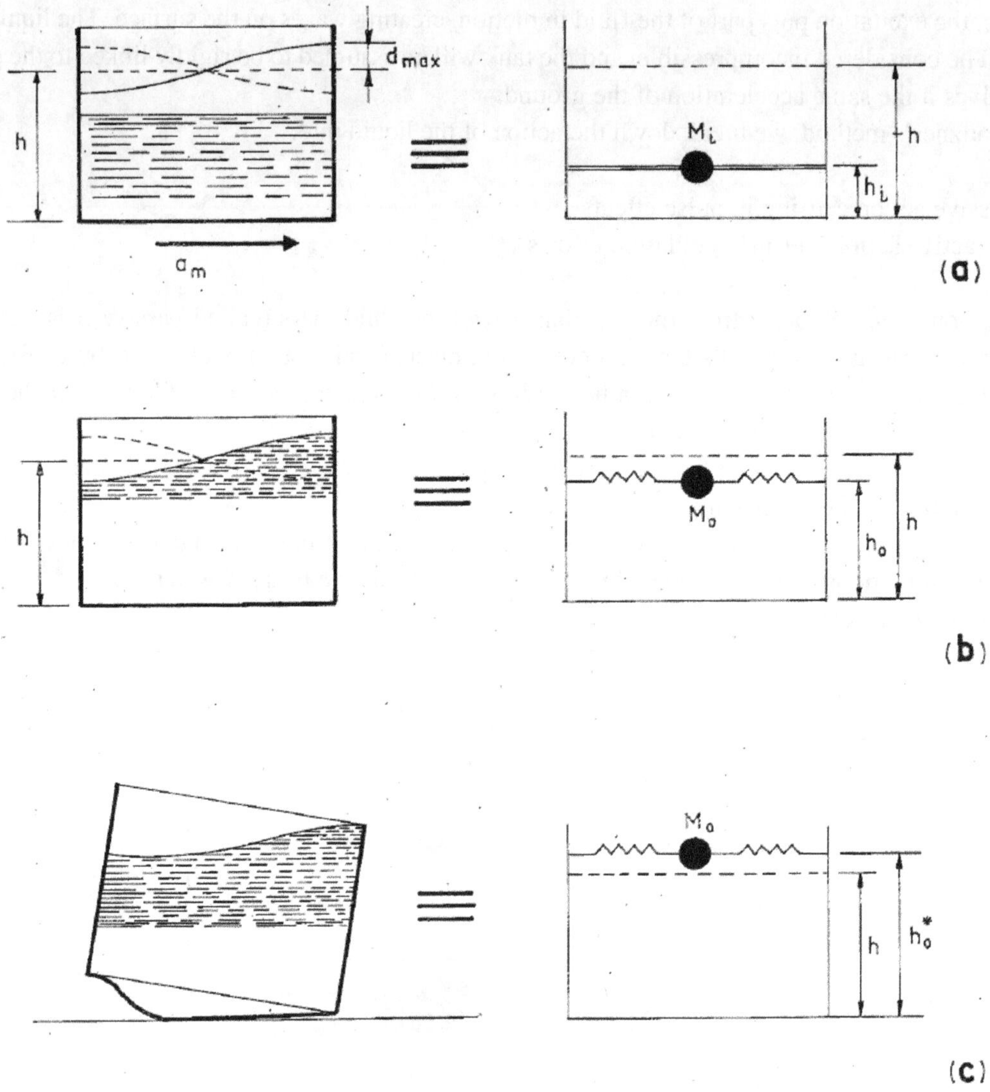

For the calculation of the bending moment, the only actions taken into account are those on the walls, in this case, the masses M_{0n} are applied at a level h_{on} (see figure 10.b)

For the calculation of the overturning moment, we take into account the action of the overpressures on the tank bottom; in this case, the masses M_{0n} are applied to a level h^*_{on} (see figure 10.c) thus the model which one will retain for the set of the two types of actions will be that of figure bellow.

Figure 11. Model has a passive mass M_i (impulse) and two active masses M_{01} and M_{02} (Oscillation)

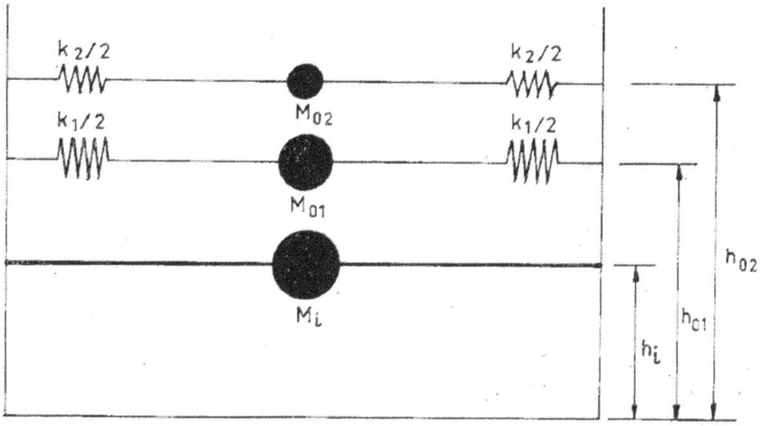

Figure 12. Cylindrical tank subjected to maximum acceleration am

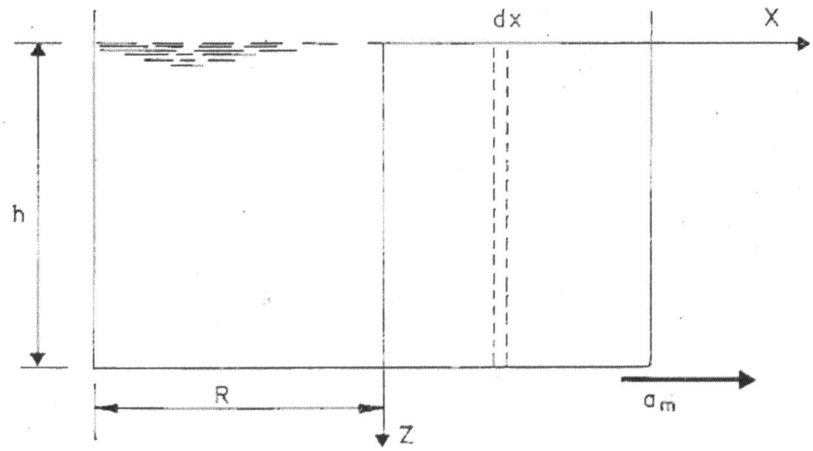

IMPULSE ACTION

Considering a cylindrical tank with horizontal base and vertical walls subjected to a maximum acceleration a_m as defined by the figure bellow.

If we neglect the compressibility of the liquid, by expressing the principle of mass conservation and the fundamental principle of dynamics, we determine the following expression of the hydrodynamic pressure exerted on the walls tank.

$$P = -\rho^2 \left[\frac{z}{h} - \frac{1}{2} \left(\frac{z}{h} \right)^2 \right] \frac{d\dot{v}}{dt} \tag{41}$$

Where V is the liquid velocity in the ox direction and

$$\dot{V} = a_m \frac{ch\sqrt{3}\frac{x}{h}}{ch\sqrt{3}\frac{R}{h}} \quad (42)$$

The pressure distribution (direction ox) is

$$P = -\rho a_m h\sqrt{3}\left[\frac{z}{h} - \frac{1}{2}\left(\frac{z}{h}\right)^2\right] th\sqrt{3}\frac{R}{h} \quad (43)$$

By integrating, we obtain the result of horizontal hydrodynamic impulse pressures

$$P_i = -\rho a_m \pi R^2 h \frac{th\sqrt{3}\frac{R}{h}}{\sqrt{3}\frac{R}{h}} \quad (44)$$

We put $\varepsilon_i = \frac{th\sqrt{3}\frac{R}{h}}{\sqrt{3}\frac{R}{h}}$ (45)

We obtain

$$P_i = -\rho a_m \pi h R^2 \varepsilon_i \quad (46)$$

Thus $P_i = a_m M_i$ (47)

OSCILLATION ACTION

On the other hand, the kinetic energy of the whole system, we establish the expression of the distribution of the hydrodynamic overpressures, according to the angular coordinate θ, in the form

$$P_i = \rho \frac{R^2}{3}\sqrt{\frac{27}{8}}\left(1 - \frac{\cos^2\theta}{3} - \frac{\sin^2\theta}{2}\right)\cos\theta \frac{ch\sqrt{\frac{27}{8}}\frac{h-z}{R}}{sh\sqrt{\frac{27}{8}}\frac{h}{R}} \varnothing_o \omega_o^2 \sin\omega_o t \quad (48)$$

The fundamental vibration pulsation of the liquid is:

$$\omega_O^2 = \frac{g}{R}\sqrt{\frac{27}{8}}th\sqrt{\frac{27}{8}}\frac{h}{R} \tag{49}$$

The oscillation angle is

$$\varnothing = \varnothing_o \frac{sh\sqrt{\frac{27}{8}}\frac{h-z}{R}}{sh\sqrt{\frac{27}{8}}\frac{h}{R}} \sin \omega_o t \tag{50}$$

ϕ_0: Maximum oscillation angle (at Z = 0)

By integrating p on Z and θ, we obtain the result of the horizontal hydrodynamic oscillation pressures

$$P_O = \rho \frac{10}{48} \pi R^4 \varnothing_O \omega_O^2 \sin \omega_O t$$

Therefore

$$P_O = 1.2 M_O g \varnothing_O \sin \omega_O t \tag{51}$$

The maximum of this equation (51) is obtained for $\sin\omega_0 t=1$

$$P_0 = 1.2\ M_0 g \phi_0 \tag{52}$$

APPLICATIONS

Example 1

The building below has two floors below, with it's two modes determine the free vibration response due to the initial displacement $U(0) = \left\langle -\frac{1}{2} \quad 2 \right\rangle^t$.

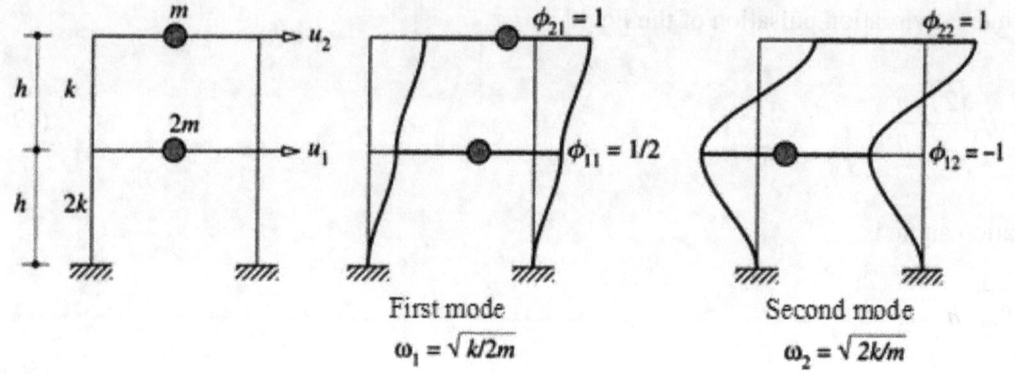

Solution

Using equation (20) we find:

$$\begin{cases} y_n(0) = \dfrac{\{\varphi_n\}^t M\{U(0)\}}{\{\varphi_n\}^t [M]\varphi_n} = \dfrac{\{\varphi_n\}^t M\{U(0)\}}{M_n} \\ \dot{y}_n(0) = \dfrac{\{\varphi_n\}^t M\{\dot{U}(0)\}}{\{\varphi_n\}^t [M]\{\varphi_n\}} = \dfrac{\varphi_n{}^t M\{\dot{U}(0)\}}{M_n} \end{cases}$$

We must determine the generalized mass $M_n = \{\phi n_j t^t M]\phi n$

$$M_1 = \{\varphi_1\}^t [M]\varphi_1 = \begin{Bmatrix} \frac{1}{2} & 1 \end{Bmatrix} \begin{bmatrix} 2m & 0 \\ 0 & m \end{bmatrix} \begin{Bmatrix} \frac{1}{2} \\ 1 \end{Bmatrix} = \frac{3m}{2}$$

$$M_2 = \{\varphi_2\}^t [M]\varphi_2 = \{1 \quad -1\} \begin{bmatrix} 2m & 0 \\ 0 & m \end{bmatrix} \begin{Bmatrix} 1 \\ -1 \end{Bmatrix} = 3m$$

$$\begin{cases} y_1(0) = \dfrac{\{\varphi_1\}^t [M]\{U_1(0)\}}{M_1} = \dfrac{\begin{Bmatrix} \frac{1}{2} & 1 \end{Bmatrix} \begin{bmatrix} 2m & 0 \\ 0 & m \end{bmatrix} \begin{Bmatrix} -\frac{1}{2} \\ 2 \end{Bmatrix}}{\frac{3m}{2}} = \dfrac{\frac{3m}{2}}{\frac{3m}{2}} = 1 \\ \dot{y}_1(0) = \dfrac{\{\varphi_1\}^t [M]\{\dot{U}_1(0)\}}{M_1} = \dfrac{\begin{Bmatrix} \frac{1}{2} & 1 \end{Bmatrix} \begin{bmatrix} 2m & 0 \\ 0 & m \end{bmatrix} \begin{Bmatrix} 0 \\ 0 \end{Bmatrix}}{\frac{3m}{2}} = \dfrac{0}{\frac{3m}{2}} = 0 \end{cases}$$

$$\begin{cases} y_2(0) = \dfrac{\{\varphi_2\}^t [M]\{U(0)\}}{M_2} = \dfrac{\{-1 \quad 1\}\begin{bmatrix} 2m & 0 \\ 0 & m \end{bmatrix}\begin{Bmatrix} -\frac{1}{2} \\ 2 \end{Bmatrix}}{3m} = \dfrac{3m}{3m} = 1 \\ \dot{y}_2(0) = \dfrac{\{\varphi_2\}^t [M]\{\dot{U}(0)\}}{M_2} = \dfrac{\{-1 \quad 1\}\begin{bmatrix} 2m & 0 \\ 0 & m \end{bmatrix}\begin{Bmatrix} 0 \\ 0 \end{Bmatrix}}{3m} = \dfrac{0}{3m} = 0 \end{cases}$$

The solution of an undamped free vibration in the modal base is as follows:

$$y_1(t) = y_1(0)\cos\omega_1 t + \frac{\dot{y}_1(0)}{\omega_1}\sin\omega_1 t \qquad y_2(t) = y_2(0)\cos\omega_2 t + \frac{\dot{y}_2(0)}{\omega_2}\sin\omega_2 t$$

By replacing the values of the initial conditions at the level of the natural coordinates:

$y_1(t)$=1cosω1$_t$ … $y2_($t)=1cosω2t

By replacing the expressions of y1 (0) and y2 (0) in equation (19) we find

$$\begin{Bmatrix} U_1(t) \\ U_2(t) \end{Bmatrix} = \begin{Bmatrix} \frac{1}{2} \\ 1 \end{Bmatrix}\cos\omega_1 t + \begin{Bmatrix} -1 \\ 1 \end{Bmatrix}\cos\omega_2 t = \begin{Bmatrix} \frac{1}{2} \\ 1 \end{Bmatrix} y_1(t) + \begin{Bmatrix} -1 \\ 1 \end{Bmatrix} y_2(t)$$

This last expression U as a function of y from which we can see that the two natural modes contributing to the answer this is due to the initial conditions.

Example 2 (Smith 2010)

The two-story building of the following figure is considered. We are interested in oscillations of the high floor. The intermediate floor have a negligible mass.

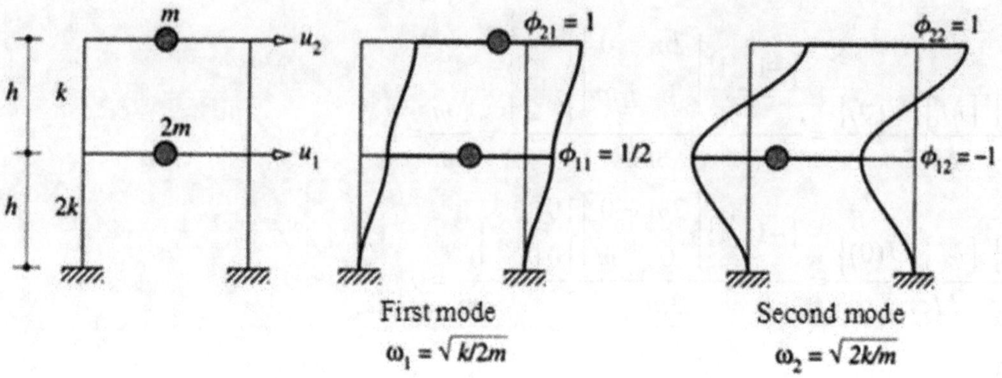

E = 200 000 MPa

I = 200 10^6 mm^4

H = 3 m

M = 2000 kg

1. We ask to draw the fundamental mass-spring system.
2. Determine the own pulsation of the structure?

Solution

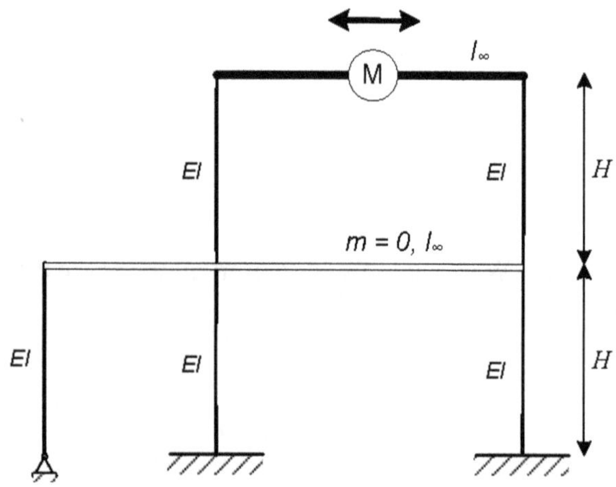

Soient:

$K_{éq-1}$: The equivalent stiffness of the three columns of the ground floor:

$$K_{éq-1} = \frac{3EI}{H^3} + 2.\frac{12EI}{H^3} = \frac{27EI}{H^3}$$

$K_{éq-2}$: The equivalent stiffness of the two columns of the first floor

$$K_{éq-2} = 2.\frac{12EI}{H^3} = \frac{24EI}{H^3}$$

$K_{éq-1}$ and $K_{éq-2}$: are in series, the equivalent stiffness of the structure is therefore written

$$\frac{1}{K_{eq}} = \frac{1}{K_{eq-1}} + \frac{1}{K_{eq-2}} = \frac{K_{eq-1}K_{eq-2}}{K_{eq-1} + K_{eq-2}} = \frac{216EI}{17H^3}$$

We obtain:

$E=200\times10^9 N/m^2$ … $I=2\times10^{-4} m^4$

$$K_{éq-1} = 3.56\ 10^7 N / m \quad K_{éq-2} = 4.0\ 10^7 N / m \quad K_{éq} = 1.88\ 10^7 N / m$$

The own pulsation of the structure: $\omega = \sqrt{\frac{K_{eq}}{M}} = 97 ras / s$

Example 3 (Smith 2010)

The following figure shows a steel frame on which a motor is placed. The engine is still supported on the rubber blocks. Determine the mass matrices and stiffness for the following degrees of freedom (U_i):

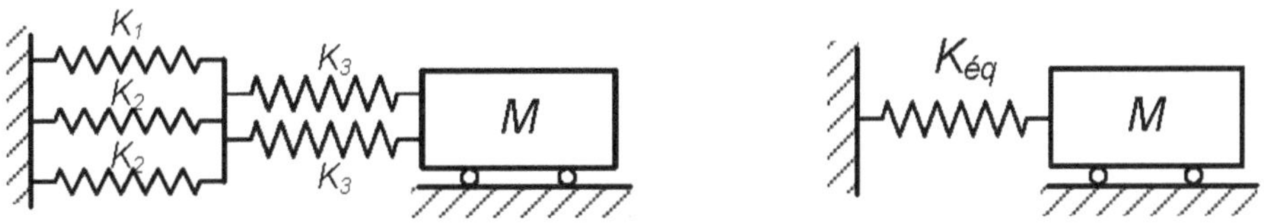

We ask to calculate Eigen frequencies and eigenmodes.

Solution

a) $M = \begin{bmatrix} 6000 & 0 \\ 0 & 10000 \end{bmatrix}$ (kg)

$$\mathrm{K} = \begin{bmatrix} K_1 + K_2 & -K_2 \\ -K_2 & K_2 \end{bmatrix} = \begin{bmatrix} 5060 & -2000 \\ -2000 & 2000 \end{bmatrix} \cdot 10^3 \,(\mathrm{N/m})$$

b) $|K - \omega 2^{M}| = 0$

$$\begin{vmatrix} 5060 \cdot 10^3 - \omega^2 6000 & -2000 \cdot 10^3 \\ -2000 \cdot 10^3 & 2000 \cdot 10^3 - \omega^2 10000 \end{vmatrix} = 0$$

$\Leftrightarrow (5060\times10^3 - \omega^2\, 6000)\,(2000\times10^3 - \omega^2 1000) - (2000\times10^3)^2 = 0$

$\Leftrightarrow 6{,}0 * 10^7\, \omega^4 - 6{,}26 * 10^{10}\, \omega^2 + 6{,}12 * 10^{12} = 0$

$$\Leftrightarrow \omega^2 = \frac{6{,}26*10^{10} \pm \sqrt{(6{,}26*10^{10})^2 - 4*6{,}0*10^7*6{,}12*10^{12}}}{2*6{,}0*10^7}$$

$\Leftrightarrow \omega 2 = 934 \rightarrow \omega 2 \,_{=}\, 30.6$ rad$s$-1

$\Leftrightarrow \omega 2 = 19 \rightarrow \omega 1 = 10.5$ rad s-1

Eigenvectors:

For $\omega 2 = 934$rad$s$-1

$$\begin{bmatrix} -544 \cdot 10^3 & -2000 \cdot 10^3 \\ -2000 \cdot 10^3 & 7340 \cdot 10^3 \end{bmatrix} \begin{Bmatrix} \Phi_{12} \\ \varphi_{22} \end{Bmatrix} = \begin{Bmatrix} 0 \\ 0 \end{Bmatrix}$$

We have (for $\phi_{12} = 1$):

$\Phi_{12} = 1$ and $\Phi_{22} = -0.27$

For $\omega 1 = 10.5$ rads^{-1}

$$\begin{bmatrix} 4406 \cdot 10^3 & -2000 \cdot 10^3 \\ -2000 \cdot 10^3 & 910 \cdot 10^3 \end{bmatrix} \begin{Bmatrix} \Phi_{11} \\ \Phi_{21} \end{Bmatrix} = \begin{Bmatrix} 0 \\ 0 \end{Bmatrix}$$

We have (for $\phi_{11} = 1$):

$\Phi_{11} = 1$ and $\phi_{21} = 2.2$

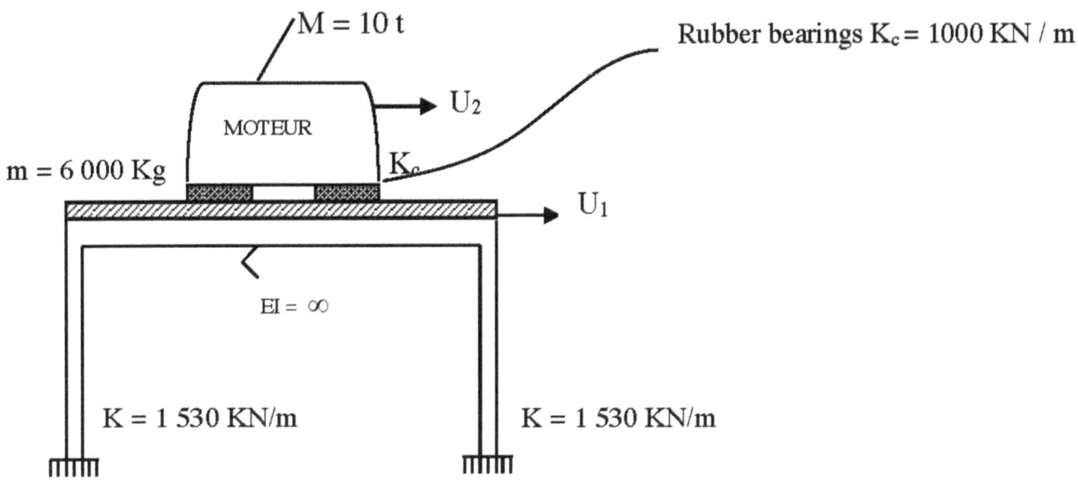

Example 4

A three mass oscillator models represent a three-story building with the features below.

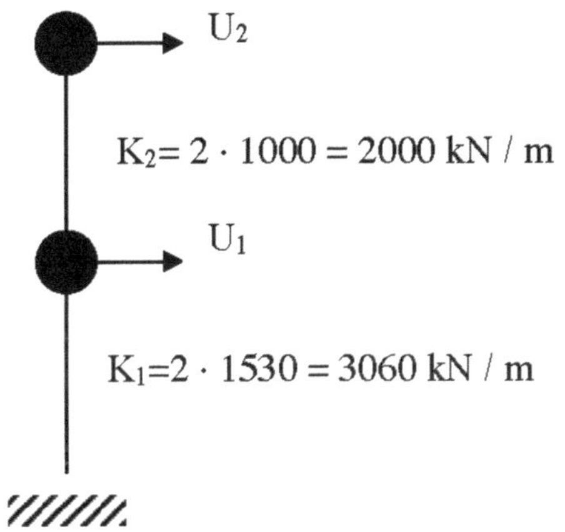

$$K = \frac{6EI}{h^3}\begin{bmatrix} 3.0769 & -1.7692 & 0.4615 \\ -1.7692 & 1.6923 & -0.6154 \\ 0.4615 & -0.6154 & 0.2692 \end{bmatrix}$$

$$\omega_{n,2} = 90.56\frac{rad}{s};\omega_{n,3} = 243.32\frac{rad}{s}$$

1. Determine the mass matrix.
2. Determine the eigenvectors as well as the generalized stiffness and masses.

Solution

A three mass oscillator models a three-story building with the features below.

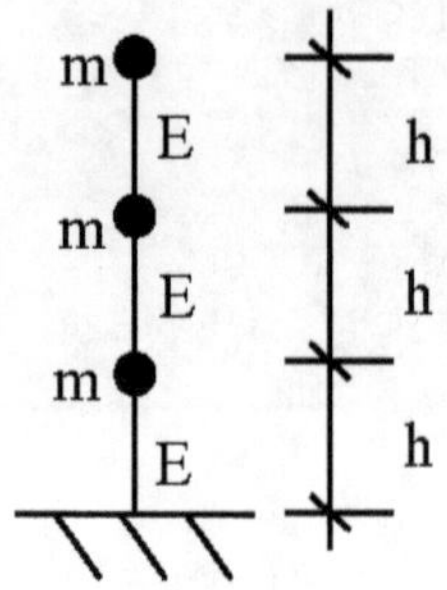

$$\underline{K} = \begin{bmatrix} 33026 & -18990 & 4954 \\ -18990 & 18164 & -6605 \\ 4954 & -6605 & 2890 \end{bmatrix} \cdot 10^3 \frac{kN}{m}$$

Determination of the mass matrix

$$\underline{M} = \begin{bmatrix} 800 & 0 & 0 \\ 0 & 800 & 0 \\ 0 & 0 & 800 \end{bmatrix} t$$

Determination of the eigenvectors as well as the rigidities and the generalized masses.

In order to determine the eigenvectors, we must solve the equation $\left(K - \omega_i^2 M\right)\phi_i = 0$ for each own frequency.

Calculation of the first eigenvector:

$$K - \omega_1^2 M = \begin{bmatrix} 32873 & -18990 & 4954 \\ -18990 & 18011 & -6605 \\ 4954 & -6605 & 2737 \end{bmatrix} \cdot 10^3$$

Choose modal deformation in level 3 $\varphi_{31} = 1$

Determine the modal deformations of level 1 and 2:

$$\text{(I)} \quad 3.2873\phi_{11} - 1.8990\phi_{21} + 0.4954 = 0$$

$$\text{(II)} \quad -1.8990\phi_{11} + 1.8011\phi_{21} - 0.6605 = 0 \Rightarrow \phi_{21} = 0.3667 + 1.054\phi_{11}$$

$$\text{(I)'} \quad 3.2873\phi_{11} - 1.8990(0.3667 + 1.054\phi_{11}) + 0.4954 = 0 \Rightarrow \phi_{11} = 0.1563$$

$$\phi_{21} = 0.3667 + 1.054 \cdot 0.1563 \Rightarrow \phi_{21} = 0.5314$$

$$\phi_1^T = [0.1563 \quad 0.5314 \quad 1]$$

Using this method for the other two eigenvectors, we find φ

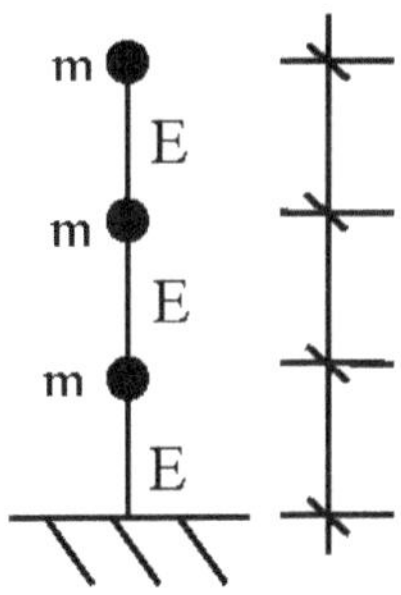

$$\phi = \begin{bmatrix} 0.16 & 0.84 & 1 \\ 0.53 & 1 & -0.70 \\ 1 & -0.66 & 0.22 \end{bmatrix} \text{ Generalized mass matrix}$$

$$M^* = \phi^T M \phi = \begin{bmatrix} 1.05 & 0 & 0 \\ 0. & 1.72 & 0 \\ 0 & 0 & 1.23 \end{bmatrix} 10^3 t \text{ Generalized stiffness matrix}$$

$$K^* = \phi^T K \phi = \begin{bmatrix} 200 & 0 & 0 \\ 0. & 14100 & 0 \\ 0 & 0 & 72700 \end{bmatrix} 10^3 \frac{kN}{m}$$

Example 5

The system that represents a car shown in the following figure the values corresponding to this system are as follows

W = 150N, L1 = 1.35m, L2 = 1.65m, K_1 = 360N / m, K_2 = 370N / m J_c = 27m4

1. Determine the equations of motion
2. Determine the eigenvalues of the system

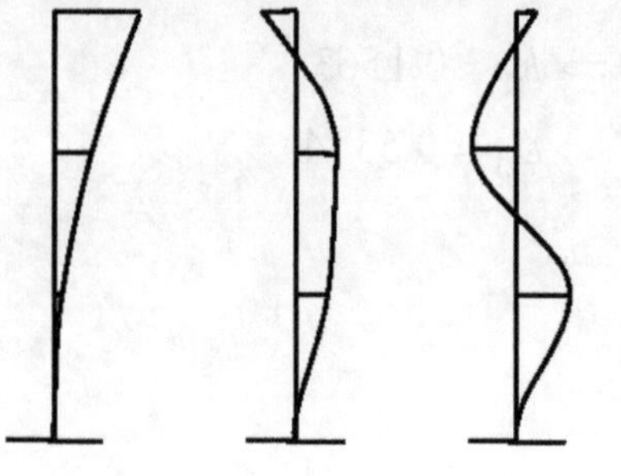

Solution

In a modest form the car is treated as a rigid body; it is connected by springs at the front and at the back; the system above will have two degrees of freedom (y & θ) we put L1 + L2 = L.

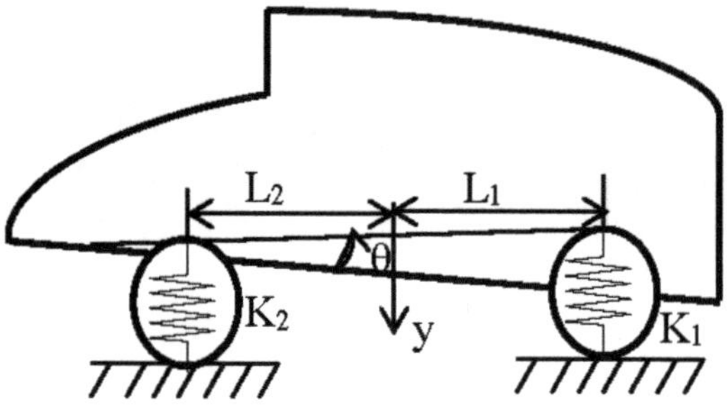

The equation of motion of the system is as follows:

$$\begin{cases} m\,\ddot{y} + k_1(y - L_1\theta) + k_2(y + L_2\theta) = 0 \\ J_c\,\ddot{\theta} - k_1(y - L_1\theta)L_1 + k_2(y + L_2\theta)L_2 = 0 \end{cases}$$

$$\begin{cases} m\,\ddot{y} + (k_1 + k_2)y + \theta(k_2L_2 - k_1L_1) = 0 \\ J_c\,\ddot{\theta} + (-k_1L_1 + k_2L_2)y + \theta(k_2L_2^2 - k_1L_1^2) = 0 \end{cases}$$

$$\begin{bmatrix} m & 0 \\ 0 & J_c \end{bmatrix}\begin{Bmatrix} \ddot{y} \\ \ddot{\theta} \end{Bmatrix} + \begin{bmatrix} k_2 + k_1 & -k_1L_1 + k_2L_2 \\ -L_1k_1 + L_2k_2 & -L_1^2\,k_1 + L_2^2\,k_2 \end{bmatrix}\begin{Bmatrix} y \\ \theta \end{Bmatrix} = \begin{Bmatrix} 0 \\ 0 \end{Bmatrix}$$

Determination of eigenvalues

$$\det[K] - \omega_i^2[M] = 0$$

$$\begin{vmatrix} \left(k_2+k_1-\omega_i^2 m\right) & -k_1L_1+k_2L_2 \\ -k_1L_1+k_2L_2 & \left(-k_1L_1^2+k_2L_2^2-\omega_i^2 J_c\right) \end{vmatrix} = \begin{vmatrix} 10-15.29\omega_i^2 & 1596 \\ -1096.5 & -1175.4-27\omega_i^2 \end{vmatrix}$$

$412.83\omega_i^4+17701.866\,\omega_i^2+1738260=412.83\lambda_i^2+17701.866\,\lambda_i+1738260$ with $\lambda_i=\omega_i^2$

Example 6 (Commend & Zimmermann 2001)

Let have the structure (shear building) defined below:

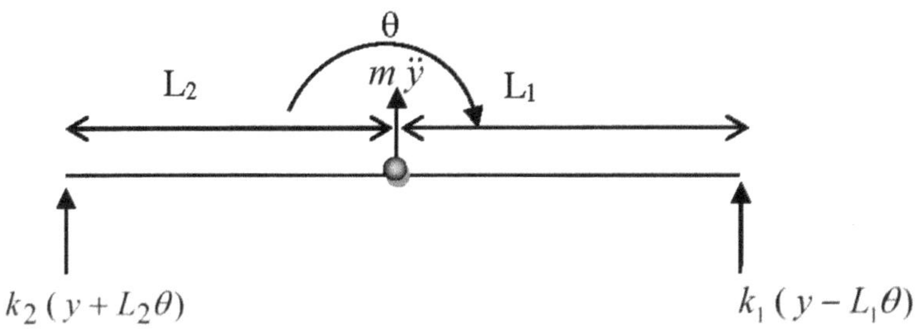

1. Determine the modes and natural frequencies of the structure?
2. Determine the normalize the vector with respect to mass?

Solution

The stiffness matrix is as follows

$$\mathbf{K}=\begin{bmatrix} 6 & -2 & 0 \\ -2 & 3 & -1 \\ 0 & -1 & 1 \end{bmatrix}$$

The mass matrix is defined by isolated masses at degrees of freedoms, namely:

$$\mathbf{M}=\begin{bmatrix} 6 & 0 & 0 \\ 0 & 2 & 0 \\ 0 & 0 & 1 \end{bmatrix}$$

The eigenvalues are the solutions of:

$$P(\omega) = \left\| \mathbf{K} - \omega \mathbf{M} \right\| = \left| \begin{bmatrix} 6-6\omega & -2 & 0 \\ -2 & 3-2\omega & -1 \\ 0 & -1 & 1-\omega \end{bmatrix} \right| = (\omega - 0.3035)\cdot(\omega - 1)\cdot(\omega - 2.1965) = 0$$

From where *:*

*ω1≅0.3rad/s; ω2=*1ras/s*; ω3≈2.2ras/s ;*

Eigenvalues are the solutions of $([K] - \omega i[M])\{\Phi i\}=0$
Take $\omega 2=1$ $([K] - \omega 2[M])\{\Phi 2\}=0$ From where:

$$\begin{bmatrix} 0 & -2 & 0 \\ -2 & 1 & -1 \\ 0 & -1 & 0 \end{bmatrix} \begin{bmatrix} \phi_{12} \\ \varphi_{22} \\ \varphi_{32} \end{bmatrix} = \begin{bmatrix} 0 \\ 0 \\ 0 \end{bmatrix}$$

Let's put $\phi_{12}=1$. The first equation gives $\phi_{22}=0$, and the second $\phi_{32}= -2\bullet\phi 12= -2$, where:

Table 1.

eignevalue (ras/s)	Period(s)	Modes
$\omega_1 = 0.3$	$T_1 = 11.405$	Mode 1 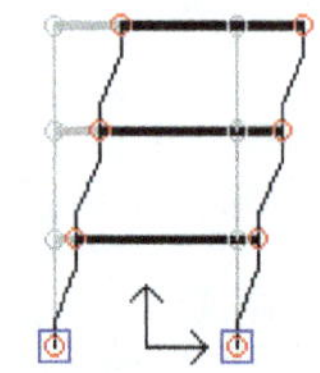
$\omega_2 = 1.000$	$T_2 = 6.283$	Mode 2 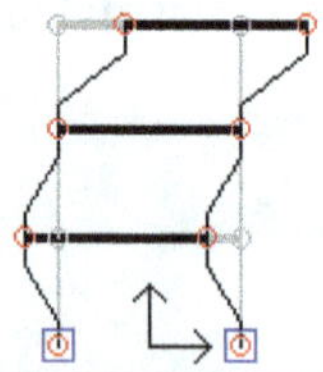
$\omega_3 = 2.2$	$T_3 = 4.240$	Mode 3 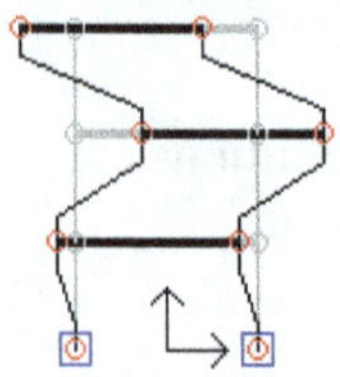

$$\varphi_2 = \begin{bmatrix} \varphi_{12} \\ \varphi_{22} \\ \varphi_{32} \end{bmatrix} = \begin{bmatrix} 1 \\ 0 \\ -2 \end{bmatrix}$$

Similarly, for $\lambda 1 \cong 0.3$ et $\lambda 3 \cong 2.2$:

$$\varphi_1 = \begin{bmatrix} \varphi_{11} \\ \varphi_{21} \\ \varphi_{31} \end{bmatrix} = \begin{bmatrix} 0.48 \\ 1 \\ 1.43 \end{bmatrix} \text{ and } \varphi_3 = \begin{bmatrix} \varphi_{13} \\ \varphi_{23} \\ \varphi_{33} \end{bmatrix} = \begin{bmatrix} 0.28 \\ 1 \\ -0.83 \end{bmatrix}$$

Normalization with respect to mass

Take $\alpha\, \Phi_1^{\,T} \mathbf{M} \alpha\, \varphi_1 = 1$, from where $\alpha \cong 0.43$, and:

$$\varphi_{n1} = \begin{bmatrix} 0.20527 \\ 0.42892 \\ 0.61583 \end{bmatrix}; \varphi_{n2} = \begin{bmatrix} 0.31624 \\ 0 \\ -0.63243 \end{bmatrix}; \varphi_{n3} = \begin{bmatrix} 0.15662 \\ -0.56217 \\ 0.46985 \end{bmatrix}$$

Example 7 (Smith 2010)

Let us present a diving board with a density of 330 Kg / m3 and a modulus of elasticity E = 80 GPa.

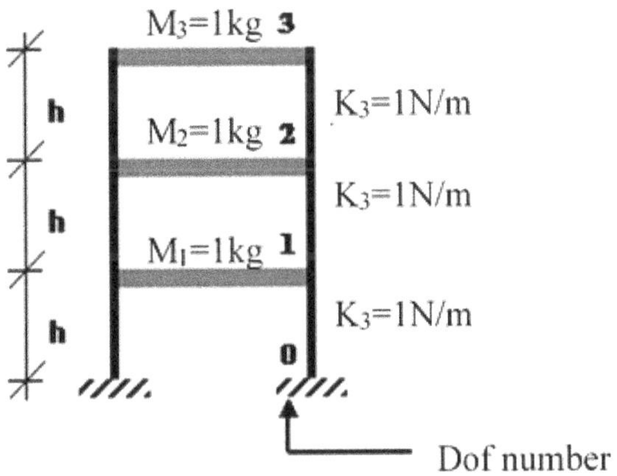

Using for the diving board the deformed function, $u(x,t) = \left(\dfrac{x^2 - 2,5x}{16,5} \right) z(t)$.

1. Draw qualitatively the deformation for z (t) = -1 m (upwards) and for z (t) = 0.5 m.
2. Without considering the 100 kg man on the board. Calculate the natural frequency of the structure.
3. Same question as b), but with consideration of the man in the board.

Solution

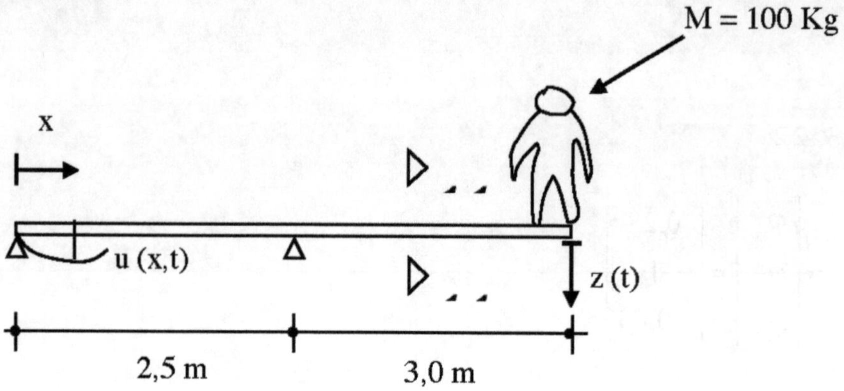

Section A-A

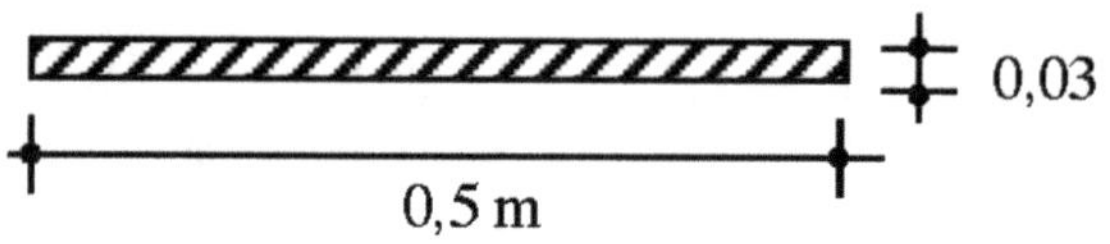

$$\Psi(x) = \frac{x^2 - 2,5x}{16,5}$$

$m = (0.5*0.03*1.0)*330 \approx 5Kg/m$

$$M^* = \int m(x)\left[\Psi(x)\right]^2 dx = \int \frac{5}{16,6^2}\left[x^2 - 2,5x\right]^2 dx$$

$$= \frac{5}{16,6^2}\int\left(x^4 - 5x^3 + 6,25x^2\right) dx = \frac{5}{16,6^2}\left(\left[\frac{x^5}{5}\right]_0^{5,5} - \frac{5}{4}\left[x^4\right]_0^{5,5} + \frac{6,25}{3}\left[x^3\right]_0^{5,5}\right)$$

$$= \frac{5}{16,6^2}\left(\frac{5,5^5}{5} - \frac{5}{4}*5,5^4 + \frac{6,25}{3}*5,5^3\right) = 3,84Kg$$

$$I = \frac{0,5*0,03^3}{12} = 1,125*10^{-6}m^4$$

$E = 80*10^9 N/m^2$

$$\Psi''(x) = \frac{2}{16,5}$$

$$K^* = EI\int\left[\Psi''(x)\right]^2 dx = 80*10^9 * 1{,}125*10^{-6}\int_0^{5,5}\left(\frac{2}{16{,}5}\right)^2 dx = 7273N/m$$

$$\omega_n = \sqrt{\frac{K^*}{M^*}} = 43{,}51\,rad\ s^{-1}\ \text{ And } T{=}0.14s$$

$$M^* = \int m(x)\left[\Psi(x)\right]^2 dx + M\left[\Psi(x=5{,}5)\right]^2 = 3{,}84+100(1)^2 = 103{,}84Kg$$

$$K^*{=}7273N/m\ \ \omega = \sqrt{\frac{7273}{103{,}84}} = 8{,}73\,rad\ s^{-1}\ \mathrm{T} = 0{,}75\ \mathrm{s}$$

Example 8

This example present equipment project study, which is composed of a single block in two floors plus a basement.

Architectural description:

The dimensions of the building are:

Total height is 13.80 m. The building includes a ground floor and a basement including:

The width is 10.00 m; the length is 33.30 m, the height of the basement 3.60 m, the height of the ground floor 3.40 m.

For current floors, we have. The width 10.00 m, the length 3.30 m, the height 3.40 m.

With numerical model, using ETABS software we need to:

1. Determine the modes of the structure?
2. Determine the vibration periods and mass participation rate?
3. Determine support reactions?

Solution

The structure 3D and side view model are represented in figure 13 and figure 14 respectively:

Figure 13. 3D model

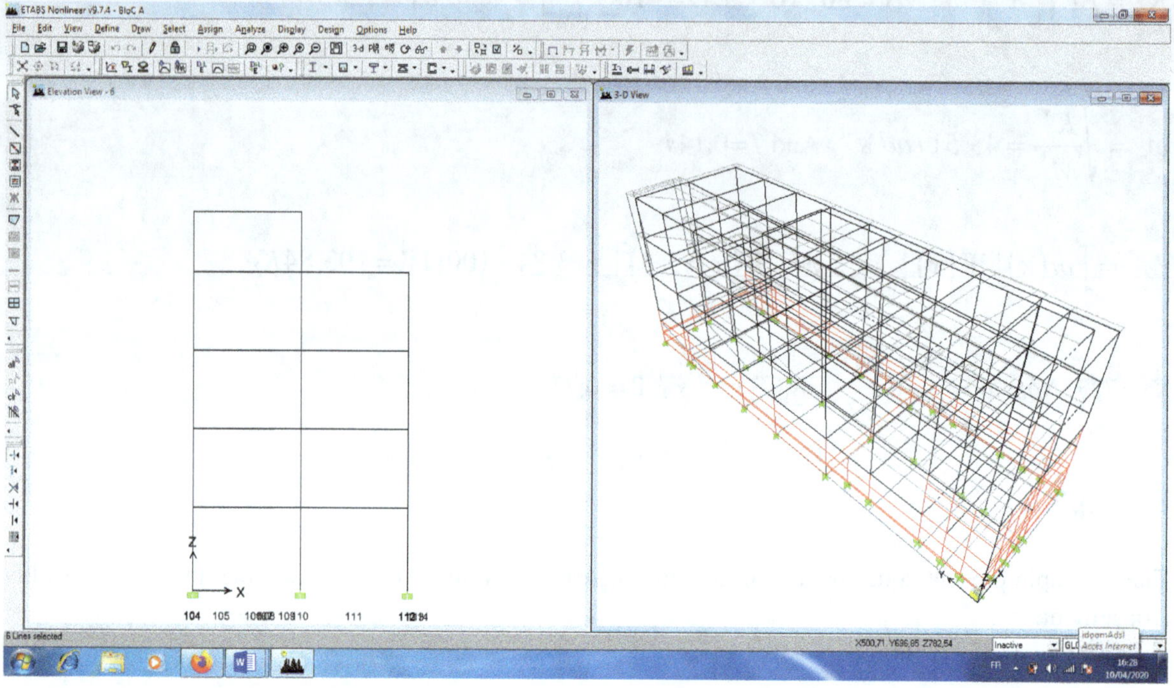

Figure 14. Side view structure model

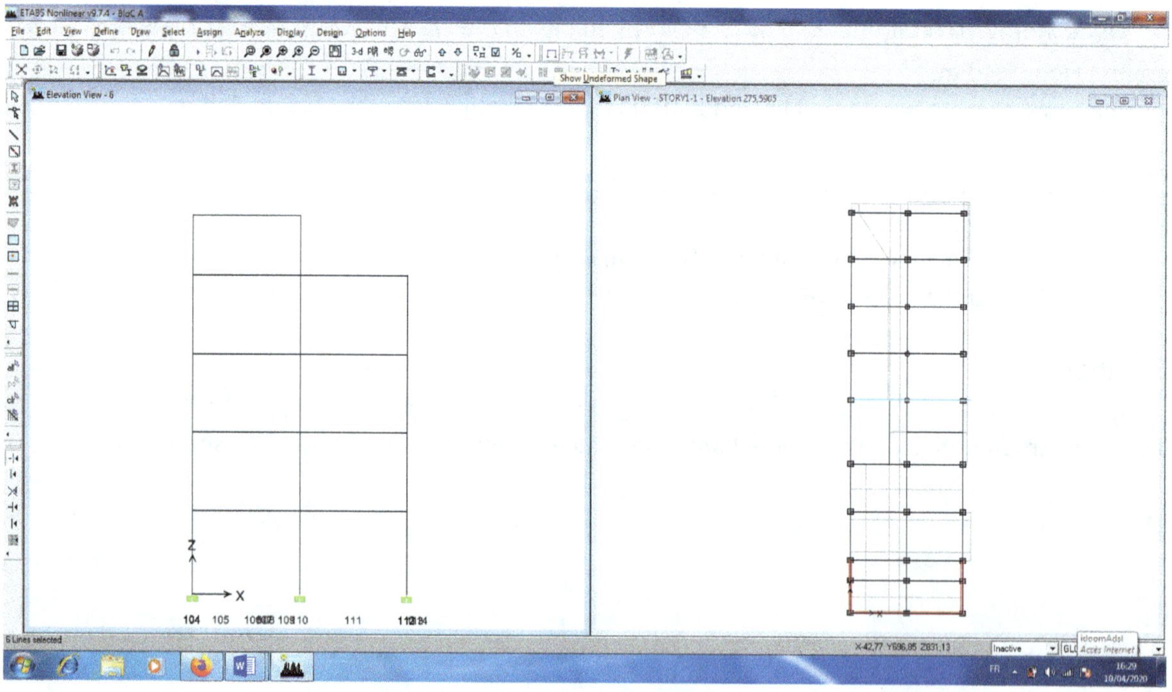

The three mode vibration as well as the period are represented in figures 15, 16 and 17 repectively:

Figure 15. First Mode vibration with a period T=0.55s

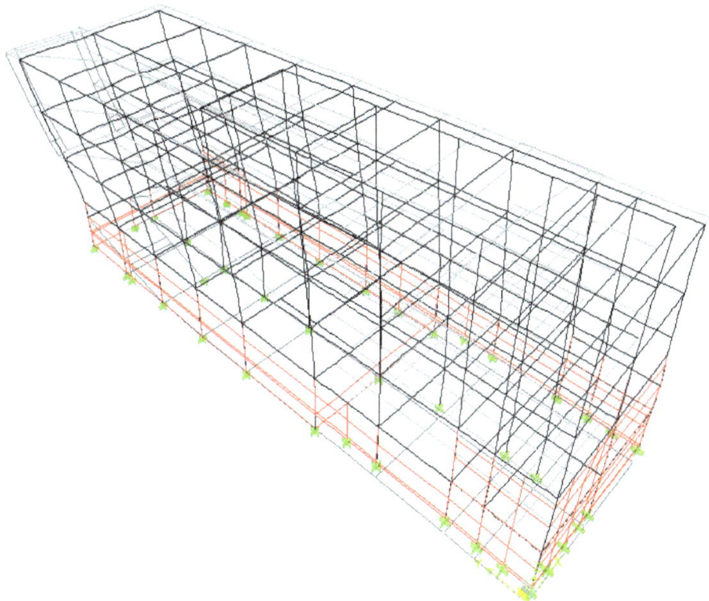

Figure 16. Second mode vibration with a period T=0.39s

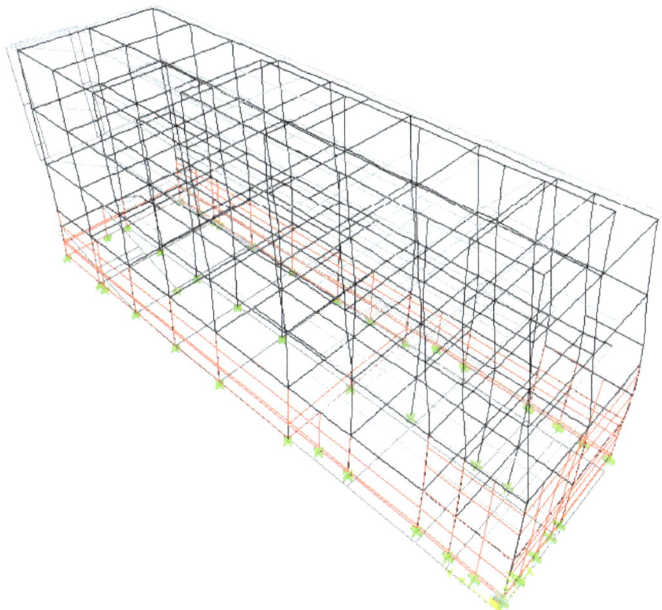

Figure 17. Third mode vibration with a period T=0.35s

Modal Participating Mass Ratios

Edit View

Modal Participating Mass Ratios

Mode	Period	UX	UY	UZ	SumUX	SumUY	SumUZ
1	0.558412	44.6872	1.2255	0.0000	44.6872	1.2255	0.0000
2	0.394195	0.9990	45.4721	0.0000	45.6862	46.6976	0.0000
3	0.357183	6.8460	0.1206	0.0000	52.5322	46.8181	0.0000
4	0.166477	8.9353	0.1383	0.0000	61.4675	46.9564	0.0000
5	0.124618	0.5046	8.2686	0.0000	61.9722	55.2250	0.0000
6	0.110937	0.0029	0.0782	0.0000	61.9750	55.3032	0.0000
7	0.100895	0.0097	0.0021	0.0000	61.9848	55.3053	0.0000
8	0.100591	1.8743	0.0000	0.0000	63.8590	55.3054	0.0000
9	0.080342	3.5920	0.0144	0.0000	67.4510	55.3197	0.0000
10	0.078273	0.1939	3.7669	0.0000	67.6449	59.0866	0.0000
11	0.068214	2.2021	10.7823	0.0000	69.8470	69.8689	0.0000
12	0.064969	0.5233	0.6387	0.0000	70.3703	70.5077	0.0000
13	0.050396	0.0000	0.0000	0.0000	70.3703	70.5077	0.0000
14	0.046812	24.4913	1.0311	0.0000	94.8616	71.5388	0.0000
15	0.043329	3.8075	0.0495	0.0000	98.6691	71.5883	0.0000
16	0.035743	0.0194	0.1588	0.0000	98.6885	71.7471	0.0000
17	0.032308	0.0878	0.0168	0.0000	98.7763	71.7639	0.0000
18	0.031937	0.0148	0.0028	0.0000	98.7911	71.7668	0.0000
19	0.019611	0.0886	26.8436	0.0000	98.8796	98.6103	0.0000
20	0.018240	1.0995	1.3867	0.0000	99.9792	99.9970	0.0000

OK

Let us give the vibration periods and mass participation rate in this table

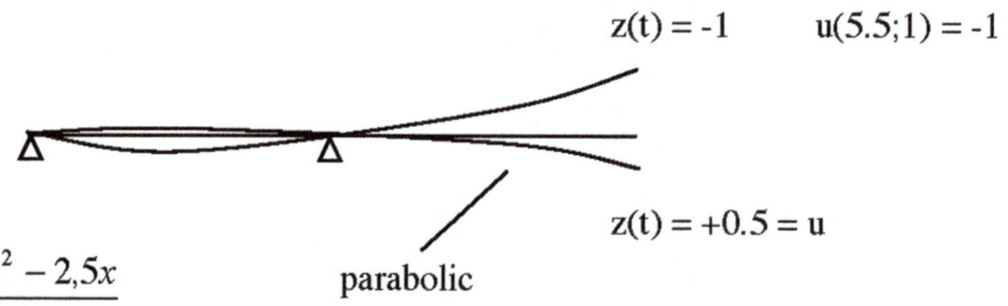

$$\Psi(x)=\frac{x^2-2,5x}{16,5}$$

Example 9

We need to do a tank numerical model which have 1000 m^3 storage capacity. The geometric characteristics of the tank are as follows:

- Interior length A = 17.00 m
- Interior width B = 15.00 m
- Storage capacity 1000 m^3
- Height of the tank including the upper belt H = 5.50 m
- Max liquid level h = 4.50m.

Table 2.

ETABS v9.1.4 File:BLOC A Units: KN-m April 08, 2020 8:43 PAGE 2								
S U P P O R T R E A C T I O N S								
STORY	POINT	LOAD	FX	FY	FZ	MX	MY	MZ
STORY1	1-6	COMB1	12.00	22.90	180.82	-2.881	1.910	0.021
STORY1	7-6	COMB1	-10.14	5.11	285.17	-4.251	0.310	0.006
STORY1	9-7	COMB1	-20.76	19.88	283.09	-3.111	-2.520	-0.070
STORY1	27-5	COMB1	15.78	2.18	430.76	-0.636	13.245	-0.062
STORY1	32-1	COMB1	-8.01	0.48	603.06	-0.379	-6.471	-0.010
STORY1	34-7	COMB1	-2.77	-4.53	431.95	-0.121	-2.245	0.007
STORY1	60-3	COMB1	20.55	-11.54	424.22	1.320	17.551	-0.491
STORY1	65-1	COMB1	-9.41	9.01	803.10	-7.461	-7.549	-0.010
STORY1	67-6	COMB1	-3.17	-16.53	438.79	1.427	-2.427	-0.033
BASE	80	COMB1	32.14	27.64	475.28	-2.546	-0.957	-0.012
BASE	85	COMB1	10.74	5.32	1004.18	-7.144	0.509	-0.040
BASE	87	COMB1	-41.49	14.92	426.46	-1.694	6.090	0.446
BASE	98	COMB1	-2.89	16.10	235.91	0.109	-2.893	0.013
BASE	104	COMB1	12.61	24.14	505.20	0.320	15.049	-0.018
BASE	110	COMB1	-4.82	-5.68	865.37	6.741	-5.680	-0.003
BASE	112	COMB1	-3.55	6.04	496.45	-0.661	-4.170	-0.492
BASE	115	COMB1	9.89	6.02	507.46	-0.031	11.962	0.008
BASE	119	COMB1	-5.46	0.45	733.66	-0.531	-6.430	-0.003
BASE	121	COMB1	-1.95	-8.70	550.29	0.727	2.134	0.037
BASE	124	COMB1	-0.32	-14.98	487.53	5.181	5.475	2.043
BASE	128	COMB1	-5.12	0.58	725.79	-0.681	-6.013	-0.003
BASE	130	COMB1	-1.98	-14.74	509.02	0.528	-2.053	-0.002
BASE	137	COMB1	3.53	-27.52	293.90	2.193	-1.312	-0.342
BASE	141	COMB1	-3.22	-6.27	720.58	7.401	-3.551	-0.007
BASE	143	COMB1	-1.38	-14.02	207.90	1.317	-1.443	-0.069
BASE	150	COMB1	15.21	-30.59	349.16	2.531	0.942	-0.327
BASE	155	COMB1	-9.13	-6.47	382.65	7.520	2.987	0.365
BASE	157	COMB1	-36.19	-24.66	350.15	3.723	-5.337	0.256

Figure18. Tank 3D model

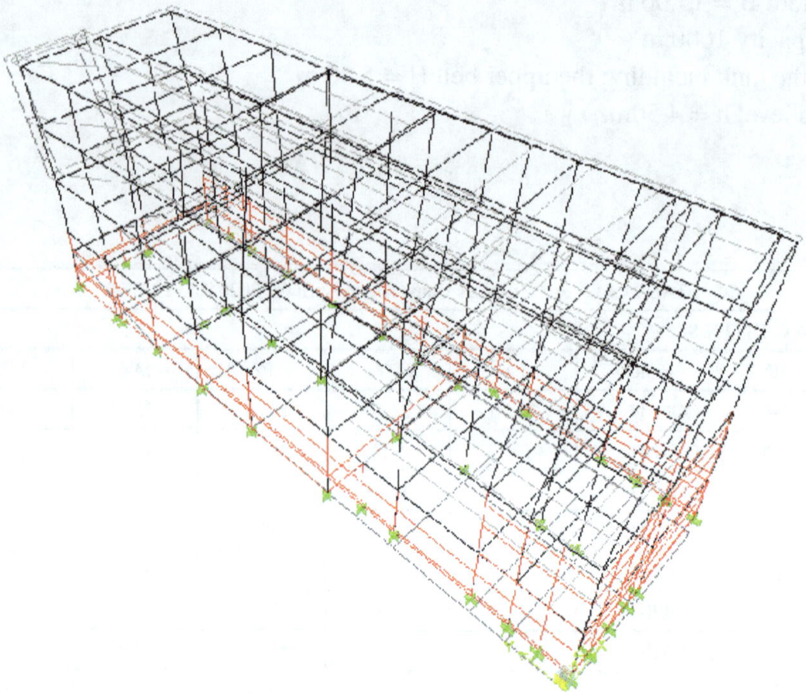

With numerical model, using SAP 2000 software we need to

1. Find the loading walls with the water pressure.
2. Find the loading walls with earth pressure.
3. Find the Moment on the small and large face.
4. do the Hydrodynamic study.

Solution

The 3D model of the tank using Sap2000 software is represented in figures bellow:

In our case, the reservoir must be calculated in two unfavorable cases, namely

1. Full tank without earth pushing.
2. Empty tank with earth pushing (before commissioning),
    a. Full tank without earth pressure case

The loading of the wall with the hydrostatic pressure is done by the method of the slices, all of them consider slices of 1ml, and the calculation of the pressure is made in relation to a height H that will be calculated between the high level of the water and the middle of the slice.

The loading walls without earth pressure is given on the following figure:

For the case of walls loading with water pressure, the results obtained as a moment on the small and large face are represented in figure 20 and 21 respectively.

Figure 19. Loading walls without earth pressure

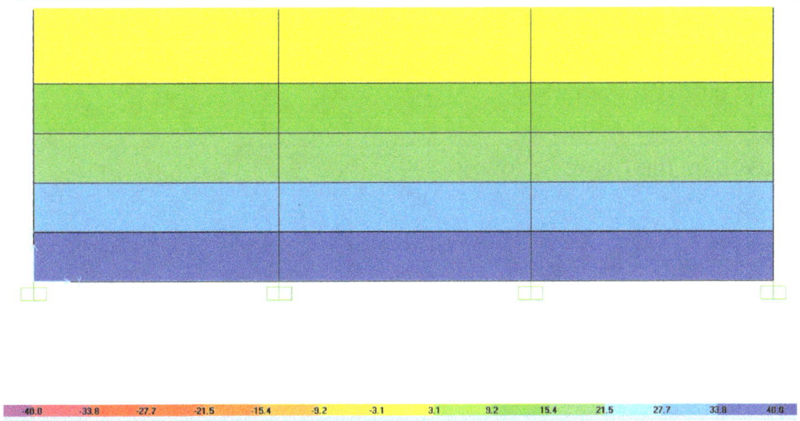

Figure 20. Moment on the small face with water pressure

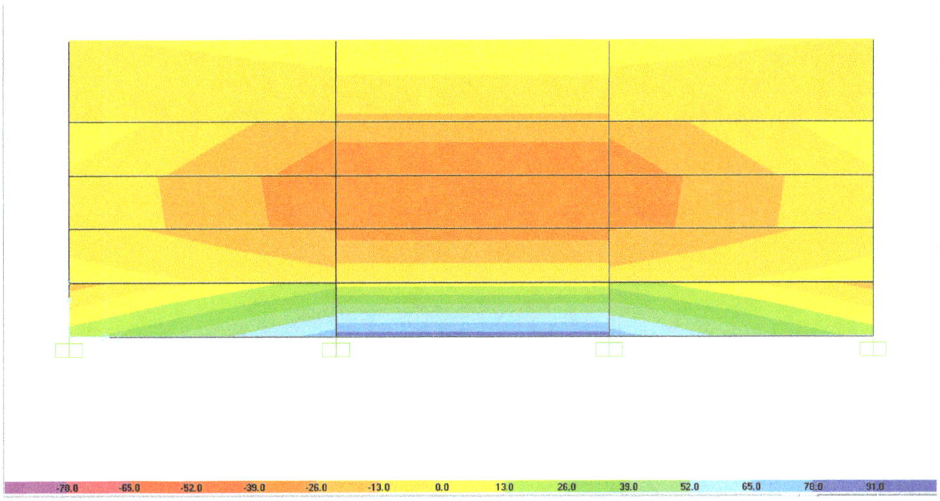

Figure 21. Moment on the large face with water pressure

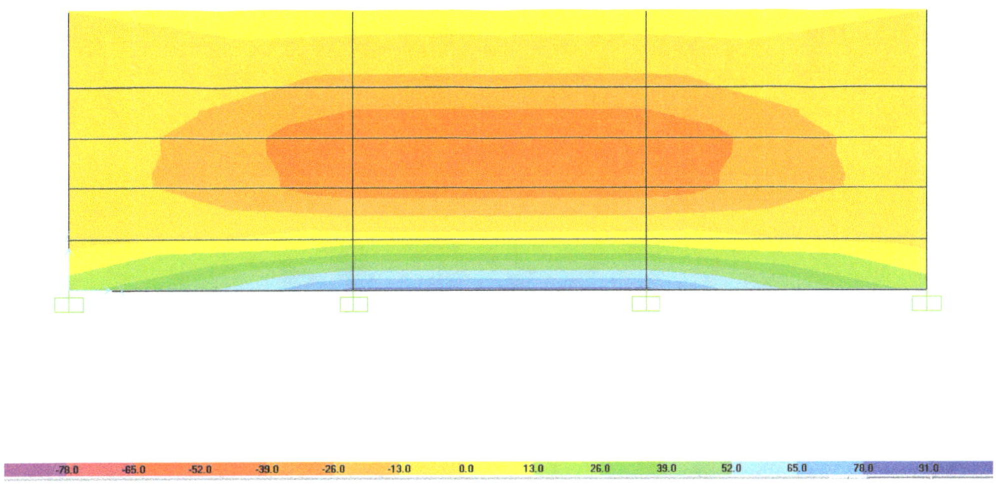

The maximum moment is: 38.53KNxm / m. and a normal effort 102.52KN/m. The reinforcement calculation is done for a 1.00x0.30 section; we find the following steel section: 2.61cm^2.

b. Empty tank with earth pressure case

The loading is given on the figure bellow:

Figure 22. Empty tank with earth pressure

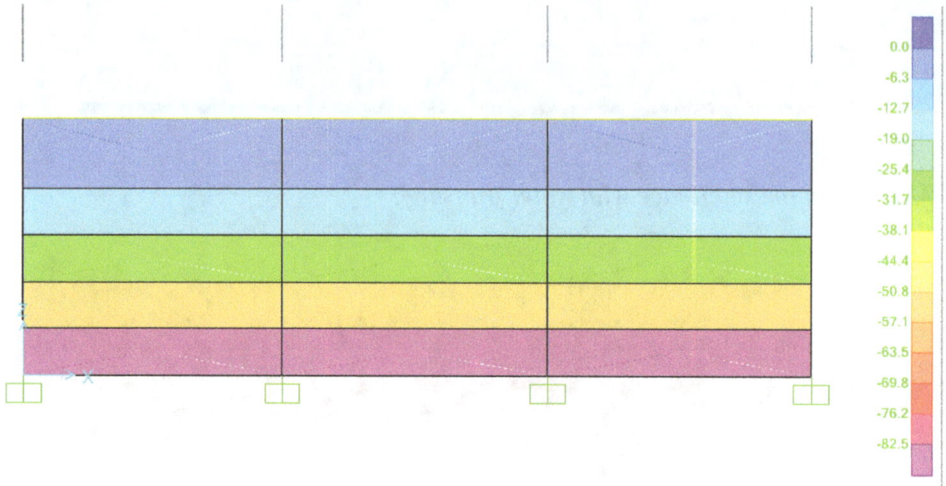

For the case of walls loading with earth pressure, the result obtained as a moment on the large face (the worst case) is presented in figure bellow:

Figure 23. Moment on the large face with earth pressure

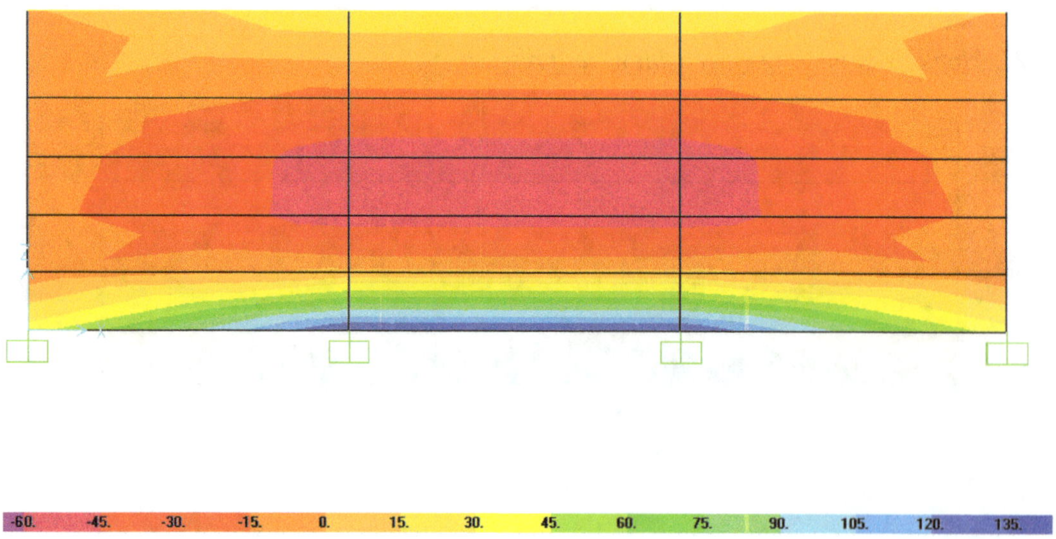

The maximum moment is 57.84KNxm / m. and a normal effort 164.57KN/m. The reinforcement calculation is done for a 1.00x0.30 section; we find the following steel section: 3.88cm^2.

Hydrodynamic Study

When a tank is completely full, there is no relative moment of the liquid with respect to the tank under an excitation. From a dynamic point of view, everything happens as if the liquid-tank assembly constituted a single mass.

On the other hand, in partially filled tanks, the excitation puts part of the fluid in motion, which leads to the formation of waves on the surface, causing the creation of stresses on the walls. In this case, the hydrodynamic calculation is necessary in order to avoid damage to the dome and the wall by the waves.

The hydrodynamic effect can significantly increase the shearing based force of the tank (by a value greater than about 20%).In our case, we used SOCOTEC software.

Before showing the results, the figure 24 shows us the 2D model made with SOCOTEC software showing the data (dimensions) of the tank, in order to prepare the hydrodynamic results.

Figure 24. 2D tank model before Hydrodynamic study

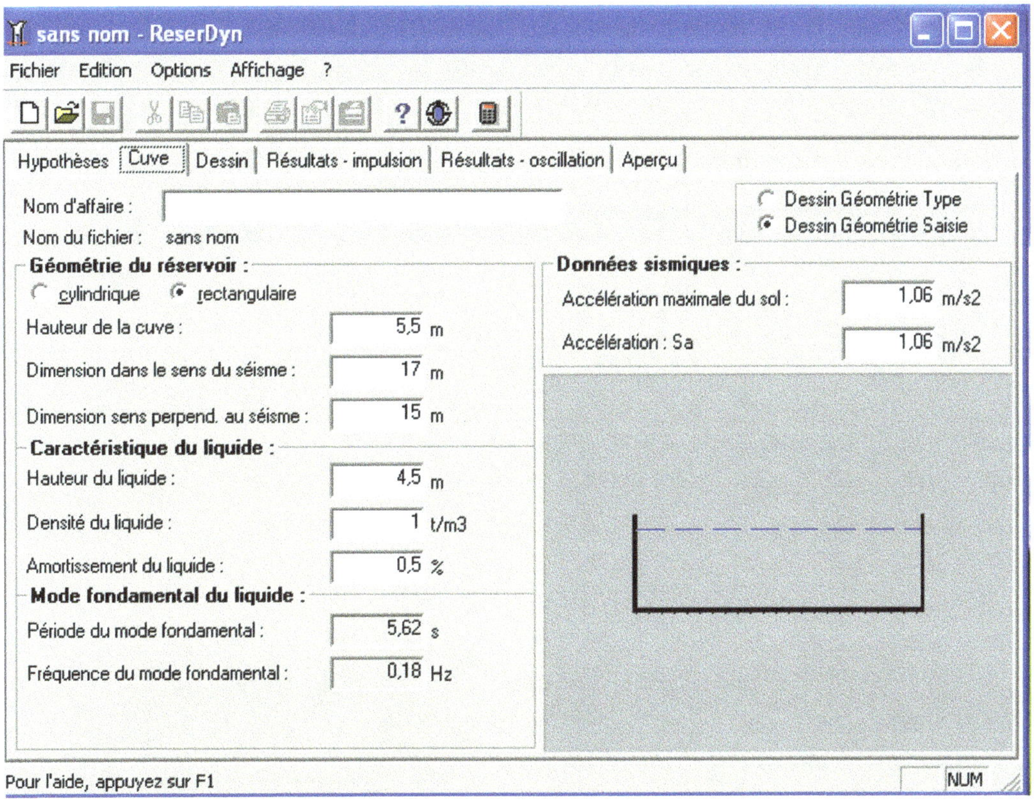

Figure 25. Impulse pressure

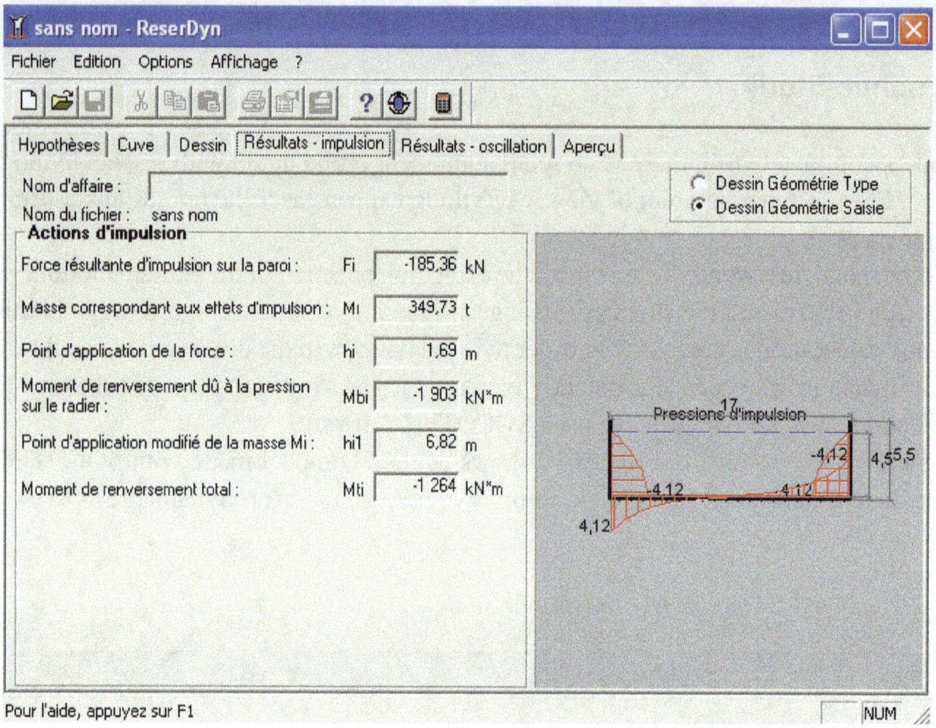

Figure 26. Maximal wave oscillation pressure

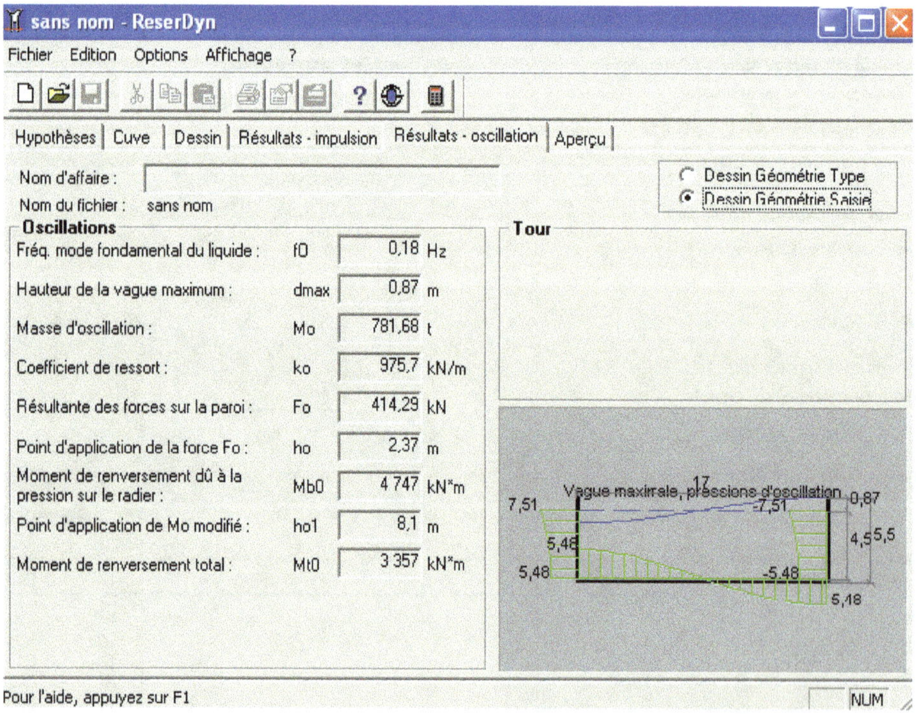

Figure 27. Maximal wave under impulse and oscillation pressure

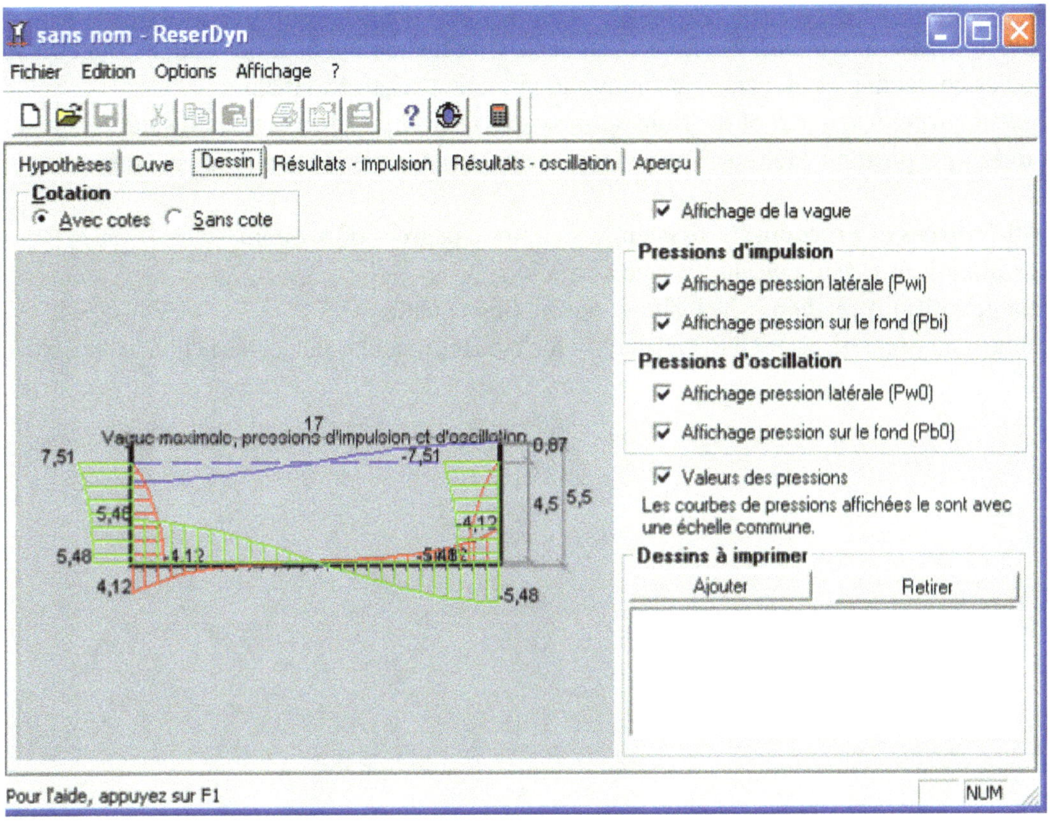

REFERENCES

Chopra, A. K. (2001). *Dynamics of Structures: Theory and Applications to Earthquake Engineering* (2nd ed.). Prentice Hall.

Clough, R.W. & Penzien, J. (1995). *Dynamic of structures.* Compute and Structure, Inc.

Commend, S., & Zimmermann, T. (2001). *LSCDYN 1DOF and nDOF: two applications for computer-assisted teaching of structural dynamics. LSC Internal Report.* Swiss Federal Institute of Technology.

SAP. (2000). Integrated structural analysis and design software. Computers and Structures, Inc.

Smith, I. (2010). *Structural dynamics, Internal course.* Swiss Federal Institute of Technology.

KEY TERMS AND DEFINITIONS

Coupled Coordinates: Translation and rotation coordinates.

Eigenvalues: Mathematically each of a set of values of a parameter for which a differential equation has a non-zero solution (an eigenfunction) under given conditions.

Eigenvectors: Mathematical vector which when operated on by a given operator gives a scalar multiple of itself.

Hydrodynamic: Is a branch of physics that deals with the motion of fluids and the forces acting on solid bodies containing or immersed in fluids and in motion relative to them.

Impulse Force: It is a part of the fluid mass, which reacts by inertia to the translation of solid.

Modal Superposition Method: Deformed structure can be expressed as a linear combination of all modes.

Multi-Degrees of Freedom: Is a system with several degrees of freedom.

Normalization: Usually means to scale a variable.

Orthogonality: A relation, which allows decoupling equations.

Oscillation Force: It a part of the fluid mass, which starts to oscillate under the action of the earthquake.

Section 3

Structural Damping and Additional Damping Devices: Influence Study on Structural Behavior

Chapter 4
Multi-Degrees of Freedom Structural Damping

ABSTRACT

Damping is the most difficult structure characteristic to evaluate because it depends on several parameters, who are in the structure itself, like support type, internal friction materials of a studied structure. In addition, parameters surrounding the structure like soil type have a direct effect on damping. These things make it difficult to evaluate properly. This chapter allows us to evaluate different types of damping, as damping proportional to stiffness matrix: Rayleigh damping, Cauchey damping, and viscous equivalent damping.

INTRODUCTION

The core of this chapter is the evaluation of all damping types namely Viscous equivalent, Rayleigh, Cauchey damping. Damping is a means of reducing speed through resistance to motion, particularly by forcing an object through a liquid or a gas, or along another body. The damping unit is often proposed as Force * length / time (N.m /s).

In practice, all structures, meets resistance during their movements. Consequently, the amplitude of the movement does not remain unchanged; but it will diminish gradually with time, in the case of free vibrations; until the motion stops. The damping forces must depend on the vibratory system itself, as well as external factors surrounding the system. In a system, damping occurs due to the internal friction forces of the materials of the structure and the external effects in which the structure is immersed.

DAMPING TYPES

An exact representation of damping phenomenon involves difficulties; however, as for each of its nature, the damping can be presented into four following types:

DOI: 10.4018/978-1-7998-4399-3.ch004

Copyright © 2021, IGI Global. Copying or distributing in print or electronic forms without written permission of IGI Global is prohibited.

1. Structural damping.
2. Viscous damping.
3. Coulomb damping.
4. Negative damping.

The structural damping is due to the internal molecular friction force of the material of the structure, or it is due to the own connections of the system.

The viscous damping appears in the case of a vibrating system in a fluid; the damping force in this case is proportional to the speed:

$$F_0 = C\dot{U}$$

Where $\dot{u}$ the velocity and C is the damping constant.

The Coulomb damping, it is also known as a dry friction; it appears when the movement of the body is in a sealed surface, the damping force in this case is given as follows:

$$F_0 = \mu N$$

Where μ is the kinematic coefficient of friction and N is the normal force of body motion on the surface.

Negative damping is a very special case, where damping adds energy to the system rather than dissipating it, resulting in amplification of the amplitude, it is the case of poor soil.

In this chapter, we will focus only on damping proportional to stiffness matrix Rayleigh damping, Cauchey damping, and Viscous equivalent damping

Just like a one-degree of freedom system, there is no analytic way of assessing damping the best way it is experimental.

As long as there is no firm form of the damping matrix [C]; we can safely assume that the matrix [C] so that the orthogonality is ensured.

$$[\varphi]^t[C][\varphi] = \underbrace{[C^*]}_{diagonal\ matrix}$$

There are five ways to choose the damping matrix [C] which are enumerated as follows (Chopra 2001) & (Clough 1995).

Proportional Mass Damping

In this case we assume that there is a proportional relationship between damping matrix and mass matrix by coefficient.

$$[C] = \alpha[M]$$

We assume

$$[\varphi]^t[C][\varphi]=[\varphi]^t\alpha[M][\varphi]$$

$$[\varphi]^t[C][\varphi]=\alpha\underbrace{[M^*]}_{\text{diagonal matrix}}$$

$$c_i^*=\alpha m_i^*$$

$$\alpha=\frac{c_i^*}{m_i^*}=2\xi_i\omega_i \quad (1)$$

Often, ξi represent the damping rate, for the first mode is called $\xi 1={}_5\%$, and then we solve the system for α.

The depreciation in the remaining modes is $\xi_i=\frac{\alpha}{2\omega_i}=\xi_1\frac{\omega_1}{\omega_i}$

Proportional Stiffness Damping

In this case we assume that there is a proportional relationship between damping matrix and stiffness matrix by β coefficient.

$[C]=\beta[K]$

We assume

$$[\varphi]^t[C][\varphi]=[\varphi]^t\beta[K][\varphi]$$

$$[\varphi]^t[C][\varphi]=\beta\underbrace{[K^*]}_{\text{diagonal matrix}}$$

$$c_i^*=\beta k_i^*$$

$$\beta=\frac{c_i^*}{k_i^*}=\frac{2\xi_i}{\omega_i} \quad (2)$$

We suppose again $\xi 1$ of the first mode $\xi 1$

$$\xi_i=\xi_1\frac{\omega_1}{\omega_i} \quad (3)$$

RAYLEIGH DAMPING

In this case we assume that damping matrix is a linear combination of mass and stiffness matrix using α and β coefficients

$$[C] = \alpha[M] + \beta[K] \tag{4}$$

We assume

$$[\varphi]^t[C][\varphi] = [\varphi]^t(\alpha[M] + \beta[K])[\varphi]$$

$$[\varphi]^t[C][\varphi] = \alpha[\varphi]^t[M][\varphi] + \beta[\varphi]^t[K][\varphi]$$

$$[\varphi]^t[C][\varphi] = \alpha\left[M^*\right] + \beta\left[K^*\right]$$

$$c_i^* = \alpha\, m_i^* + \beta\, k_i^*$$

$$2\xi_i\omega_i m_i^* = \alpha\, m_i^* + \beta\, k_i^*$$

$$2\xi_i\omega_i = \alpha + \beta\,\omega_i^2$$

In a similar way we have

$$2\xi_j\omega_j = \alpha + \beta\,\omega_j^2$$

As matrix form we have

$$\frac{1}{2}\begin{bmatrix} \frac{1}{\omega_i} & \frac{1}{\omega_i} \\ \frac{1}{\omega_j} & \frac{1}{\omega_j} \end{bmatrix}\begin{Bmatrix} \alpha \\ \beta \end{Bmatrix} = \begin{Bmatrix} \xi_i \\ \xi_j \end{Bmatrix} \tag{5}$$

In this case, we assume the damping for two modes we solve the system for α, β; and once we have α and β the solution for ξk is $\xi_k = \frac{\alpha}{2\omega_k} + \frac{\beta\,\omega_k}{2}$

Considering this three cases (the three relations) in graphical form we have.

CAUCHEY DAMPING (Jerome 2012)

For this, case the damping based on more than two modes. From the problems of the eigenvalues (which will be detailed in chapter 7). In addition, for the mathematical writing considerations we change Rayleigh coefficients by proposing $\alpha 0 = \alpha$ and $\beta = \alpha 1$

$$[K]\{\varphi_j\} = \omega_j^2[M]\{\varphi_j\} \tag{6}$$

Figure 1. (a) Proportional mass damping (b) Proportional stiffness damping (c) Rayleigh damping

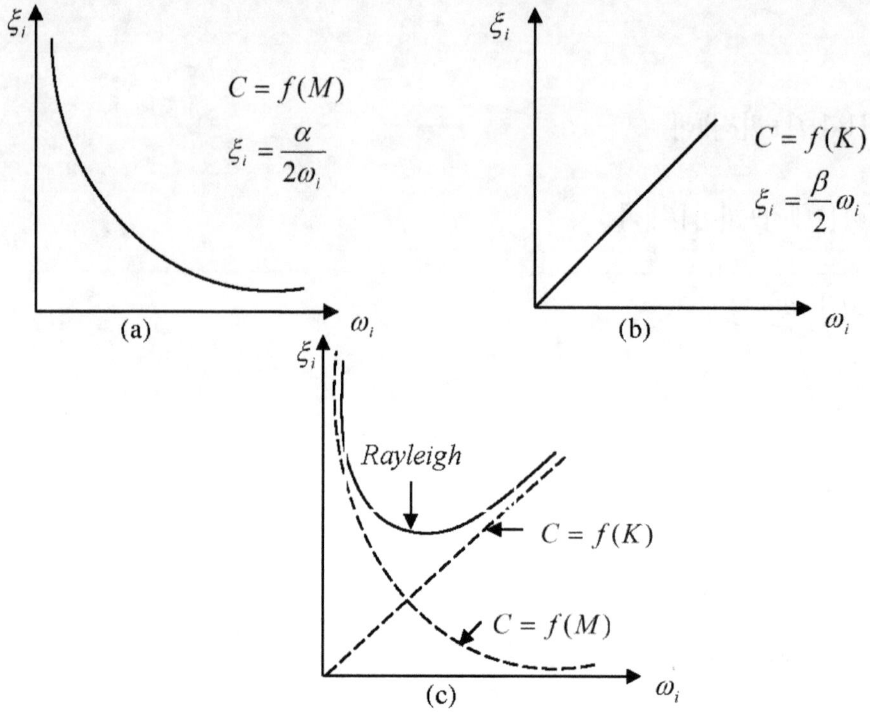

Pre-multiplying the equation (6) by

$$\{\varphi_i\}^t[M][M]^{-1}$$

$$\{\varphi_i\}^t[M][M]^{-1}[K]\{\varphi_j\} = \omega_j^2\{\varphi_i\}^t[M][M]^{-1}[M]\{\varphi_j\}$$

$$\{\varphi_i\}^t[M][M]^{-1}[K]\{\varphi_j\} = \omega_j^2\{\varphi_i\}^t[M]\{\varphi_j\} = \begin{cases}\omega_j^2 m_j^* & i = j\\ 0 & i \neq j\end{cases} \tag{7}$$

Pre-multiplying again the equation (7) by

$$\{\varphi_i\}^t[K][M]^{-1}$$

$$\{\varphi_i\}^t[K][M]^{-1}[K]\{\varphi_j\} = \omega_j^2\{\varphi_i\}^t[K][M]^{-1}[M]\{\varphi_j\}$$

$$\{\varphi_i\}^t[K][M]^{-1}[K]\{\varphi_j\} = \omega_j^2\{\varphi_i\}^t[K]\{\varphi_j\} = \begin{cases}\omega_j^2 k_j^* & i = j\\ 0 & i \neq j\end{cases} = \begin{cases}\left(\omega_j^2\right) m_j^* & i = j\\ 0 & i \neq j\end{cases} \tag{8}$$

We assume

$$[C]=\sum_k \alpha_k [M]\left([M]^{-1}[K]\right)^k$$

If k=0 case of Proportional mass damping
If k=1 case of Proportional stiffness damping

$$[\varphi]^t[C][\varphi]=[\varphi]^t\left(\sum_k \alpha_k [M]\left([M]^{-1}[K]\right)^k\right)[\varphi]$$

$$[\varphi]^t[C][\varphi]=\sum_k \alpha_k [\varphi]^t[M]\left[\omega^2\right]^k[\varphi]$$

$$[\varphi]^t[C][\varphi]=\sum_k \alpha_k \left[\omega^2\right]^k\left[M^*\right]$$

With $[\varphi]^t[M][\varphi]=\underbrace{\left[M^*\right]}_{\text{diagonal matrix}}$

$$\begin{bmatrix} c_1^* & 0 & 0 & 0 \\ 0 & c_2^* & 0 & 0 \\ 0 & 0 & \ddots & 0 \\ 0 & 0 & 0 & c_n^* \end{bmatrix}=\sum_k \alpha_k \begin{bmatrix} \left(\omega_1^2\right)^k m_1^* & 0 & 0 & 0 \\ 0 & \left(\omega_2^2\right)^k m_2^* & 0 & 0 \\ 0 & 0 & \ddots & 0 \\ 0 & 0 & 0 & \left(\omega_1^2\right)^k m_1^* \end{bmatrix} \tag{9}$$

$$c_i^*=m_i^*\sum_k \alpha_k \omega_j^{2k}$$

$$\xi_i=\frac{1}{2}\sum_k \alpha_k \omega_j^{2k-1}$$

With k is the number of mode to keep in account
Assuming all damping coefficients for all modes
We assume $\xi_1, \xi_2, \xi_3 \ldots\ldots \xi_n$

$$[\varphi]^t[C][\varphi]=\begin{bmatrix} 2\xi_1\omega_1 m_1^* & 0 & 0 & 0 \\ 0 & 2\xi_2\omega_2 m_2^* & 0 & 0 \\ 0 & 0 & \ddots & 0 \\ 0 & 0 & 0 & 2\xi_n\omega_n m_n^* \end{bmatrix}=\left[C^*\right]$$

$$[C]=\left([\varphi]^t\right)^{-1}\left[C^*\right][\varphi]^{-1}$$

Note

The damping is actually measured after installing a vibrating table, in order to swept a wide frequency band.

Figure 2. Damping Measurement

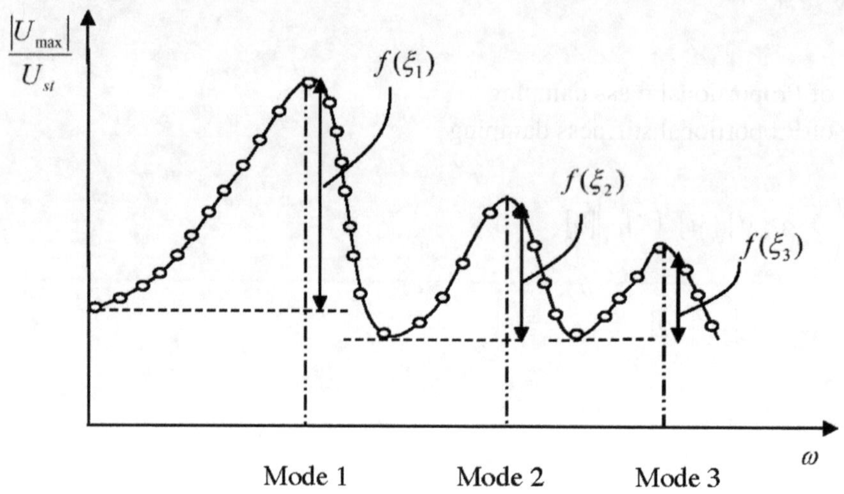

Some damping rate values depending on the type of structure

welded structure → ξ= 2- 3%
Reinforced concrete structure with cracking → ξ= 3 - 5%
Reinforced concrete structure without cracking → ξ= 2%
Wooden structure → ξ= 6 - 7%
Bolted steel structure → ξ= 5%

EQUIVALENT VISCOUS DAMPING

Part of the elastic energy stored in the spring, is dissipated over time; this phenomenon is called damping. Any internal damping mechanism can be grouped under viscous damping; example internal friction in materials (sliding between grains in a particle assembly), resistance of a fluid. On the other hand, loss of energy from a damped system observed from the decrease of the amplitude example $U(t)=u_0\sin(\Omega t - \theta)$

Figure 3. Deformation energy of a spring

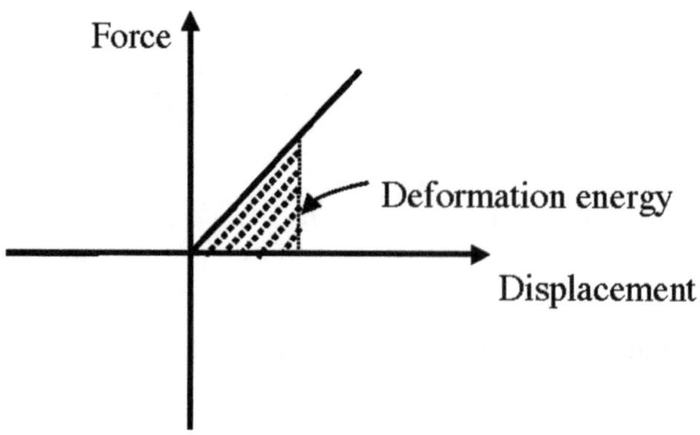

Figure 4. Dissipated energy E_D in a harmonic vibration cycle

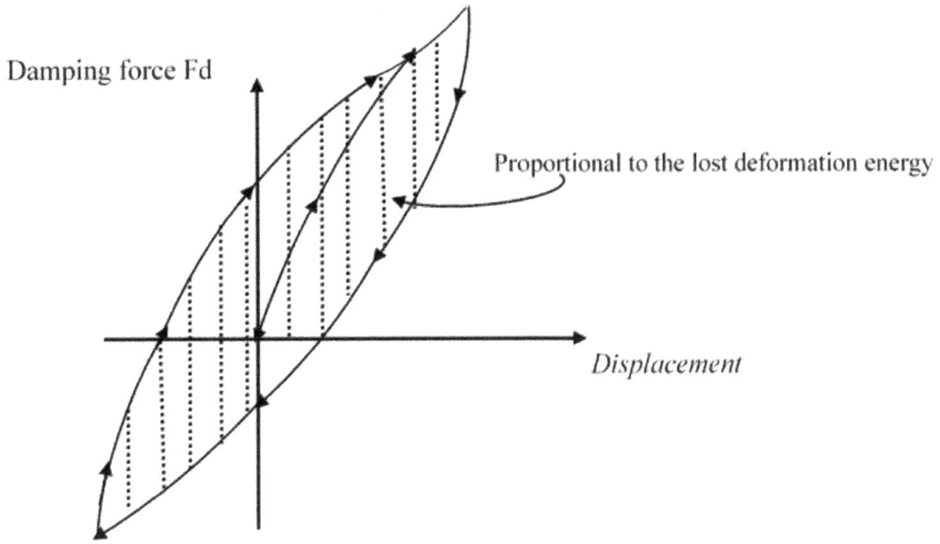

Under a harmonic loading, the curve of forces displacement system, encloses an area called hysteresis curve.

DETERMINATION OF DAMPING COEFFICIENT AS A FUNCTION OF DISSIPATED ENERGY

Let have the following system:

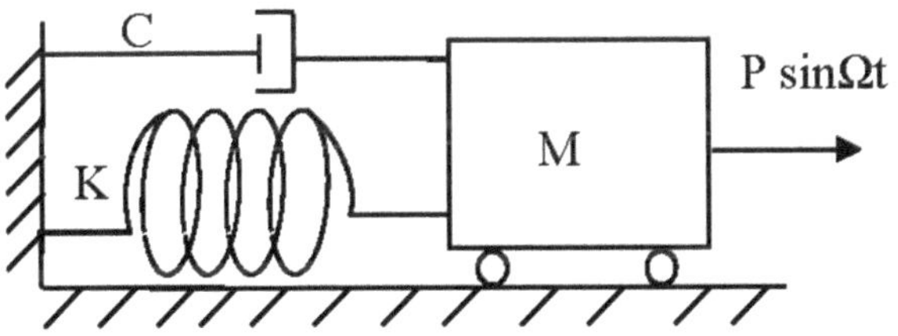

The stationary response $U(t)=u_0\sin(\Omega t - \theta)$

The velocity Is expressed as follows $\dot{U}(t) = u_0\Omega\cos(\Omega t - \theta)$

Knowing that cosα2+sinα2=1

We have $U(t) = \pm u_0\Omega\sqrt{1-\sin^2(\Omega t - \theta)}$

$$U = \pm u_0\Omega\sqrt{1-\frac{1}{\Omega}u^2}$$

$$U = \pm\Omega\sqrt{u_0^2 - u^2} \tag{10}$$

The damping force $f_D = C\dot{u} = \pm C\Omega\sqrt{u_0^2 - u^2}$

$$f_D^2 = C^2\Omega^2\left(u_0^2 - u^2\right)$$

$$\left(\frac{f_D}{C\Omega u_0}\right)^2 + \left(\frac{u}{u_0}\right)^2 = 1 \quad (11)$$

Equation (10) is an ellipsoid equation

Figure 5. Hysterisis curve for viscous damping

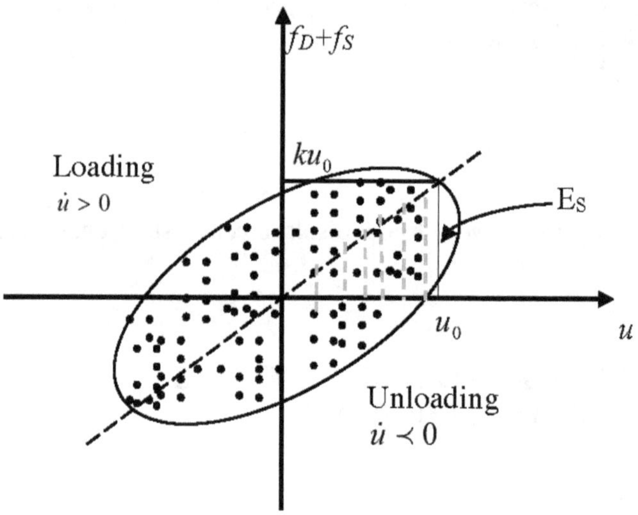

$$E_D = \int_0^{T_p} C\dot{u}\,du = \int_0^{T_p} C\dot{u}\left(\dot{u}\,dt\right) = \int_0^{T_p} C\dot{u}^2 dt = \int_0^{\frac{2\pi}{\Omega}} C\Omega^2 u_0^2 \cos(\Omega t - \theta)^2\,dt$$

$$E_D = \pi\, C\Omega u_0^2$$

$$C_{equivalent} = \frac{E_D}{\pi\,\Omega u_0^2}$$

$$\xi_{equivalent} = \frac{C_{equivalent}}{C_{critical}} = \frac{E_D\,\omega}{2\pi\,\Omega u_0^2 k}$$

We assume that $E_s = \frac{1}{2}kU_0^2$ is the elastic energy

$$\xi_{equivalent} = \frac{1}{4\pi}\left(\frac{E_D}{E_s}\right)\frac{\omega}{\Omega} \quad (12)$$

Figure 6. Hysterisis curve for a spring and viscous damping

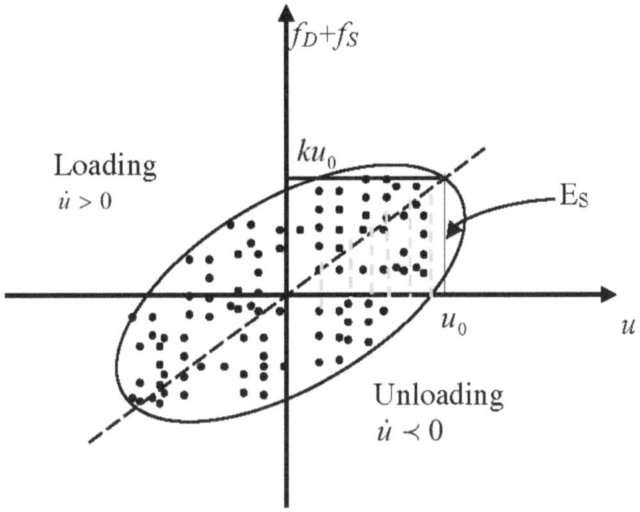

NON-ORTHOGONAL DAMPING

We always simplify things by assuming that the damping is orthogonal. With this assumption, we benefit from the decoupling of the equations governing the motion.

However, in the more general case of a damping matrix; one will not be able to use the decoupling of the equation. Therefore, a simplifying property of the mathematical computation is lost; in this case, we must use the complex mode method (Pecker 2006).

The principle is to rewrite the motion equation of for example free damped system equation (12):

$$[M]_{(nxn)}\{\ddot{U}\}_{(n)}+[C]_{(nxn)}\{\dot{U}\}_{(n)}+[K]_{(nxn)}\{U\}_{(n)}=0 \tag{12}$$

by adding a complementary system

$$\begin{cases}[M]_{(nxn)}\{\dot{V}\}_{(n)}+[C]_{(nxn)}\{V\}+[K]_{(nxn)}\{U\}_{(n)}=0\\ [M]_{(nxn)}\{\dot{U}\}_{(n)}=[M]_{(nxn)}\{V\}_{(n)}\end{cases} \tag{13}$$

For the system of equation (13) its dimension changes it goes from N*N to 2N * 2N, that can be written in matrix form as follows:

$$\begin{bmatrix}0 & [M]\\ [M] & [C]\end{bmatrix}\begin{Bmatrix}\{\dot{V}\}\\ \{\dot{U}\}\end{Bmatrix}+\begin{bmatrix}-[M] & 0\\ [0 & [K]\end{bmatrix}\begin{Bmatrix}\{V\}\\ \{U\}\end{Bmatrix}=\begin{Bmatrix}0\\ 0\end{Bmatrix} \tag{14}$$

The resolution of equation 14 leads us to the resolution of a classic problem of eigenvalues, which will be detailed in chapter 7.

APPLICATIONS

Example 1

A three mass oscillator models represent a three-story building with the features below.

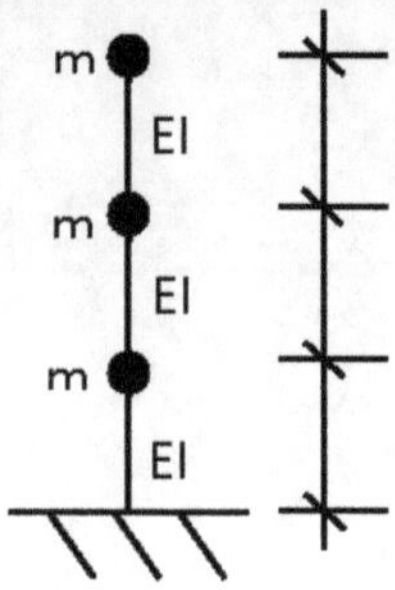

h = 3.00 m
m = 800 t
$EI=4.83*10^7$ kNm^2

With

$$K=\frac{6EI}{h^3}\begin{bmatrix}3.0769 & -1.7692 & 0.4615\\ -1.7692 & 1.6923 & -0.6154\\ 0.4615 & -0.6154 & 0.2692\end{bmatrix}$$

and

$$M=\begin{bmatrix}800 & 0 & 0\\ 0 & 800 & 0\\ 0 & 0 & 800\end{bmatrix}t$$

We ask for a graphic representation of three degree of freedoms structure response Under El Centro earthquake, taking into account Rayleigh damping coefficients as follows: α=0.0241 and β =0.2

With a damping rate $\xi = 0.0149$.

Solution

After developing a code using Matlab language that determines the response of a multi degree of freedom system by respecting the coefficients given in the problem and uploading El Centro earthquake, the graphical response is as follows.

Figure 7. Structure response to El Centro earthquake with Rayleigh damping coefficients α=0.0241 and β =0.2

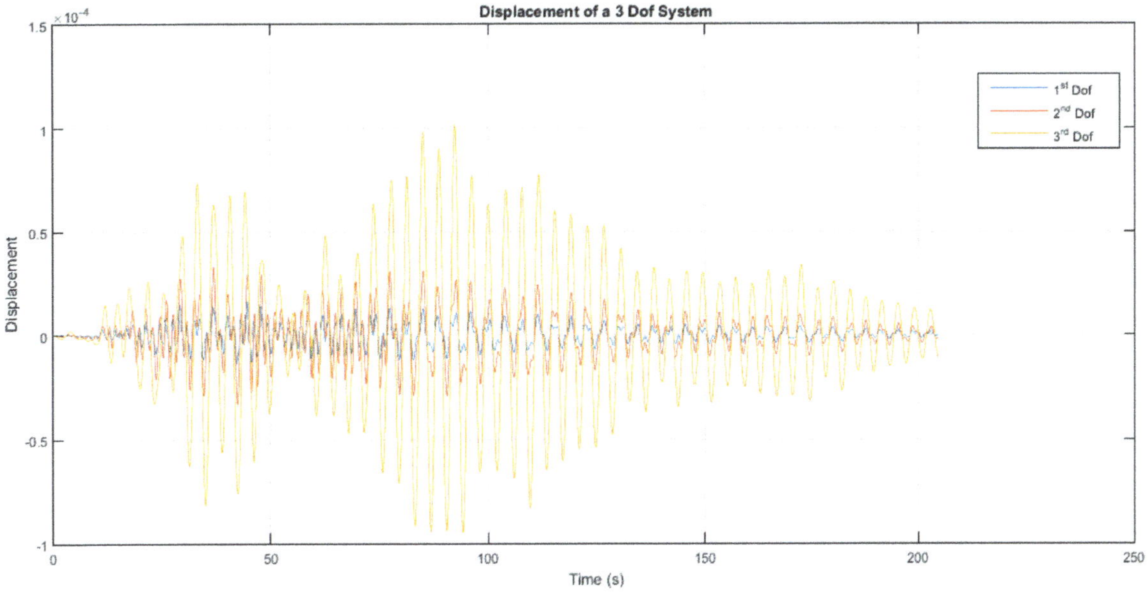

Example 2

We consider a step load ($F = 0$ for $t < 0$ and $F = F_0$ for t ³ 0) applied to single degree of freedom oscillator. The oscillator, modeled as a bar, has the following characteristics: stiffness K=1 N/m (modulus of elasticity E =1 N/m^2, section A=1 m^2, length L = 1 m), mass M =1 kg, force $F_0 = 1$.

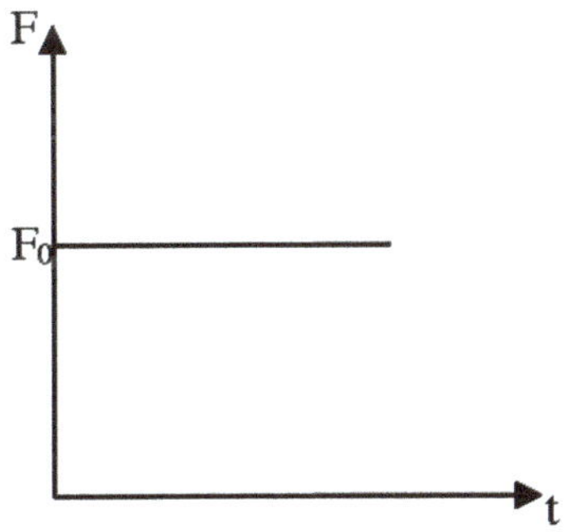

We ask for:

1. graphic representation of single degree of freedoms structure response axial load, taking into account two cases the first one is Rayleigh damping coefficients as follows: α=0. 1 and β =0.2, with a daming rate ξ = 0.01The second case is ; damping proportional the mass matrix with α =0.1 and β =0, with a daming rate ξ = 0.01.
2. what we can conclude ?

Solution

Linear response of single degree of freedom system in the case of Rayleigh damping (α=0.1 and β =0.2) with a daming rate ξ = 0.01, under axial load is as follows.

Figure 8. Oscillator response to axial step load force with Rayleigh damping coefficients α=0.1 and β =0.2

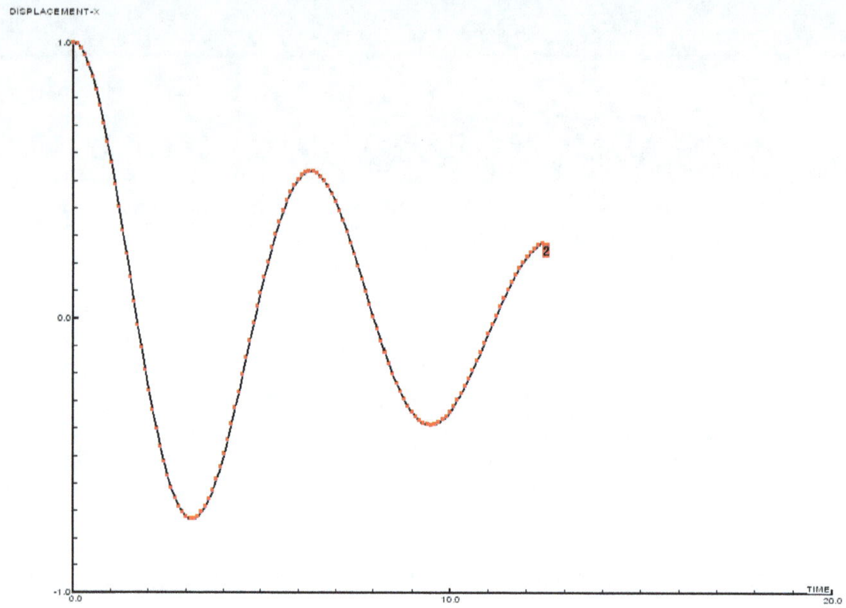

Figure 9. Oscillator response to axial step load force with Rayleigh damping coefficients α=0.1 and β =0

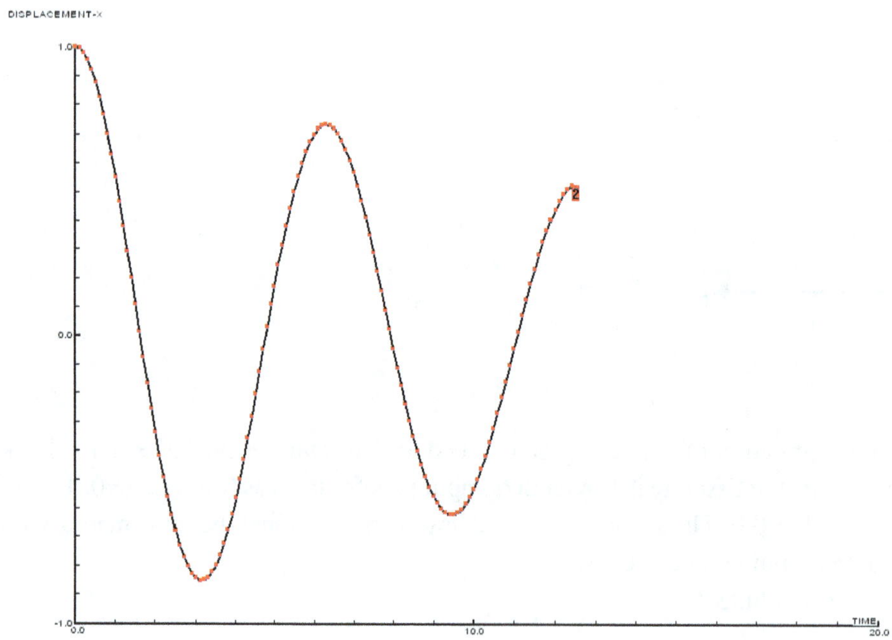

The second case is ; damping proportional to the mass matrix with α =0.1 and β =0, the daming rate ξ = 0.01, under axial step load. In the figure bellow you find the graphic representation of oscillator response.

We can conclude from this example that in the case of Rayleigh damping the decrease in motion magnitude of the oscillator, is more pronounced than in the case of damping proportional the mass matrix. Which can be seen in figures 8 and 9.

REFERENCES

Chopra, A. K. (2001). *Dynamics of Structures: Theory and Applications to Earthquake Engineering* (2nd ed.). Prentice Hall.

Clough, R.W. & Penzien, J. (1995). *Dynamic of structures.* Compute and Structure, Inc.

Jerome, P. L. (2012). *Structure dynamics course note.* Civil and Environmental Engineering University of Michigan.

Pecker, A. (2006). *Dynamique des structures et des ouvrages. Ecole National des ponts et chaussées.*

KEY TERMS AND DEFINITIONS

Cauchey Damping: damping is based on more than two modes.

Dissipated Energy: A dissipative process is a process in which energy (internal, bulk flow kinetic or system potential) is transformed from some initial form to some final form; the capacity of the final form to do mechanical work is less than that of the initial form.

Non-Orthogonal Damping: It is a non-diagonal damping matrix.

Rayleigh Damping: Damping matrix is a linear combination of mass and stiffness matrix.

Viscous Damping: Is the dissipation of energy that occurs when a particle in a vibrating system is resisted by a force the magnitude of which is a constant.

Chapter 5
Passive Dampers:
Fluid Viscous Dampers

ABSTRACT

Civil structures are subjected to various types of loading, which induce severe damage to the structures. Many techniques have been developed for structural rehabilitation; one of the emerging technologies is the use of energy dissipation systems such as fluid viscous dampers (referred to hereafter by FVD). In this chapter, the effect of these devices on the dynamic behavior of an RC building is investigated, with an optimal choice of the linear FVD parameter (i.e., damping coefficient), using a simplified and effective approach. It was found that the maximum inter-story drift of the analyzed retrofitted structures can be significantly reduced compared to the original ones.

INTRODUCTION

In this chapter, our efforts are focused on the effect of Fluid Viscous Dampers (FVD); which is a case of a passive dampers; in the dynamic behavior of an RC building.

As we know civil structure are usually designed to resist a moderate to mild seismic intensity; these structures may sustain significant damage and deformation during their functionality. In order to limit seismic induced structural displacement and peak accelerations to acceptable levels many rehabilitation techniques are used to extend their lifetime and ensure the functionality. Two main options are available for engineers to reduce seismic effects; enhance the building's lateral stiffness either by increasing the structural size and material quantity, or by increasing the level of damping by means of supplemental devices (Duflot et al., 2008). Retrofitting consists in the addition of supplemental devices in the lateral motion of a structural system without a major modification in the building (Rodrigo et al., 2003). Fluid Viscous Dampers (FVD) are energy dissipation systems, their use for civil applications have emerged after their proven efficiency in various military branches. Many researchers (Canstantinou et al., 1993) (Soong et al., 2002) proceeded in their testing and proved their efficiency for improving the seismic behavior of full-scale civil structures.

DOI: 10.4018/978-1-7998-4399-3.ch005

Copyright © 2021, IGI Global. Copying or distributing in print or electronic forms without written permission of IGI Global is prohibited.

Their ability to dissipate large quantities of the input energy and reducing the deformation demand is why they are used more frequently for seismic design and retrofitting (Pollini et al., 2016). Since they have proved their efficiency in such applications, this work focuses mainly on the design of linear FVDs with optimum damping coefficient in such way the computation time is reduced, along with an illustration that studies an RC building, where the reduction amount will be shown and facilitating the comprehension of their preliminary design method.

DIFFRENCE BETWEEN ACTIVE AND PASSIVE CONTROL SYSTEMS

Active control systems are considered as a more advanced technology, their concept is based on a real time processing that allows them to act accordingly to the excitation, thanks to the sensors integrated within the device. These sensors collect information about the excitation and the structural response, then adapts the device's behavior based on the collected data (Saaed & Nikolakopoulos, 2015, Soong & Spencer, 2002).

Passive control systems consists in materials and devices that enhance damping, strength and stiffness, the most common devices of this class operate generally on principles such as deformation of fluids and fluid orificing (the case of this chapter), deformation of viscoelastic solids, frictional sliding and yielding of metals (Soong & Spencer, 2002). A passive dissipation device utilizes the motion of the structure to generate the control force (no external power source needed). In the following, it will be mainly focused on fluid viscous dampers i.e. FVD, and their design optimization, which is seen as intensive task, all illustrated with an example. Figure 1 shows some of the passive dissipation devices followed by a brief definition and a summary of their advantages and disadvantages.

Figure 1. Passive energy dissipation systems.

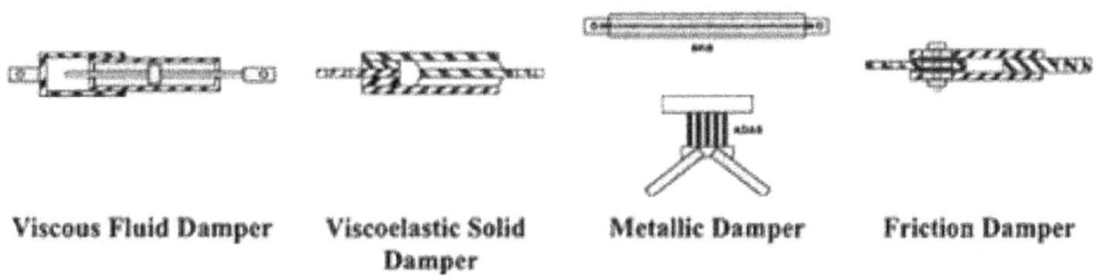

Friction damper consists of four links located at the intersection of cross bracings (tension brace and compression brace) (Canstantinou & Spencer, 1992). A slippage is permitted for one of the braces then the other slips, which allow the device to dissipate energy in both braces. Filiatrault & Cherry (1990) tested and confirmed that this type of devices can increase substantially the capacity of dissipation in one cycle and reduce the amount of drifts. However Symans & al (2008), cited some reliability issues that concerns this sliding option, which can change the interface condition with time, and since their behavior is highly nonlinear, it may provoke undesirable structural behavior by exciting higher modes.

Metallic damper or also called yielding steel elements, many concepts are proposed, it can consist on round steel bar integrated in the bracing frame (Canstantinou & Symans, 1992, Tyler, 1985 and

Whittaker & al, 1991), with their stable hysteretic behavior and reliability for long periods, they also present an insensitivity to temperature. One must mention that after a severe excitation, this device may be severely damaged and must be replaced.

Viscoelastic damper initially developed for wind protection, consisting of bonded viscoelastic layer, suitable also for earthquake resistance (Canstantinou & Symans, 1992, Chang & al, 1991 and Aiken & al, 1990), which mean that they are activated at low displacements and easily modeled due to their linear behavior. Some of the inconvenient when using this type of dampers is their temperature and frequency dependency, along with their low capacity for deformation and some reliability concerns (Symans & al, 2008).

Fluid viscous damper, which will make a subject for this chapter, have a proven record of performance in the military applications and present a large independency to frequency and temperature, also activated at low displacements with the only disadvantage of a possible fluid seal leakage (Symans & al, 2008).

FLUID VISCOUS DAMPERS

Fluid viscous dampers (FVDS) operate on the principle of fluid flow through orifices. A stainless steel piston travels through chambers that are filled with silicone oil (inert, non-flammable, non-toxic and stable). The seismic energy is transformed into heat when the silicone oil flows through orifices between the two chambers (Duflot et al., 2008). In (Canstantinou & al., 1993) it was demonstrated that energy dissipation systems are capable of producing significant reduction of inter-story drift to structures and column bending moments to which they are installed. Thus, they are all suitable in seismic retrofit of existing buildings.

Many manufacturers supply a wide variety of devices operating at high or low fluid pressure having a force-velocity relationship as follows (Paola & al., 2007):

$$F = C\,V^{\alpha}$$

Where V is the relative velocity across the damper, C is the damping coefficient and α is a damping exponent that takes a value in the range of 0.3 – 1.95. When α is equal to one the damper is said to be a linear damper which is the case of this work.

Figure 2. Typical FVD (Canstantinou & al., 1993).

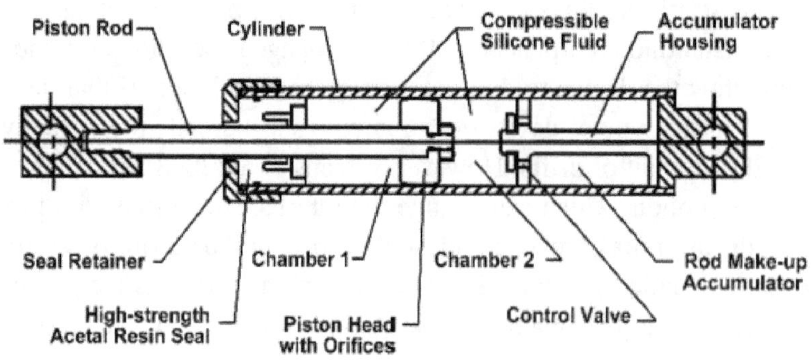

The FVDs can be introduced to the building and having different inclination possibilities, the engineer may choose which is suitable for any given structure, respecting the architectural requirements, and some of these inclination possibilities are presented in the next figure.

Figure 3. FVD inclination: a- diagonal; b- chevron; c- toggle; d-scissor. (Guo & al. 2014).

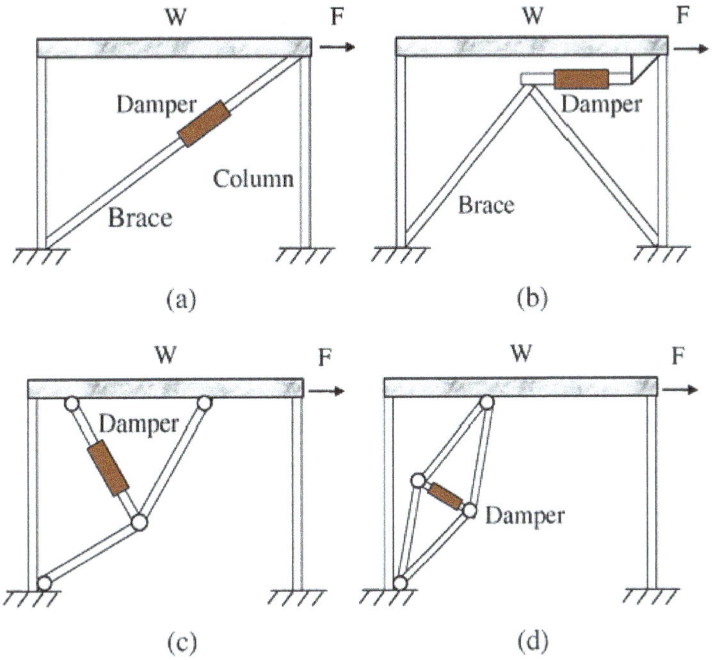

The design of FVDs i.e. the choice of optimal damping values, can be seen as an intensive and time consuming task, many researchers were faced to this difficulty, (Rodrigo & al. 2003; Gherbi & al. 2017; Mcnamara & al. 2003). It has been shown that the choice of an optimal value of the damper's parameters can be done after numerous analysis based on finite element based software. In the following section, a brief representation of simple methodology for the preliminary design of FVDs, which was proposed by (Duflot & al., 2017) simplifies the design process for determining a suitable damping coefficient, initially used for wind excitation; a slight modification concerning the aerodynamic damping, which will be neglected in this work.

METHODOLOGY

For a an MDOF structure, the governing equation of motion can be written in the matrix form as follows

$$[M]\ddot{U}+[C]\dot{U}+[K]U=F(t) \tag{1}$$

[M], [C] and [K] are the mass, damping and stiffness matrices of the MDOF structure, F(t) is the dynamic excitation. Using Finite Element methods, system's characteristics can be determined by assembling element matrices; in the case of added damping devices, the damping matrix can be expressed as:

$$C = C_s + C_d \tag{2}$$

C_s is the structural damping and C_d is the damping in the additional FVDs, the assembling of each added device's matrix constitutes the C_d matrix, in a local frame:

$$C_d = \sum_{j=1}^{nd} l_j \begin{bmatrix} c_j & -c_j \\ -c_j & c_j \end{bmatrix} l_j \tag{3}$$

c_j is the viscosity of damper j, n_d is the number of additional dampers and l_j is an $N \times 2$ vector symbolizing the rotation localization. It should be mentioned that the only non-zero elements in C_d matrix corresponds to the added dampers ends. A projection in the modal basis is necessary in the preliminary design phase, since the appearing of the mode shape matrix ϕ simplifies the upcoming steps (diagonalized matrices M*, K*, and C*). The diagonal element of the added dampers matrix is:

$$C^*_{d_{n,n}} = \sum_{j=1}^{nd} c_j \Delta\phi^2_{n,j} \tag{4}$$

Where $\Delta\phi^2_{n,j}$ is the relative longitudinal displacement of both ends of damper j in mode n, and it is more naturel to express this quantity as a function of the inter-story drift $\Delta\phi = \varphi\Delta u_{n,j}$, with φ is called an efficiency factor (depending on the inclination of the FVD). The non-diagonal elements of this matrix could be non-zero especially when mode shapes are similar. The damping ratio of the added dampers can be expressed as follows:

$$\zeta_d = \frac{C^*_{d_{n,n}}}{2M^*_{n,n}\Omega_{n,n}} \tag{5}$$

Where $\Omega_{n,n}$ is the matrix of the natural circular frequencies. The modal damping ratio in mode n can be expressed $\zeta=\zeta_s+\zeta_d$ and since the target damping ζ_t is initially known (required by the client of the building for example) it can be deduced that $\zeta_d=\zeta_t-\zeta_s$, the minimal viscosity for all dampers can be obtained:

$$c_j \geq \frac{2M^*_{n,n}\Omega_{n,n}\left(\zeta_t - \zeta_s\right)}{\sum_{j=1}^{nd} c_j \Delta\phi^2_{n,j}} \tag{6}$$

This formula gives a flexibility in the design from different points of view, number of the dampers, their location and their inclination. All these parameters are left for the designer to consider, and some iterations maybe needed to obtain the optimal values after comparing few scenarios.

APPLICATIONS

Example 1

A three mass oscillator models represent a three-story building with the features below.

h= 3.00 m

m= 800 t

EI=4.83*10^7 kNm^2

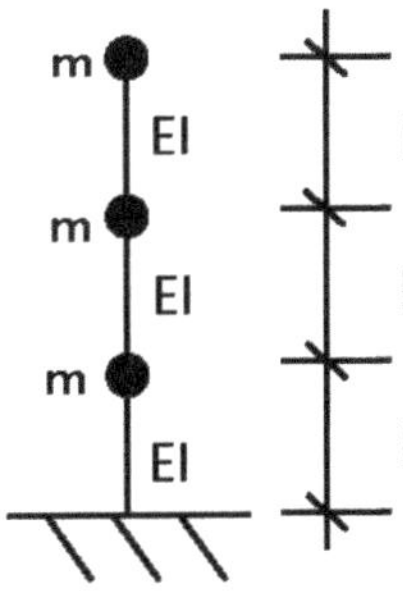

With

$$K = \frac{6EI}{h^3}\begin{bmatrix} 3.0769 & -1.7692 & 0.4615 \\ -1.7692 & 1.6923 & -0.6154 \\ 0.4615 & -0.6154 & 0.2692 \end{bmatrix}$$

and

$$M = \begin{bmatrix} 800 & 0 & 0 \\ 0 & 800 & 0 \\ 0 & 0 & 800 \end{bmatrix} t$$

We ask for a graphic representation of three degree of freedoms structure response Under Al Centro earthquake, taking into account Rayleigh damping coefficients (in the case of structural damping) as follows: α=0.0241 and β =0.2 with a damping rate ξ = 0.0149. For additional damping coefficient Cd= 50Kn s/m

Solution

After developing a code using Matlab language that determines the response of a multi degree of freedom system by respecting the coefficients given in the problem and uploading El Centro earthquake.

As the response amplitude of the three degree of freedom is greater compared to the two others degree of freedom of the studied structure. We decide to put the dashpot (damper) at the level of the third level, in order to decrease the response amplitude at this level.

The two graphs bellow represents the three degree of freedom response of the studied structure with structural damping (without additional damping) see example of chapter 5; and the comparaison between the third floor response when we have structural damping only and when we add dashpot (damper) at this floor level. All this for α which is a damping exponent is equal to one in this case that means that the damper is linear.

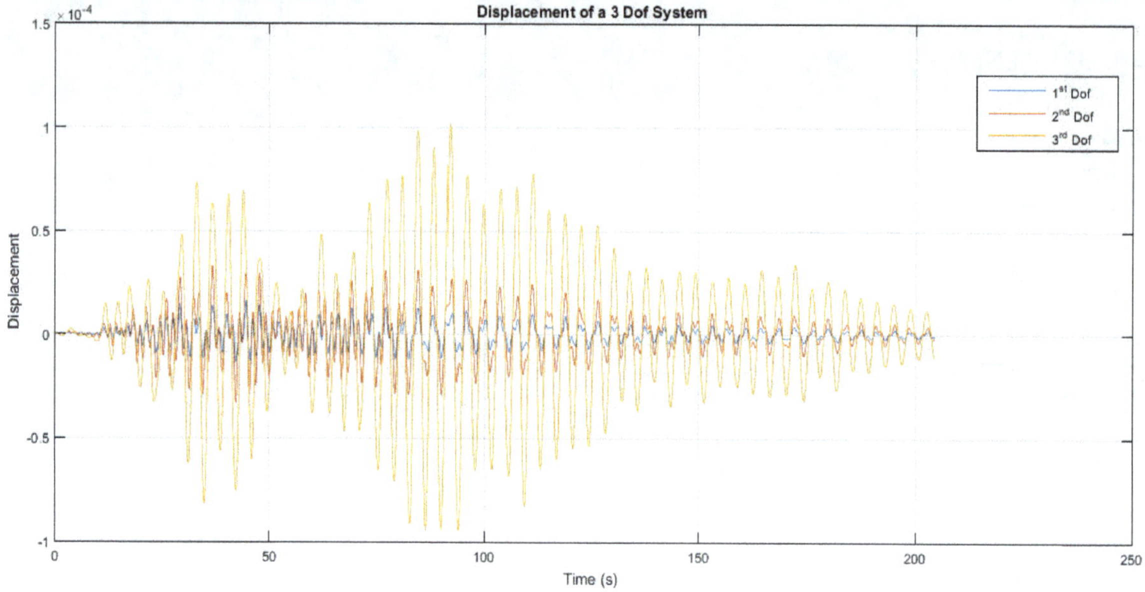

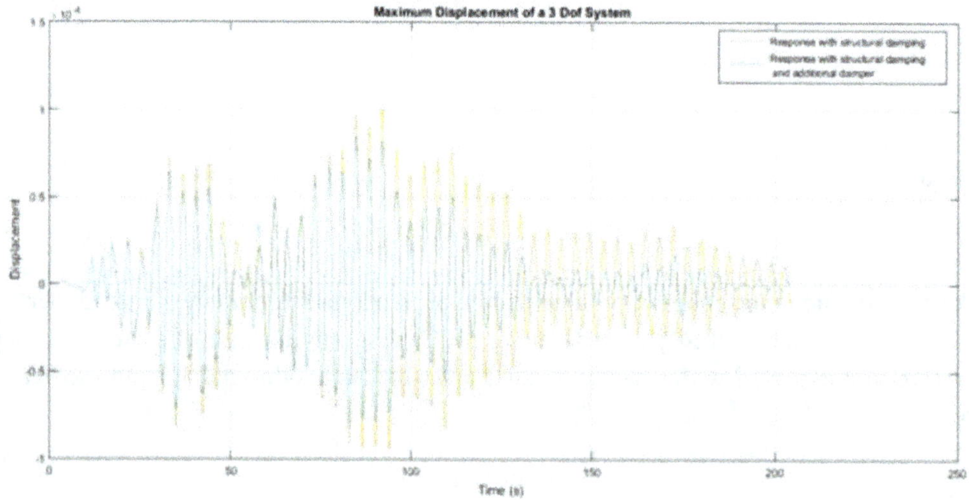

Example 2 (Gherbi & al.2019)

In this example, the preliminary design method will be applied on a 5-story RC building, the structural scheme consists in a column-beam system, figure 13 presents a plan view. With a total height of 18.28 meters and concrete grade of 25 MPA, this structure can be designed using commercially available finite element-based software package. In order to apply a time history analysis, and for example wind loads changes along the height of the building (we can imagine seismic load), the structure was divided into three representative section along its height, where the wind load is applied as a dynamic force. Using the method explained above for a multivariate time history generation. Based on the inter-story drift, the additional dampers are installed in the building; few scenarios will be proposed to finally settle with the optimal design.

Figure 4. Plan view of the analyzed structure

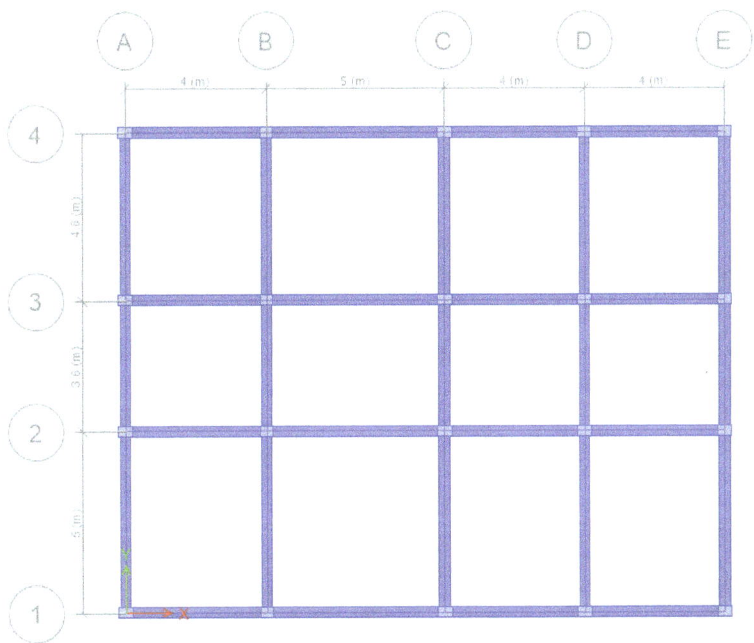

Solution

Since the approach adopted in this chapter is to simplify the comprehension, and in order to illustrate the flexibility obtained by the described method, two scenarios will be considered:

1. Additional dampers will be installed in every floor, in the middle bay in lines (A) and (E).
2. In the second scenario, the dampers will be installed where the structure presents higher values of interstory drift, since their use is to reduce the drift.

Figure 5. Inter-story drift for the structure (no additional dampers)

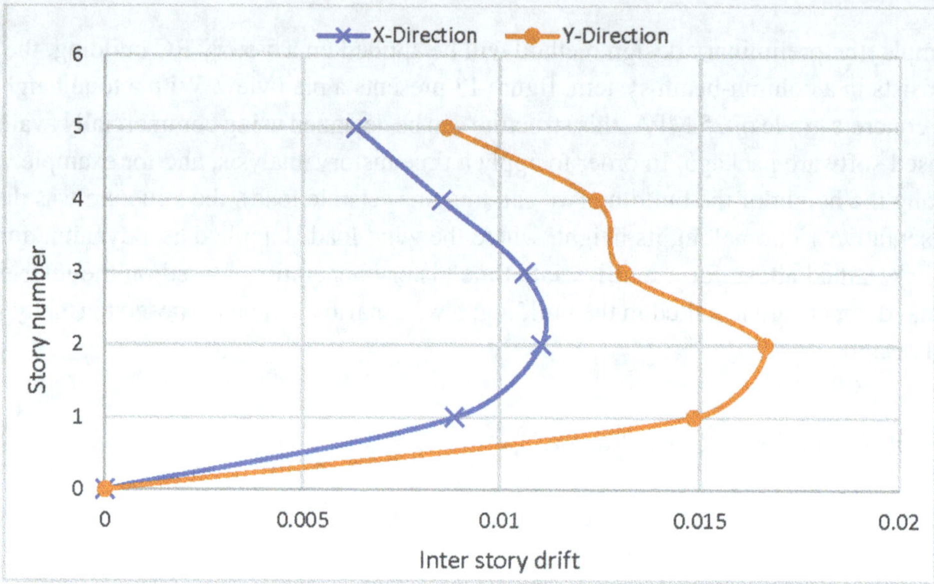

To illustrate the design methodology, table.1 summarizes the properties used for this purpose, choosing a target damping ratio for this structure to be 8%, also, the structural damping ratio is supposed to be 2%, then, $\zeta_d=\zeta_t-\zeta_s=6\%$ (these values where taken arbitrarily for illustration purpose). As for the aerodynamic damping, it is usually taken equal to 1%, however, due to a certain reduction of the damper's contribution that can be caused during their installation, the aerodynamic damping is neglected, and thus insuring that the target damping is reached.

Table 1. Structural properties for target damping

Mode	Frequency (Hz)	Modal Mass (ton.)	Additional damping ζd	Target viscosity (KN.s / m)
1 - Y direction	0.694	1371.7	6%	720
2 – X direction	0.833	1340	6%	840

Before proceeding with the proposed methodology, some assumption must be made in order to restrain the various design possibilities.

- The dampers are installed in the building diagonally (see figure 3). This means that the efficiency factor φ in equation (17) is equal to φ=cosθ, with the inclination of θ=45°, φ is assumed to be equal to $\sqrt{2}/2$.
- For the first scenario, the damper's viscosity is assumed to be the same for all dampers (10 dampers).

- As stated before in equation (17), the relative longitudinal displacement is expressed as a function of the inter-story drift, $\Delta\phi = \varphi \Delta u_{n,j}$.

After the modal analysis, the required modal viscosity is easily defined using equation (18). It was found that for the first mode (Y direction), the target modal viscosity needed to achieve the target damping ratio is equal to a total of 720 KN.s / m. Dividing the target viscosity by the number of added dampers, we find that the average contribution of a single damper is 71 KN.s / m. From figure 14, the inter-story drift is $\Delta u_n^2 = 0.016$, and φ=cos45 (damper inclination). From equation (17), the obtained value is $0.51.6.10^{-3} c_j$, it is found that to reach a damping ratio of 6%, each damper should have a viscosity approximately equal to 9000 KN.s / m.

In the same manner for the X direction, and for a value of inter-story drift of $\Delta u_n^2 = 0.011$, with ten dampers diagonally installed in the middle bay (between (B) and (C)) of lines (1) and (4). The average contribution of each damper is 83 KN.s/m; following the same procedure, it is found that the viscosity of each damper is 13000 KN.s/m.

Figure 6 shows that the maximum inter-story drift was reduced by more than 30% when adding FVDs, which is considered as a major improvement and proves the efficiency of these devices.

Figure 6. Inter-story drift reduction after the first scenario

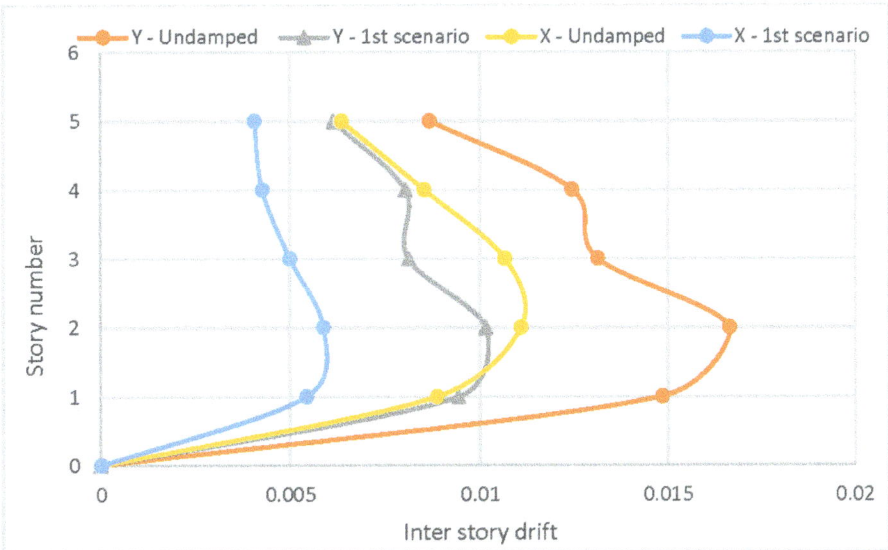

For the second scenario, the focus will be on the Y direction, since it represents higher values of drift and the same procedure can be followed for the other direction. The designer can obviously see that floors from first to fourth present high inter-story drift, and even after the addition of dampers, it may be demanded or for better behavior to reduce the drift in first and second floor. Thus, the flexibility given by this method permits to adjust the viscosity of the dampers installed in those floors, or increase the number of dampers. The latter proposition may be seen as a costly solution; so, the damping coefficient is increased for this case.

Figure 7 illustrates the optimized solution, after a few iteration where the viscosity of the dampers installed in the first and second floors is adjusted, respecting the total target viscosity. In addition dampers installed in the fifth story where removed, and an optimization of time, damping value and cost is achieved. Many other optimization proposition can be adopted for this case, but a reduction of over 50% in inter-story drift is a major improvement.

Figure 7. Inter-story drift reduction after the second scenario

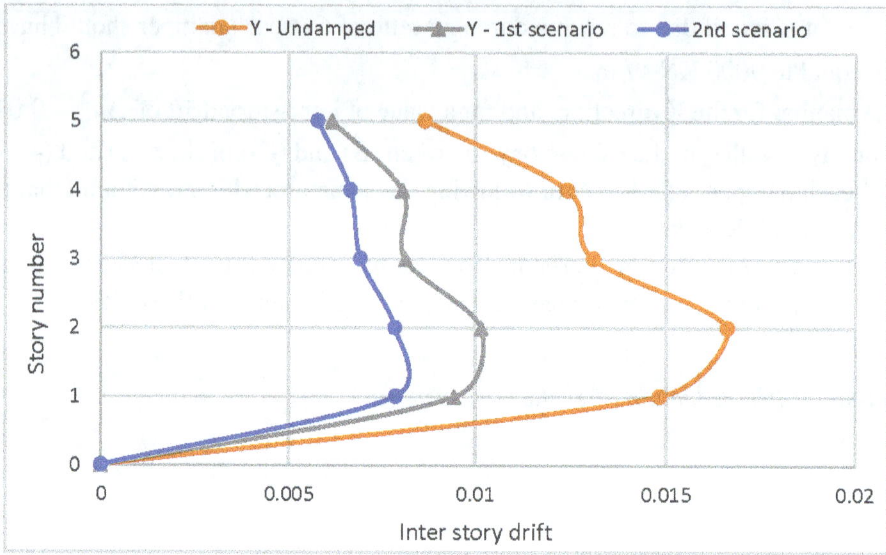

REFERENCES

Aiken, I. D., Kelly, J. M., & Mahmoodi, P. (1990, May). The application of viscoelastic dampers to seismically resistant structures. In *Proceedings of Fourth US National Conference on Earthquake Engineering* (*Vol. 3*, pp. 459-468). Academic Press.

Chang, K., Soong, T. T., Oh, S. T., & Lai, M. L. (1991). *Seismic response of a 2/5 scale steel structure with added viscoelastic dampers*. Earthquake Engineering to Extreme Events.

Constantinou, M. C. & Symans, M. D. (1992). *Experimental and analytical investigation of seismic response of structures with supplemental fluid viscous dampers*. Buffalo, NY: National Center for Earthquake Engineering Research.

Constantinou, M. C., Symans, M. D., Tsopelas, P., & Taylor, D. P. (1993). Fluid viscous dampers in applications of seismic energy dissipation and seismic isolation. *Proceedings ATC*, *17*(1), 581–592.

Duflot, P., & Taylor, D. (2008). Experience and Practical Considerations in the Design of Viscous Dampers. In *Third international conference*. Academic Press.

Duflot, P., Vigano, M. G., & Denoël, V. (2017). Method for Preliminary Design of a System of Viscous Dampers Applied to a Tall Building. *European African Conference on Wind Engineering*.

Filiatrault, A., & Cherry, S. (1990). Seismic design spectra for friction-damped structures. *Journal of Structural Engineering*, *116*(5), 1334–1355. doi:10.1061/(ASCE)0733-9445(1990)116:5(1334)

Gherbi, A., & Belgasmia, M. (2017). Dynamic Response of an RC High-Rise Building under Wind load, and Eneregy Dissipation Systems Effect. *European African Conference on Wind Engineering*.

Gherbi, A., & Belgasmia, M. (2019). *Wind Loads on Structures, and Energy Dissipation Systems Optimization, book*. IGI Global.

Guo, T., Xu, J., Xu, W., & Di, Z. (2014). Seismic Upgrade of Existing Buildings with Fluid Viscous Dampers: Design Methodologies and Case Study. *Journal of Performance of Constructed Facilities*, *29*(6), 04014175. doi:10.1061/(ASCE)CF.1943-5509.0000671

Mcnamara, R. J., & Taylor, D. P. (2003). Fluid viscous dampers for high-rise buildings. *Structural Design of Tall and Special Buildings*, *12*(2), 145–154. doi:10.1002/tal.218

Paola, M. D., Mendola, L. L., & Navarra, G. (2007). Stochastic seismic analysis of structures with nonlinear viscous dampers. *Journal of Structural Engineering*, *133*(10), 1475–1478. doi:10.1061/(ASCE)0733-9445(2007)133:10(1475)

Pollini, N., Lavan, O., & Amir, O. (2016). Towards realistic minimum-cost optimization of viscous fluid dampers for seismic retrofitting. *Bulletin of Earthquake Engineering*, *14*(3), 971–998. doi:10.100710518-015-9844-9

Rodrigo, M. M., & Romero, M. L. (2003). An optimum retrofit strategy for moment resisting frames with nonlinear viscous dampers for seismic applications. *Engineering Structures*, *25*(7), 913–925. doi:10.1016/S0141-0296(03)00025-7

Saaed, T. E., Nikolakopoulos, G., Jonasson, J. E., & Hedlund, H. (2015). A state-of-the-art review of structural control systems. *Journal of Vibration and Control*, *21*(5), 919–937. doi:10.1177/1077546313478294

Soong, T. T., & Spencer, B. Jr. (2002). Supplemental energy dissipation: State-of-the-art and state-of-the-practice. *Engineering Structures*, *24*(3), 243–259. doi:10.1016/S0141-0296(01)00092-X

KEY TERMS AND DEFINITIONS

Active Control Systems: Is based on a real time processing that allows system to act accordingly to the excitation.

Fluid Viscous Damping: Is energy dissipation systems.

Passive Control Systems: It consists in materials and devices that enhance damping, strength and stiffness.

Retrofitting: Means to put new parts or new equipment to improve an existing.

Time History Analysis: Is a step-by- step analysis of the dynamic response of a structure to a specified loading that may vary with time.

Section 4
Numerical Methods

Chapter 6
Extracting Eigenvalues and Eigenvectors Technique

ABSTRACT

In the process of solving dynamics problems at a given stage, one needs to have the dynamic characteristic of the structure in order to be able to push the calculation (resolution). Further, these characteristics are the eigenvalues and eigenvectors. In this chapter, some important points are dealt with, namely problem with standard eigenvalues, property of eigenvalues and eigenvectors, Rayleigh quotient, offset of eigenvalues, and different methods for extracting eigenvalues and eigenvectors are presented. At the end, some examples are exposed.

INTRODUCTION

This chapter describes the different methods of extracting eigenvalues and eigenvectors as well as their classifications and their characteristics. Calculating the eigenvalues and the eigenvectors of matrices is one of the most important problems in linear numerical analysis. Techniques requiring knowledge of the spectrum of matrices are used in fields as varied as quantum mechanics, structural analysis. Knowledge of the vectors and eigenvalues provides key information on the linear application considered. There are moreover many cases where this knowledge totally characterizes the linear application. The naive method of ðnding the eigenvalues of a matrix involves ðnding the roots of the characteristic polynomial of the matrix.

BRIEF REMINDER

The equation of motion, case of free vibrations for example (without load term) is written

$$[M]\{\ddot{U}(t)\} + [K]\{U(t)\} = 0 \tag{1}$$

DOI: 10.4018/978-1-7998-4399-3.ch006

Copyright © 2021, IGI Global. Copying or distributing in print or electronic forms without written permission of IGI Global is prohibited.

There is a solution of the form

$$U(t) = \phi\sin(\omega t + \theta) \qquad (2)$$

$$\ddot{U}(t) = -\varphi\,\omega^2 \sin(\omega t + \theta)$$

with ϕ is a vector, ω the vibration frequency, θ the phase angle.

By replacing $U(t)$ and $\ddot{U}(t)$ in equation (1) we obtain

$$([K] - \omega 2^{[}M]) \,\phi\sin(\omega t + \theta) = 0 \qquad (3)$$

Which must be verified whatever time t, hence the condition

$$([K] - \lambda[M])\,\phi = 0 \qquad (4)$$

Because $\sin(\omega t + \theta)\neq 0$

Equation (4) defines a eigenvalues problem, where [M] is a defined-positive mass matrix, $\lambda_i = \omega_i^2$ and ωi Ar_e the eigen frequencies expressed in rad / s.

The condition for existence of a non-trivial solution of equation (4) is:

$$P(\lambda) = \det([K] - \lambda[M]) = 0 \qquad (5)$$

Where P (λ) is the characteristic polynomial of order n with n is the number of degrees of freedom.

Note

1. If the matrix [K] and [M] are positive, the eigenvalues are real and positive
2. If the matrix [M] is singular, there are infinite eigenvalues.
3. If the matrix [K] is singular, it exits zero eigenvalues.

STANDARD EIGENVALUES PROBLEM

The problem with the standard eigenvalues is defined by the equation:

$$(A - \lambda I)\,\phi = 0 \qquad (6)$$

Where I is the identity matrix, ϕ is an eigenvector of A, λ is an eigenvalue of A and equation (6) expresses that by multiplying a matrix by one of its eigenvectors we obtain n proportional vectors.

If A = [K]e (the stiffness matrix of a finite element), equation (6) will define the Eigen frequencies of the element. The solution of the whole is written as follows:

$$[K]^e[\Phi] = [\Phi]\Lambda \qquad (7)$$

From where $[\Phi] = [\varphi 1 \ldots \varphi n]$

$$[\Lambda] = \begin{bmatrix} \lambda_1 & & & & 0 \\ & \lambda_2 & & 0 & \\ & & \ddots & & \\ & 0 & & \ddots & \\ 0 & & & & \lambda_n \end{bmatrix}$$

$[\Lambda]$: Spectral matrix
$[K]^e$: Stiffness matrix of order n, definite-positive or semi-definite-positive.

Note

- If $[K]^e$ is positive-definite: $\lambda i > 0$ i = 1 ... n, and n is the number of degrees of freedom of matrix [K].
- If $[K]^e$ is semi-definite-positive: $\lambda i, 0$ and the number of λi nulls defines the number of rigid modes.
- (λ_i, φ_i): defines a proper value-vector pair and we admit that λ_i is such that: $0 \le \lambda 1 \le \lambda 2 \le \lambda 3 \ldots \le \lambda n$
- From the values λi, we obtain the corresponding eigenvectors by the equation: $([K] - \lambda i[M]) \phi i = 0$
- The system being singular, it will be necessary to give oneself a priori at least one component of ϕi
- If the matrices [K] and [M] are symmetric and real and one is positive-definite, then we can transform the problem to the initial eigenvalues into a standard problem, as follows:

Case 1. Let [M] be positive-definite, [M] = [S] [S] T, [S] is non-singular (via cholesky); we obtain:

$[K]\phi = \lambda[S][S]T\phi$,

let's then assume $\varphi' = [S]^T \varphi$ and pre multiplied by [S]-1 we obtain:

$$[K]' \varphi' = [S]^T \varphi$$

With $[K]' = [S]^{-1}[K][S]^{-T}$

Case 2. If defined-positive, then the same approach is applied to $[M]\varphi = \frac{1}{\lambda}[K]\varphi$

The advantage of such an approach is that it allows the use of eigenvalue extraction techniques adapted to the standard problem.

EIGENVALUES AND EIGENVECTORS PROPERTY

The eigenvectors of the eigenvalue problem resulting from the equation of motion have properties of orthogonality.

Each pair (λ_k, φ_k) is such that

$$[K]\phi_i = [M]\lambda_i\phi_i \tag{8}$$

$$[K]\phi_j = [M]\lambda_j\phi_i \tag{9}$$

By pre multiplying (8) by φ_j^t one gets

$$\varphi_j^t[K]\varphi_i = \lambda_i\varphi_j^t[M]\varphi_i \tag{10}$$

By transposing the two members of equation (10) taking into account $[M]=[M]^T$ and $[K]=[K]^T$ we have

$$\varphi_i^t[K]\varphi_j = \lambda_i\varphi_i^t[M]\varphi_j \tag{11}$$

By pre multiplying equation (9) by φ_i^t one gets

$$\varphi_i^t[K]\varphi_j = \lambda_j\varphi_i^t[M]\varphi_j \tag{12}$$

Subtract from equation (11) to obtain equation (13)

$$(\lambda_i - \lambda_j)\phi_i^t[M]\phi_j = 0_j \tag{13}$$

Let's $\varphi_i^t[M]\varphi_j = 0$ if $\lambda i_, \lambda j$ (14)

Equation (14) expresses that the eigenvectors corresponding to distinct eigenvalues. They are orthogonal to the mass matrix, under certain conditions, namely the symmetry of [M] and [K]. Equation (14) can be generalized as follows:

$$\varphi_i^t[M]\varphi_j = 0 \text{ if } i^{1}j$$

Which expresses orthogonality extended to multiple eigenvalues.

By substituting equation (15) in equation (11), we obtain:

$$\varphi_i^t[K]\varphi_j = 0$$

Which expresses the orthogonality of eigenvectors with respect to rigidity.

Moreover, in pre multiplying equation (8) by ϕ_i^t we have

$$\lambda_i = \frac{\varphi_i^t [K] \varphi_j}{\varphi_i^t [M] \varphi_j} = 0 \tag{16}$$

Equation (16) allows eigenvalues to be calculated from eigenvectors.

STIFFNESS AND MASS MATRIX TRANSFORMATION

The objective of this part of the chapter is to see the influence of a transformation of the matrix and stiffness and mass, caused for example by a rotation of a bar element on the vectors and eigenvalues.

Let us have:

$$[K]^* = [Q]^T[K][Q] \tag{17}$$

$$[M]^* = [Q]^T[M][Q] \tag{18}$$

$[Q]$ Is the rotation matrix

The characteristic equation becomes

$$\det\left[[K]^* - \lambda^*[M]\right] = 0$$

Where

$$\left|[Q]^T \left([K] - \lambda^*[M]\right)[Q]\right| = 0$$

$$\left|[Q]^T\right| \left|[Q]\right| \left|\left([K] - \lambda^*[M]\right)\right| = 0$$

From where $\lambda^* = \lambda$

As a conclusion by rotating the bar, there is no influence on the frequencies.

RAYLEIGH QUOTIENT

The calculation of the eigen frequencies of complex systems with several degrees of freedom makes use of laborious numerical methods, which can only be used with the help of a computer. However, in many dynamic problems, determining the fundamental frequency is sufficient to describe the dynamic response of a structure. The Rayleigh method is a quick and easy way to obtain an approximation of the natural frequency of a system from the selection of a function approaching the shape of the fundamental mode of vibration. This method is based on the principle of energy conservation during free undamped oscillations (Clough & Penzien, 1995) & (James 1991).

Consider the N-stage frame of Figure 1. Suppose that the beams are infinitely rigid and that the equivalent stiffness of the columns of the floor i is equal to k_i.

Figure 1. Deformed structure.

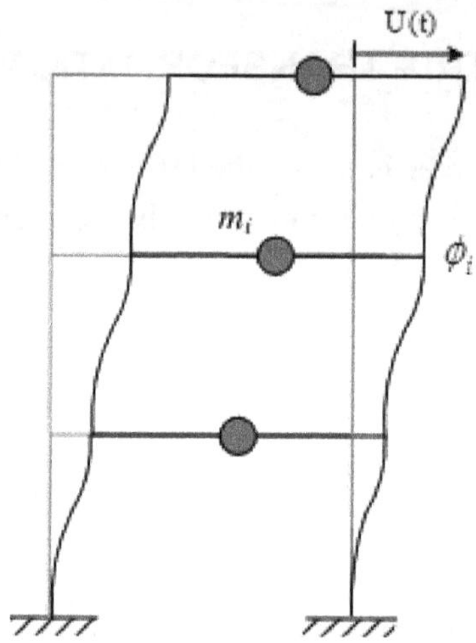

We can express the vectors of displacement and velocity by the dimensionless vector-representing mode of vibration:

$U(t) = \phi \sin(\omega t + \theta)$

$\dot{U}(t) = \varphi \omega \sin(\omega t + \theta)$

The elastic potential energy of the system is given by:

$$E_p = \sum_{i=1}^{N} \frac{1}{2} k_i \left(U_i - U_{i-1}\right)^2 \tag{19}$$

From where (U_i- U_{i-1}) represents the relative displacement of the level i.
The maximum potential energy is obtained by considering the maximum of the displacement:

$$E_{p,\max} = \frac{1}{2} \sum_{i=1}^{N} k_i \left(\varphi_i - \varphi_{i-1}\right)^2 \tag{20}$$

On the other hand, the kinetic energy of the structure in free vibration is given by:

$$E_C = \sum_{i=1}^{N} \frac{1}{2} m_i \dot{U}_i^2 \tag{21}$$

The kinetic energy reaches its maximum value when the speeds reach their maximum this appear when the structure passes through the equilibrium position:

$$E_{C,\max} = \frac{1}{2} \omega^2 \sum_{i=1}^{N} m_i \varphi_i^2 \tag{22}$$

The proper pulsation of the structure is obtained by equalizing the maximum potential energy and the maximum kinetic energy according to the principle of energy conservation. Thus, we obtain:

$$E_{p,\max} = E_{C,\max} \Rightarrow \frac{1}{2} \sum_{i=1}^{N} k_i \left(\varphi_i - \varphi_{i-1}\right)^2 = \frac{1}{2} \omega^2 \sum_{i=1}^{N} m_i \varphi_i^2$$

$$\omega^2 = \frac{\sum_{i=1}^{N} k_i \left(\varphi_i - \varphi_{i-1}\right)^2}{\sum_{i=1}^{N} m_i \varphi_i^2} \tag{23}$$

The equation above is called Rayleigh Quotient.
In Matrix form

$$\omega^2 = R(v) = \frac{\varphi^T [K] \varphi}{\varphi^T [M] \varphi}$$

RAYLEIGH QUOTIENT PROPERTIES

For any vector v, the Rayleigh quotient is written:

$$R(v) = \frac{\varphi^T [K] \varphi}{\varphi^T [M] \varphi} \tag{24}$$

Let's express the arbitrary vector v as a linear combination of N modal vectors:

$$v = \sum_{j=1}^{N} \beta_j \varphi_j$$

If the eigenvalues are distinct then the eigenvectors are orthogonal in the matrices K and M.

$$v_i^T M v_i = \begin{cases} 0 & if\ i \neq j \\ \beta_i & if\ i = j \end{cases};$$

$$v_i^T K v_i = \begin{cases} 0 & if\ i \neq j \\ \omega_i \beta_i & if\ i = j \end{cases}$$

The orthogonality property of the modal vectors makes it possible to write the Rayleigh quotient in the following form:

$$R(v) = \frac{\omega_1^2 \beta_1^2 + \omega_2^2 \beta_2^2 + \cdots + \omega_N^2 \beta_N^2}{\beta_1^2 + \beta_2^2 + \cdots + \beta_N} \tag{25}$$

EIGENVALUES SEPARATION

Let's have:

1- $\lambda i(i=1,n)$ The n eigenvalues of ([K] – λ[M])=0 in which [K] and [M] are matrices n x n and
2- Ii(i=1, (n – 1)) The eigenvalues of ([K]1 – λ[M]1)=0 in which [K]1 and [M]1 are identical to [K] and [M] minus their last row and column, then

$$\lambda_1 \leq I_1 \leq \lambda_2 \ldots \leq \lambda_{n-1} \leq I_{n-1} \leq \lambda_n$$

More generally by deleting the r last rows and columns

$$\lambda_1 \leq I_1 \leq \lambda_2 \ldots \leq \lambda_{n-r-1} \leq I_{n-r-1} \leq \lambda_{n-r}$$

In addition, the polynomial sequence $P_r(\lambda)$ constitutes a sequence of Sturm.

EIGENVALUES OFFSET (SHIFTING)

Let's have $[K]^*=[K] - a[M]$ (transformation of the stiffness with respect to the mass)
The characteristic equation becomes (Bhatt, 2001)

$$\left|[K]^{*}-\lambda^{*}[M]\right|=\left|[K]-(\lambda-a)[M]\right|=0 \tag{26}$$

Therefore, the eigenvalues are shifted compared to that of $|[K]-\lambda[M]|=0$ of $\lambda^{*}=\lambda-a$ *and* the eigenvectors are identical.

Note: It can be shown that decomposing the characteristic matrix as follows:

$$[K]-\mu[M]=LDLT \tag{27}$$

In which μ is not an eigenvalue, the number of negative terms in D corresponds to the number of eigenvalues less than μ.

EIGENVALUES AND EIGENVECTORS METHODS CLASSIFICATION

The methods for extracting eigenvalues and eigenvectors can be divided into categories according to the fundamental property of the eigenvalues and eigenvectors on which they are based. Some methods also use several fundamental properties.

Table 1. Extraction methods extraction classification

Eigenvalues and eigenvectors extraction methods classification			
N°	Description	invoked Property	Method
1	Eigenvectors Iteration	$K\phi_i=\lambda_i M\phi_i$	Reverse iteration Direct iteration Othogonalisation Gram-Schmidt
2	Transformation	$\Phi^T K\Phi=\Lambda$ $\Phi^T M\Phi=I$	Jacobi Householder-QR
3	Polynomial iterations	$p(\lambda_i)=0$ $p(\lambda_i)=\det(K-\lambda M)$	Explicit iteration Implicit iteration
4	Sturm sequence	$p(\lambda)=\det(K-\lambda M)$ $p^{(v)}(\lambda^{(v)})=\det(K^{(v)}-\lambda^{(v)}M^{(v)})$	
5	Rayleigh Quotient	$R(v)=(V^T KV)/(V^T MV)$	Iteration on a Rayleigh quotient
6	Combined		Sous-espace Lanczos

Note

1- All eigenvalues extraction techniques is iterative since it returns, in one way or another, to solve a polynomial of order n where n is the dimension of the matrix K

2- Once one of the elements of the pair (λ_i, ϕ_i) got, we get the other as follows:

for a given λ_i, ϕ_i results from the resolution of $(K-\lambda_i M)\,\phi_i=0$

for a given ϕ_i, we use the Rayleigh quotient since $\lambda_i=\phi_i{}^T K\,\phi_i$ and $\phi_i{}^T M\,\phi_i=1$ thus $\lambda_i=R(\phi_i)$

REDUCTION PROBLEM

The lower modes, corresponding to the low frequencies are generally the most interesting from structure dynamic response point of view. Moreover, the high frequencies of finite element discretized systems result directly from the discretization process and generally have nothing to do with the actual eigen frequencies. We are therefore interested in algorithms that can economically extract low frequencies, and work on a typical reduced system.

$$(K^{*} - \lambda^{*} M)\,\boldsymbol{\phi} = 0 \tag{28}$$

Where K * and M * are full matrices of small size; K * is semi-definite-positive, and M * and definite-positive.

STATIC CONDENSATION

Often in structural mechanics, we have systems like:

$$\left\{\begin{bmatrix} K_{11} & K_{12} \\ K_{21} & K_{22} \end{bmatrix} - \lambda \begin{bmatrix} M_{11} & 0 \\ 0 & 0 \end{bmatrix}\right\}\begin{Bmatrix} \varphi_1 \\ \varphi_2 \end{Bmatrix} = \begin{Bmatrix} 0 \\ 0 \end{Bmatrix} \tag{29}$$

This situation arises in particular when a light structure supports a heavy structure whose mass is concentrated in a few degrees of freedom.

Static condensation consists of the determination of j_2 from the first equation of equation system (29) and replacing it in the third equation of equation system (29) to obtain.

$$(K_{11}^{*} - 1\, M_{11}) f_1 = 0$$

With

$$K_{11}^{*} = K_{11} - K_{12} K_{22}^{-1} K_{21}$$

Note

1- Matrix K_{11}* tends to be full. This method is therefore appropriate, especially when K_{11}* is small.
2- K_{22}^{-1} Is not generally reversed, we proceed rather by factoring

RAYLEIGH-RITZ REDUCTION

In this approach, sometimes called projection method, we choose some cases of charge P (Hyghes,2000), in number of n_p and we calculate the corresponding displacements R_{np}.

$K\,R = P$ (30)

The displacements R_{np} it is called test vectors, one uses them to form the reduced system:

$K^* = R^T K R$

$M^* = R^T M R$

K and M are of dimension $n_{eq} * n_{eq}$
K* and M * are of dimensions $n_p * n_p$
P and R are of dimension $n_{eq} * n_p$

Note

1. We define P in order to activate the low frequencies: the columns of P must be linearly independent and chosen to activate the large masses and the most flexible zones of the structure
2. If n_p is small, K * and M * are small.
3. We must perform n_p solution of the equation (30).
4. We obtain an approximation of eigenvectors by $\varphi = R\varphi^*$.
5. Since it is necessary to be able to solve the equation (30), it is necessary that K is not singular, that is to say does not have eigenvalues null. If it is, the case one can apply an offset.

ITERATION METHOD ON EIGENVECTORS

This type of method aims to solve the following equation:

$K\phi = \lambda M\phi$ (31)

By successive approximation on ϕ as follows :

*X1: F*irst approximation of ϕ
$\lambda_1^1 = 1$: First approximation of $\lambda 1$

The right-hand side of equation (31) is calculated as follows:

$R_1 = 1MX_1$ Normally R1 will be different from the left-hand member:
$KX_1{}^1R_1$ We can therefore calculate a second approximation of ϕ by solving the following equation:
$KX_2 = R_1$ And so on

INVERSE ITERATION METHOD

The hypothesis in this method is that K is defined-positive and M is semi-definite-positive, but if K is semi-positive we will use the shift technique.

The inverse algorithm is described in the following steps

1- Initialisation

X_1 at each iteration

2- Solving

$$K\tilde{X}_{k+1} = MX_k \tag{32}$$

$$X_{k+1} = \frac{\tilde{X}_{k+1}}{\left(\tilde{X}_{k+1}^T M\tilde{X}_{k+1}\right)^{0.5}} \tag{33}$$

3- Test the convergence

$$\lambda^{K+1} = R\left(X_{k+1}\right)$$

$$\frac{\left|\lambda_1^{K+1} - \lambda_1^K\right|}{\lambda_1^{K+1}} \leq TOL \tag{34}$$

Example

Let's have the mass and stiffness matrix that corresponds to the following structure

$$K = \begin{bmatrix} 3 & -1 \\ -1 & 1 \end{bmatrix} \quad M = \begin{bmatrix} 2 & 0 \\ 0 & 1 \end{bmatrix}$$

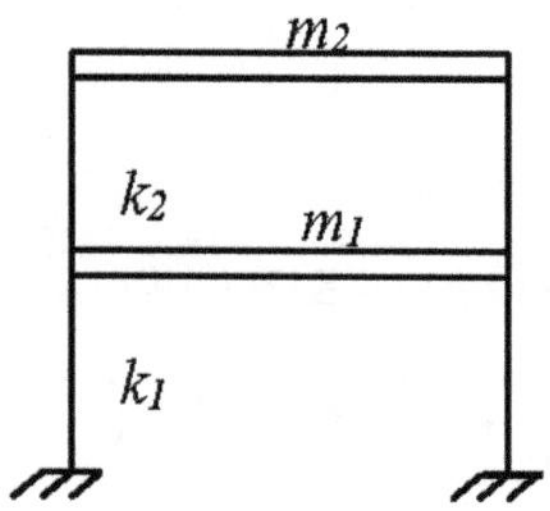

Let's have the initial approximation of the first eigenvector: X_1=[1 1]^t,

$$\tilde{X}_2 = K^{-1} M\, X_1 = \frac{1}{2}\begin{bmatrix} 1 & 1 \\ 1 & 3 \end{bmatrix}\begin{bmatrix} 2 & 0 \\ 0 & 1 \end{bmatrix}\begin{Bmatrix} 1 \\ 1 \end{Bmatrix} = \frac{1}{2}\begin{Bmatrix} 3 \\ 5 \end{Bmatrix}$$

$$\tilde{X}_2^T M\, \tilde{X}_2 = \frac{1}{2}\begin{bmatrix} 3 & 5 \end{bmatrix}\begin{bmatrix} 2 & 0 \\ 0 & 1 \end{bmatrix}\frac{1}{2}\begin{Bmatrix} 3 \\ 5 \end{Bmatrix} = \begin{bmatrix} \frac{3}{4} & \frac{5}{4} \end{bmatrix}\begin{Bmatrix} 6 \\ 5 \end{Bmatrix} = \frac{43}{4} = 10.75$$

$$X_2 = \tilde{X}_2\left(\tilde{X}_2^T M\, \tilde{X}_2\right)^{0.5} = \frac{0.5}{\sqrt{10.75}}\begin{Bmatrix} 3 \\ 5 \end{Bmatrix} = \begin{Bmatrix} 0.457 \\ 0.762 \end{Bmatrix} \text{etc...}$$

Note: The unit vector is not always the best starting choice.

JACOBI METHOD (TRANSFORMATION METHOD)

It is a transformation method based on the following:

$$\begin{aligned} \varphi^T K \varphi &= \Lambda \\ \varphi^T M \varphi &= I \end{aligned} \tag{35}$$

The matrix j diagonalizing equation (35) being unique, we try to construct it iteratively.

This method makes it possible to calculate the n value and eigenvector pairs for a system whose matrices M and K are symmetric and definite-positive. It consists in making K and M diagonals through the following successive transformations.

$$K_1 = K \quad \begin{aligned} K_1 &= Q_1^T K_1 Q_1 \\ K_{k+1} &= Q_k^T K_k Q_k \end{aligned}$$

$$M_1 = M \quad \begin{aligned} M_2 &= Q_1^T M_1 Q_1 \\ M_{k+1} &= Q_k^T M_k Q_k \end{aligned} \tag{36}$$

The matrices Kk+1 and Mk+1 become progressively diagonal. We should therefore tend towards

K=Λ

M=I

And f=Q_1 Q_2 Q_3… Q_{k+1}

In practice, it is not necessary that M_{k+1} converges to I and K_{k+1} to Λ. It's enough that they become diagonal, so

$$\lambda_i = \frac{K_{ii}^d}{M_{ii}^d} \quad \varphi = Q_1 \cdots Q_{k+1} \begin{bmatrix} \ddots & & 0 \\ & \frac{1}{\sqrt{M_{jj}^d}} & \\ 0 & & \ddots \end{bmatrix} \tag{37}$$

The transformation matrices Q_k are chosen to cancel a non-zero (non-diagonal) term of K_k and M_k after transformation. Q_k has the form:

$$Q_k = \begin{bmatrix} 1 & & & & 0 \\ & 1 & & a & \\ & & \cdot & & \\ & b & & 1 & \\ 0 & & & & 1 \end{bmatrix} \begin{matrix} \\ i \\ \\ j \\ \\ \end{matrix}$$

$$\begin{matrix} i & j \end{matrix}$$

We compute a and b by forcing $(M_{ij}) = (K_{jj})_{k+1} = 0$, so for $(k + 1)$

$a\,K_{ii}+(1+a.b)K_{ij+b}K_{jj}=0$

$a\,M_{ii}+(1+a.b)M_{ij+b}M_{jj}=0$

Whose solution is

$c_1= K_{ii}\,M_{ij}- M_{ii}K_{ij}$

$c_2= K_{jj}\,M_{ij}- M_{jj}K_{ij}$

$c_3= K_{ii}\,M_{jj}- M_{ii}K_{jj}$

$$d = \frac{c_3}{2} + signe(c_3)\sqrt{\left(\frac{c_3}{2}\right)^2 + c_1 c_2} \tag{38}$$

From where

$$a = \frac{c_2}{d} \;\cdots\cdots\; b = \frac{c_1}{d}$$

Note

1- $\sqrt{\left(\frac{c_3}{2}\right)^2 + c_1 c_2}$ is real if M is positive definite

2- if d=0 then a=0 and $b = -\frac{K_{ij}}{K_{jj}}$

CLASSIC JACOBI METHOD

The classical method applies to the standard form (M = I). In that case:

$$Q_k = \begin{bmatrix} 1 & & & & 0 \\ & \cos\theta & & -\sin\theta & \\ & & \bullet & & \\ & \sin\theta & & \cos\theta & \\ 0 & & & & 1 \end{bmatrix} \begin{matrix} \\ i \\ \\ j \\ \\ \end{matrix}$$
$$\qquad\qquad i \qquad\qquad j$$

With

$$\begin{cases} tg(2\theta) = \frac{2K_{ij}}{K_{ji} - K_{jj}} & Si\ K_{ii} \neq K_{jj} \\ \theta = \frac{\pi}{4} & Si\ K_{ii} = K_{ij} \end{cases}$$

SUB-SPACE METHOD

The subspace method combines a technique of reduction with an iterative technique in order to obtain an exact solution for the lower modes. This method is commonly used as part of the finite element method. The method follows the following steps: choice of p initial loads step, and calculation of the corresponding Ritz vectors, definition of the new load case from the inertial forces generated by the eigenvectors of the reduced problem and iteration with these new cases load. See the steps below (Hyghes,2000).

$$X_k^i = \begin{Bmatrix} 0 \\ 0 \\ \vdots \\ 1 \\ \vdots \\ 0 \end{Bmatrix}$$

APPLICATIONS

Example 1

Determine the corresponding eigenvalues and eigenvector of a bar whose stiffness and mass matrix are as follows:

$$K = k\begin{bmatrix} 1 & -1 \\ -1 & 1 \end{bmatrix} \cdots\cdots\ M = \frac{m}{2}\begin{bmatrix} 1 & 0 \\ 0 & 1 \end{bmatrix}$$

Solution

Calculation of eigenvalues:

$$\left[[K]-\lambda[M]\right] = \begin{bmatrix} k-\lambda\dfrac{m}{2} & -k \\ -k & k-\lambda\dfrac{m}{2} \end{bmatrix}$$

$$\det\left[[K]-\lambda[M]\right] = \det\begin{bmatrix} k-\lambda\dfrac{m}{2} & -k \\ -k & k-\lambda\dfrac{m}{2} \end{bmatrix} = \left(k-\lambda\frac{m}{2}\right)^2 - k^2 = 0$$

$$\lambda^2\frac{m^2}{4} - \lambda km + k^2 - k^2 = 0$$

$$\lambda\left(\frac{m^2}{4}\lambda - km\right) = 0$$

$\lambda_1 = 0 \quad \Rightarrow \quad \omega_1 = 0$

It means that there is a rigid body movement

$\lambda_2 = \frac{4k}{m} \quad \Rightarrow \quad \omega_2 = 2\sqrt{\frac{k}{m}}$

Calculation of the corresponding eigenvectors

1- $\lambda_1 = 0 \quad \Rightarrow \quad \omega_1 = 0$

$$[[K]-\lambda[M]]\begin{Bmatrix}\phi_{11}\\ \phi_{12}\end{Bmatrix} = \begin{bmatrix} k-\lambda\frac{m}{2} & -k \\ -k & k-\lambda\frac{m}{2}\end{bmatrix}\begin{Bmatrix}\phi_{11}\\ \phi_{12}\end{Bmatrix} = \begin{bmatrix} k & -k \\ -k & k\end{bmatrix}\begin{Bmatrix}\phi_{11}\\ \phi_{12}\end{Bmatrix} = \begin{Bmatrix}0\\ 0\end{Bmatrix}$$

As it is singular, I chose an arbitrary value $\varphi 1_{1=}1$ and I determine $\varphi 12$

$$\begin{bmatrix} k & -k \\ -k & k\end{bmatrix}\begin{Bmatrix}1\\ \phi_{12}\end{Bmatrix} = \begin{Bmatrix}0\\ 0\end{Bmatrix}$$

From where φ_{12}=1. Thus $\varphi_1 = \begin{bmatrix}\phi_{11} & \phi_{12}\end{bmatrix}^T = \begin{bmatrix}1 & 1\end{bmatrix}^T$

The eigenvector $\varphi 1$ represents a rigid translation it is a rigid body movement

2- $\lambda_2 = \frac{4k}{m} \quad \Rightarrow \quad \omega_2 = 2\sqrt{\frac{k}{m}}$

$$[[K]-\lambda[M]]\begin{Bmatrix}\phi_{21}\\ \phi_{22}\end{Bmatrix} = \begin{bmatrix} k-\frac{4k}{m}\frac{m}{2} & -k \\ -k & k-\frac{4k}{m}\frac{m}{2}\end{bmatrix}\begin{Bmatrix}\phi_{21}\\ \phi_{22}\end{Bmatrix} = \begin{bmatrix} -k & -k \\ -k & k\end{bmatrix}\begin{Bmatrix}\phi_{21}\\ \phi_{22}\end{Bmatrix} = \begin{Bmatrix}0\\ 0\end{Bmatrix}$$

$$\begin{bmatrix} -k & -k \\ -k & -k\end{bmatrix}\begin{Bmatrix}1\\ \phi_{22}\end{Bmatrix} = \begin{Bmatrix}0\\ 0\end{Bmatrix}$$

From where φ_{22}= -1. Thus $\varphi_2 = \begin{bmatrix}\phi_{21} & \phi_{22}\end{bmatrix}^T = \begin{bmatrix}1 & -1\end{bmatrix}^T$

In this second case $\phi 2$ represents a deformation mode

Example 2

It is proposed to calculate the eigenvectors using Rayleigh-Ritz Reduction of this building

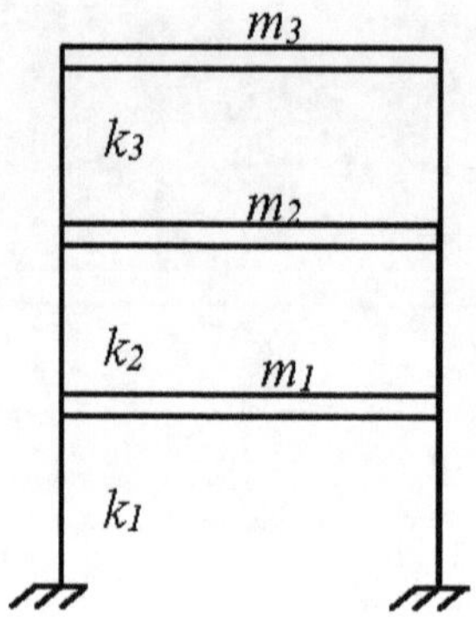

From where $\begin{bmatrix} 6 & -2 & 0 \\ -2 & 3 & -1 \\ 0 & -1 & 1 \end{bmatrix}$ and $\begin{bmatrix} 6 & 0 & 0 \\ 0 & 2 & 0 \\ 0 & 0 & 1 \end{bmatrix}$

The eigenvalues and eigenvectors of this example are:

λ_1=0.3, λ_2=1, λ_3=2.2

$$\varphi_1^T = \begin{bmatrix} 0.205 & 0.429 & 0.616 \end{bmatrix}$$

$$\varphi_2^T = \begin{bmatrix} -0.316 & 0 & 0.632 \end{bmatrix}$$

And

$$\varphi_1^T = \begin{bmatrix} -0.157 & 0.562 & -0.470 \end{bmatrix}$$

Rayleigh-Ritz Reduction

Considering the cases of charges P_1, P_2 applied to the structures defined above; we have:

$k_{11}/m_{11}=1$

$k_{22}/m_{22}=1.5$

$k_{33}/m_{33}=1$

The smaller values correspond to the lower modes

$$P_1 = \begin{Bmatrix} 1 \\ 0 \\ 0 \end{Bmatrix} ......P_2 = \begin{Bmatrix} 0 \\ 0 \\ 1 \end{Bmatrix}$$

From where

$$r_1 = \begin{Bmatrix} 1/4 \\ 1/4 \\ 1/4 \end{Bmatrix} ......r_2 = 1/4 \begin{Bmatrix} 1 \\ 3 \\ 7 \end{Bmatrix}$$

$k_{11}/m_{11}=1$ $\qquad$ $k_{22}/m_{22}=1.5$ $\qquad$ $k_{33}/m_{33}=1$

r_i Is such that $Kr_i = P_i$

From where $R = \begin{bmatrix} r_1 & r_2 \end{bmatrix} = \frac{1}{4}\begin{bmatrix} 1 & 1 \\ 1 & 3 \\ 1 & 7 \end{bmatrix}$

$$K^* = R^T K\, R = \frac{1}{16}\begin{bmatrix} 1 & 1 & 1 \\ 1 & 3 & 7 \end{bmatrix}\begin{bmatrix} 6 & -2 & 0 \\ -2 & 3 & -1 \\ 0 & -1 & 1 \end{bmatrix}\begin{bmatrix} 1 & 1 \\ 1 & 3 \\ 1 & 7 \end{bmatrix}$$

$$K^* = \frac{1}{4}\begin{bmatrix} 1 & 1 \\ 1 & 7 \end{bmatrix}$$

$$M^* = R^T M\; R = \begin{bmatrix} 0.56 & 1.19 \\ 1.19 & 4.56 \end{bmatrix}$$

From where

$$P(\lambda^*) = \det\left(K^* - \lambda^* M^*\right) = 0$$

And

$$\lambda_1^* = 0.32$$
$$\lambda_2^* = 1.62$$

(To compare with previous results λ_1=0.3 and λ_2=1)
Similarly

$$\varphi_1^* = \begin{bmatrix}0.587 & 0.295\end{bmatrix}^T \ldots\ldots \varphi_2^* = \begin{bmatrix}1.914 & -0.637\end{bmatrix}^T$$

Hence the approximations of the eigenvectors of the initial problem

$$\tilde{\varphi}_1 = R\varphi_1^* = \begin{Bmatrix}0.22\\0.37\\0.66\end{Bmatrix} \cong \varphi_1 = \begin{Bmatrix}0.205\\0.429\\0.616\end{Bmatrix}$$

$$\tilde{\varphi}_2 = R\varphi_2^* = \begin{Bmatrix}0.32\\10^{-3}\\-0.64\end{Bmatrix} \cong \varphi_1 = \begin{Bmatrix}-0.316\\0\\0.632\end{Bmatrix}$$

Example 3

It is proposed to calculate the fundamental frequency of the building using the Rayleigh quotient and assuming the initial linear form shown in Figure bellow with:

$m_1 = m_2 = m_3 = 400t$

$k_1 = 240\,10^6\ N/m;\ k_2 = 160\,10^6\ N/m;\ k_3 = 80\,10^6\ N/m$

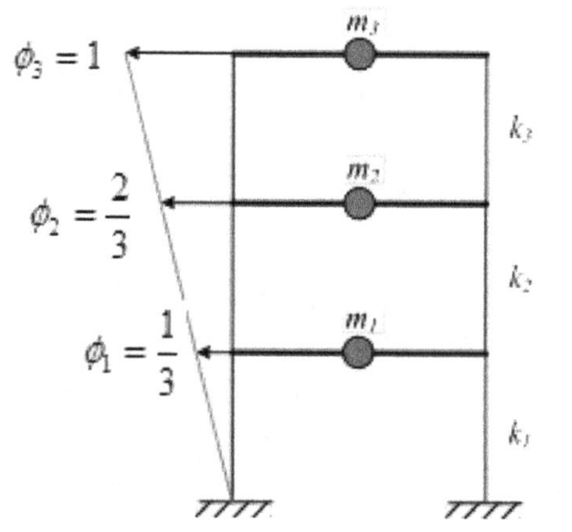

Solution

The Rayleigh quotient is written as follows:

$$\omega^2 = \frac{\sum_{i=1}^{N} k_i \left(\varphi_i - \varphi_{i-1}\right)^2}{\sum_{i=1}^{N} m_i \varphi_i^2}$$

$$\omega^2 = \frac{k_1\varphi_1^2 + k_2\left(\varphi_2 - \varphi_1\right)^2 + k_2\left(\varphi_3 - \varphi_2\right)^2}{m_1\varphi_1^2 + m_2\varphi_2^2 + m_3\varphi_3^2}$$

Which give $\omega 2^{=}85.714 \Rightarrow \omega=9.26$ rad/s

Example 4

Considering the following problem determine the eigenvalues and vector and check the orthogonality relation knowing that the stiffness and mass matrix are as follows:

$$[K] = \begin{bmatrix} 1 & 1 \\ 1 & 3 \end{bmatrix} \quad [M] = \begin{bmatrix} 1 & 1 \\ 1 & 2 \end{bmatrix}$$

Use Rayleigh quotient for frequency estimation from eigenvectors (Eigen modes) Separation of eigenvalues

Solution

The matrices [K] and [M] are defined-positive, so the eigenvalues are real and positive, they are the roots of the polynomial

$$\det\left[[K] - \lambda[M]\right] = \det\left(\begin{bmatrix} 1 & 1 \\ 1 & 3 \end{bmatrix} - \lambda\begin{bmatrix} 1 & 1 \\ 1 & 2 \end{bmatrix}\right) = (1-\lambda)(2-\lambda) = 0$$

From where $\lambda 1_{=}1$, $\lambda 2{=}_2$

Calculation of the corresponding eigenvectors

1- $\lambda 1_{=}1$ Þ $\omega 1{=}_1$ rad/s

$$\left[[K] - \lambda[M]\right]\begin{Bmatrix} \phi_{11} \\ \phi_{12} \end{Bmatrix} = \left(\begin{bmatrix} 1 & 1 \\ 1 & 3 \end{bmatrix} - 1\begin{bmatrix} 1 & 1 \\ 1 & 2 \end{bmatrix}\right)\begin{Bmatrix} \phi_{11} \\ \phi_{12} \end{Bmatrix} = \begin{Bmatrix} 0 \\ 0 \end{Bmatrix} \Rightarrow \{\phi_1\} = \begin{Bmatrix} x_1 \\ 0 \end{Bmatrix}$$

$$\left[[K] - \lambda[M]\right]\begin{Bmatrix} \phi_{21} \\ \phi_{22} \end{Bmatrix} = \left(\begin{bmatrix} 1 & 1 \\ 1 & 3 \end{bmatrix} - 2\begin{bmatrix} 1 & 1 \\ 1 & 2 \end{bmatrix}\right)\begin{Bmatrix} \phi_{11} \\ \phi_{12} \end{Bmatrix} = \begin{Bmatrix} 0 \\ 0 \end{Bmatrix} \Rightarrow \{\phi_2\} = \begin{Bmatrix} x_2 \\ -x_2 \end{Bmatrix}$$

The values of x_1 and x_2 are obtained using the condition of M-orthonormality of the vectors $\{\varphi 1\}$ and $\{\varphi 2\}$, the resulting modal matrix is:

$$\{\phi_1\}^T[M]\{\phi_1\}=1 \Rightarrow [x_1\ 0]\begin{bmatrix}1 & 1\\1 & 2\end{bmatrix}\begin{Bmatrix}x_1\\0\end{Bmatrix}=x_1^2=1$$

$$\{\phi_2\}^T[M]\{\phi_2\}=1 \Rightarrow [x_2\ -x_2]\begin{bmatrix}1 & 1\\1 & 2\end{bmatrix}\begin{Bmatrix}x_2\\-x_2\end{Bmatrix}=x_2^2=1$$

From where

$$\{\phi_1\}=\begin{Bmatrix}1\\0\end{Bmatrix}\ldots\ldots\{\phi_2\}=\begin{Bmatrix}1\\-1\end{Bmatrix}$$

Thus $\varphi=\begin{bmatrix}1 & 1\\0 & -1\end{bmatrix}$

For any vector v, the Rayleigh quotient is written:

$$R(v)=\frac{\varphi^T[K]\varphi}{\varphi^T[M]\varphi}=\frac{\{1\ \ 1\}\begin{bmatrix}1 & 1\\1 & 3\end{bmatrix}\begin{Bmatrix}1\\1\end{Bmatrix}}{\{1\ \ 1\}\begin{bmatrix}1 & 1\\1 & 2\end{bmatrix}\begin{Bmatrix}1\\1\end{Bmatrix}}=\frac{6}{5}=1.2$$

We check that $\lambda_1 \leq R(v) \leq \lambda 2$

With $\lambda 1=1$, $\lambda 2=2$

By removing the last line and the last column of stiffness [K] and [M] mass matrix:

[K]1=1 [M]1=1

The eigenvalue of

$(1-I1)\phi=0$ Is I=1

We check the property

$\lambda_1 \leq I \leq \lambda 2$

Moreover by performing the decomposition LDLT with the offset $\mu=3$

$$[K]-\mu[K]=[K]-3[K]=\begin{bmatrix}-2 & -2\\ -2 & -3\end{bmatrix}=\begin{bmatrix}1 & 0\\ 1 & 1\end{bmatrix}\begin{bmatrix}-2 & 0\\ 0 & -1\end{bmatrix}\begin{bmatrix}1 & 1\\ 0 & 1\end{bmatrix}$$

Since there are two negative terms in the diagonal matrix, we check that all values are less than 3.

REFERENCES

Bhatt, P. (2001). Programming the Dynamic Analysis of Structures. Academic Press.

Clough, R. W. & Penzien, J. (1995). *Dynamic of structures.* Compute and Structure, Inc.

Hyghes, T. J. R. (2000). *The Finite Element Method Linear Static and dynamic Finite Element Analysis.* Dover Publications, Inc.

James, F. D. (1991). *Static and Dynamique Analysis of Structures: with an Emphasis on Mechanic and Computer Matrix Method.* Kluwer Academic Publishres.

KEY TERMS AND DEFINITIONS

Eigenvalues: Are a special set of scalars associated with a linear system of equations.:

Eigenvalues Offset: Is a shifting of eigenvalues by transformation of the stiffness with respect to the mass for ease of calculation.

Eigenvectors: Is a nonzero vector that changes at most by a scalar factor when that linear transformation is applied to it.

Rayleigh Quotient: Is a scalar, which depends on the matrices [K] and [M] as well as {φ}.

Rayleigh-Ritz Reduction: Is a numerical method of finding approximations to eigenvalue equations that are difficult to solve analytically.

Static Condensation: Is a dimensionality reduction method which reduces the number of degrees of freedom by ignoring the inertial terms of the equilibrium equations and expressing the unloaded degrees of freedom in terms of the loaded degrees of freedom.

Chapter 7
Numerical Methods for Solving Dynamic Equilibrium Equations

ABSTRACT

Once the number of degrees of freedom exceeds a certain number, it would be impossible to solve the dynamic equilibrium equation manually, hence the need to switch to a numerical resolution, whose general principle is to convert a dynamic equation into a static one. We are interested, for the dynamic analysis of the structures and the continuous media, in "one-step" algorithms rather than "multi-step" one. It is mainly because the systems to be solved are of large size and that it is important to minimize the number of operations and value to be memorized to the detriment, if necessary, of precision. A "one-step" algorithm, like that of Newmark, makes it possible to calculate the solution at time t_{n+1}, starting from the solution at time t_n. In addition to the disadvantage of requiring the storage of several steps, the "multi-step" algorithms such as that of Houbolt requires a startup procedure. This chapter allows the reader to enumerate and understand different numerical method with different examples.

INTRODUCTION

This chapter describes the different numerical resolution method of dynamic equilibrium equation, which all have the same principle, which is the conversion of a dynamic equation into a static one.

Most of the known dynamics methods, the frequency domain analysis, the Duhamel integral, presuppose the validity of the superposition principle and therefore do not apply to non-linear problems.

Direct integration, step by step with a constant or variable time step Dt, evaluates the response during the time step from the initial conditions to the beginning of the time step $U_0 , \dot{U}_0$ and the evolution of the load during the step.

The response at each time step is an independent analysis that does not require the validity of the superposition principle. Step by step integration methods provide a general approach to the problem of linear and non-linear dynamics

DOI: 10.4018/978-1-7998-4399-3.ch007

Copyright © 2021, IGI Global. Copying or distributing in print or electronic forms without written permission of IGI Global is prohibited.

Before talking about numerical methods, we allow ourselves to give a brief summary in the form of a table on what has been done so far for the resolution of one degree of freedom system under different loads types.

DIFFERENT STRUCTURAL RESPONSES OVERVIEW

The table below gives an overview of different responses of the structure corresponding to the different types of solicitations; these later can be enumerated as follows: harmonic, periodic, impulsive and any type of solicitations.

For Duhamel integral there are limitations, which are:

- The assumption of the linear system
- The solutions are not always possible example earthquake
- The integral is not generalized; for each load there is a separate solution that can't be scaled.

In order to overcome these limitations, numerical methods are used which offer several advantages, namely:

- The use of these methods in software (seismic simulation)
- Treatment of non-linear systems
- Possibility of generalizing these methods
- Programmability of methods

EXACT METHOD BY INCREMENTS

This method consists in dividing the history of the charge into intervals, generally constant, and in admitting that the charge varies linearly over the interval. There is therefore a linear approximation of the load per interval, and the response for the linearized load is evaluated for a continuously variable load, the solution is improved by reducing the time step h = Dt (constant or variable).

Considering a step Dt = t1-t0 and setting t = t- t0, we then have the load:

$$P(\tau) = \mathrm{P}0 + \alpha\tau \tag{1}$$

The equation of motion becomes:

$$M\ddot{U} + kU = P_0 + \alpha\tau \tag{2}$$

The response comprises a free vibration component U_h and the particular solution U_p corresponding to the load:

$$U(\tau) = \mathrm{U}h_(\tau) + \mathrm{UP}(_\tau) \tag{3}$$

Table 1. Different solicitation structural response

Excitation	*Structure*	*Response*
Free vibration	Single degree of freedom 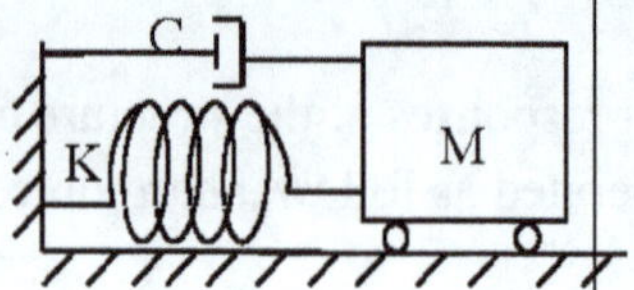	Transient 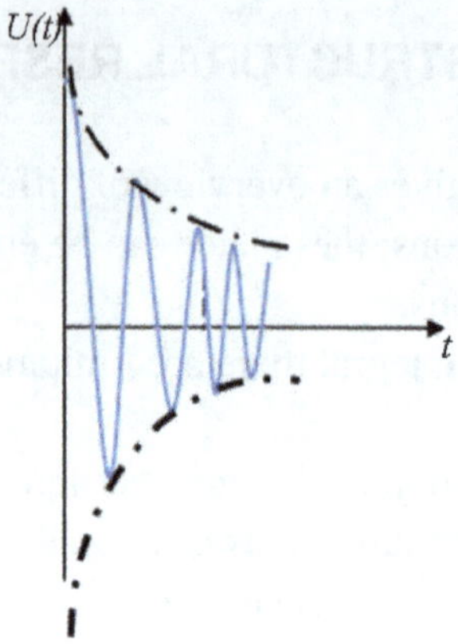
Harmonic		Transient + Stationary 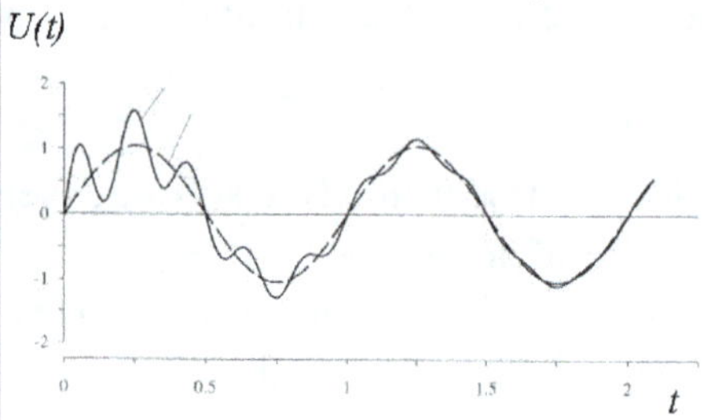
Periodic load		Stationary
Impulsive load 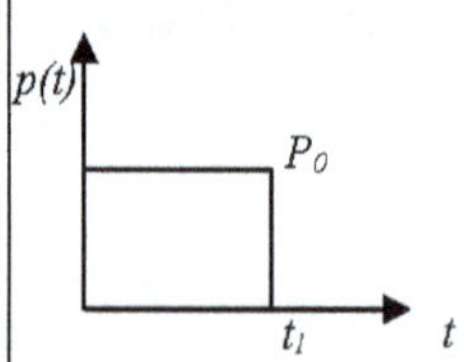		Two-phase solution 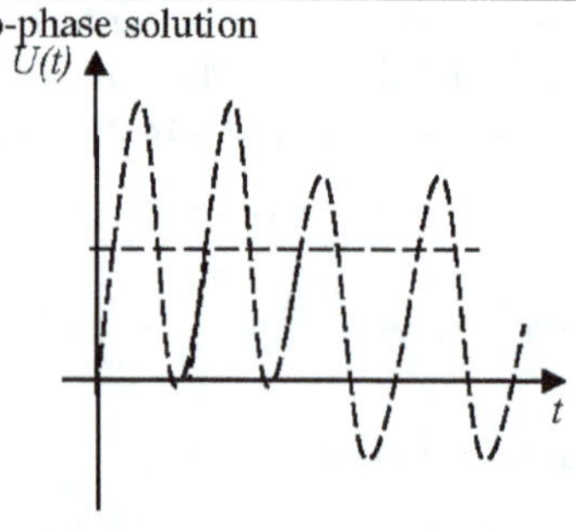
Any load type	Duhamel Integral →	Any response type

Numerical Integration

Figure 1. Loading history (Hyghes,2000)

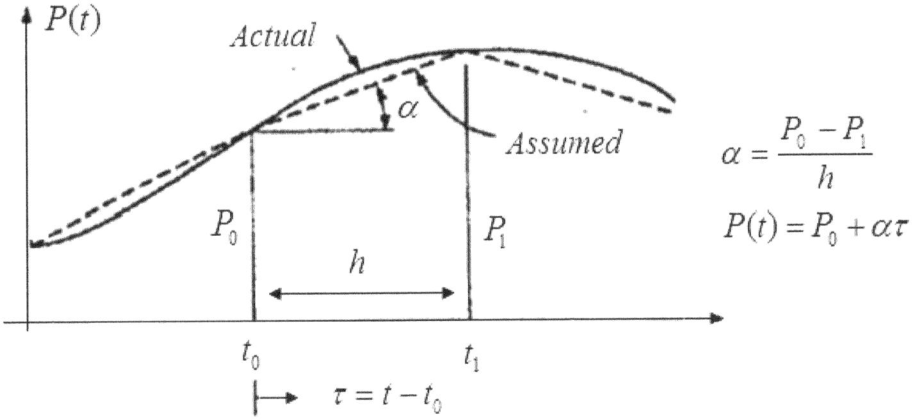

Figure 2. Response History (Hyghes,2000)

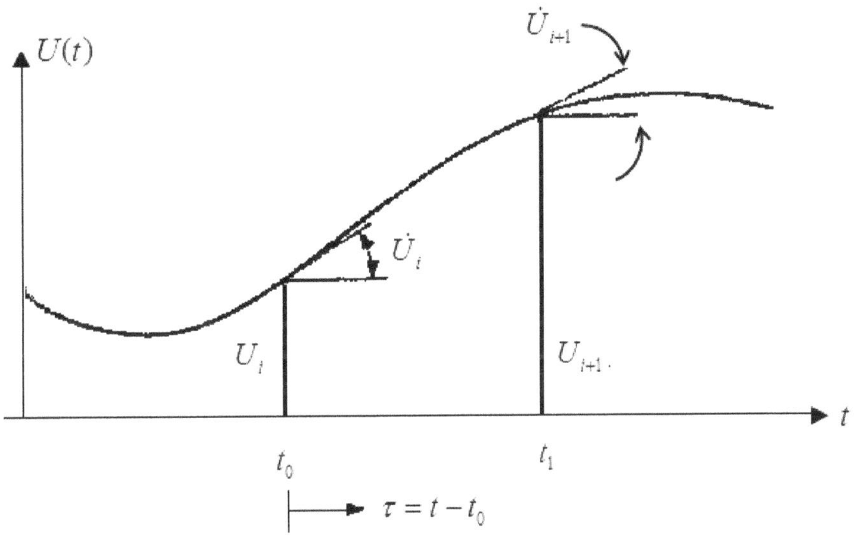

The solution of the homogeneous equation (free vibration) has been established previously:

$$U_h(\tau) = U_0 \cos\omega\tau + \frac{\dot{U}_0}{\omega}\sin\omega\tau \tag{4}$$

The particular solution has two components, the first corresponding to P0, the second to the variation of the load on Dt:

$$U_p(\tau) = \frac{P_0}{k}\left(1 - \cos\omega\tau\right) + \frac{\alpha}{k}\left(\tau - \frac{\sin\omega\tau}{\omega}\right) \tag{5}$$

The solutions represented in equation (4) and (5) make it possible to calculate the displacement and the velocity at the end of the time interval knowing the displacement and the velocity at the beginning; the operation can then be repeated for the next time interval.

The resulting algorithm therefore seems perfectly adept at one degree of freedom system (damped or not), but hardly generalizable to a multi degrees of freedom system.

Let the multi-DDL system represented by equation (6)

$$[M]_{(nxn)}\{\ddot{U}\}_{(n)}+[C]_{(nxn)}\{\dot{U}\}_{(n)}+[K]_{(nxn)}\{U\}_{(n)}=\{f\}_{(n)} \tag{6}$$

This system can be written as a force system as follows:

$$F_I(t) + F_A(t) + F_E(t) = f(t)$$

Mathematically Equation (6) is a set of differential equations that can be solved with standard methods. However, when 'n' is very large the use of the notion of differential equation with second member becomes very heavy so we use other methods to solve this system of equation, among these methods we have:

1. Direct integration method
2. Modal superposition method

DIRECT INTEGRATION METHODS

Although such methods can be used for linear systems, one of the difficulties of these methods is that the damping matrix must be explicitly defined; a formulation of the Rayleigh damping type is generally retained. This is not a requirement because the resolution method does not require the decoupling of the modal responses and the matrix C can have any form.

Many procedures are applicable for the numerical integration of motion equations (Hyghes,2000). In these direct integration methods, use the step-by-step solution; instead of solving the equation of motion (6), we will divide the time into a time increment and we will consider equilibrium of the equation only in the interval Δt. The use of direct integration can be useful when the integral of the excitatory charge is difficult to determine. Therefore, the idea in this case is to divide the surface into an increment of time with $N=\frac{t}{\Delta t}$, which is the number of time steps or increment, it is not necessary to take Δt constant.

Among the most used integration methods in structural dynamics, we have:

- Linear acceleration method
- Average acceleration method
- Wilson's method
- Newmark b method
- Houbolt method

Figure 3. Variation arbitraire de la charge dans le temps

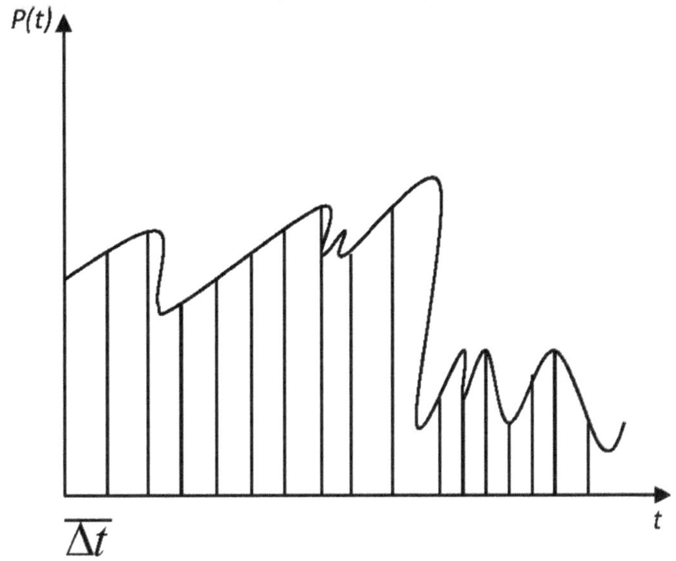

LINEAR ACCELERATION METHOD

The assumption in this method is that the acceleration varies linearly in time.

By similar triangles, we can write:

$$\frac{(\ddot{U}_{t+\Delta t}-\ddot{U})}{\Delta t}=\frac{x}{\tau}\Rightarrow x=\frac{\tau}{\Delta t}(\ddot{U}_{t+\Delta t}-\ddot{U})$$

Knowing that $\ddot{U}_{t+\tau}=\ddot{U}+x$

We will have:

$$\ddot{U}_{t+\tau}=\ddot{U}_{t}+\frac{\tau}{\Delta t}(\ddot{U}_{t+\Delta t}-\ddot{U}_{t}) \tag{7}$$

The acceleration represented in equation (7) is integrated with respect to τ we have:

$$\dot{U}_{t+\tau}=\dot{U}_{t}+\ddot{U}_{t}\tau+\frac{\tau^{2}}{2\Delta t}(\ddot{U}_{t+\Delta t}-\ddot{U}_{t}) \tag{8}$$

By integrating a second time we will have

$$U_{t+\tau}=U_{t}+\dot{U}_{t}\tau+\frac{\ddot{U}_{t}}{2}\tau^{2}+\frac{\tau^{3}}{6\Delta t}(\ddot{U}_{t+\Delta t}-\ddot{U}_{t}) \tag{9}$$

Figure 4. Linear variation of acceleration

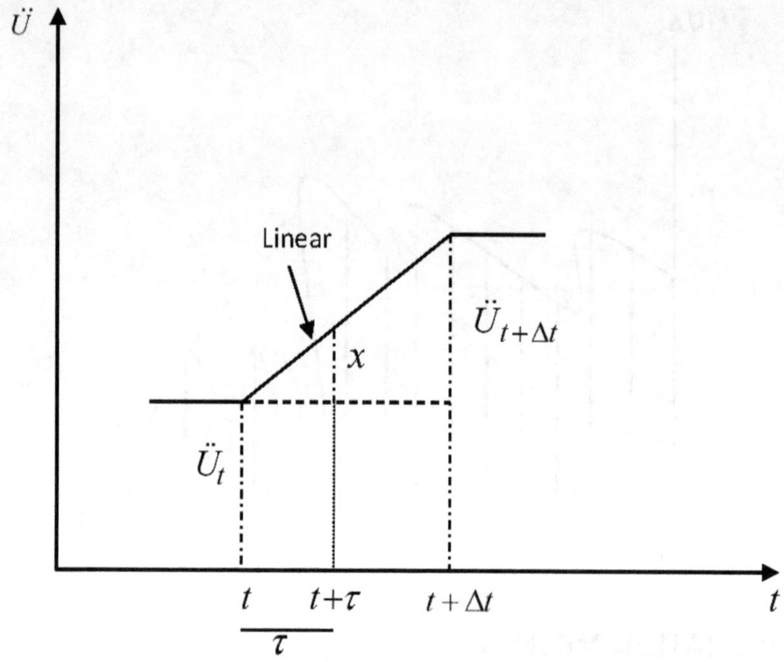

At the end of the interval τ=Δt Therefore $\ddot{U}_{t+\tau} = \ddot{U}_{t+\Delta t}$
By replacing the expression of $\ddot{U}_{t+\tau}$ in equations (8) and (9) we have

$$\dot{U}_{t+\Delta t} = \dot{U}_t + \ddot{U}_t \Delta t + \frac{\Delta t}{2}(\ddot{U}_{t+\Delta t} - \ddot{U}_t) \tag{10}$$

$$U_{t+\Delta t} = U_t + \dot{U}_t \Delta t + \frac{\ddot{U}_t}{2}\Delta t^2 + \frac{\Delta t^2}{6}(\ddot{U}_{t+\Delta t} - \ddot{U}_t) \tag{11}$$

We determine the expression of $\ddot{U}_{t+\Delta t}$ from equation (11) we find

$$\ddot{U}_{t+\Delta t} = \frac{6}{\Delta t^2}(U_{t+\Delta t} - U_t) - \frac{6}{\Delta t}\dot{U}_t - 2\ddot{U}_t \tag{12}$$

By replacing equation (12) in equation (10), we find:

$$\dot{U}_{t+\Delta t} = (\frac{3}{\Delta t})(U_{t+\Delta t} - U_t) - 2\dot{U}_t - (\frac{\Delta t}{2})\ddot{U}_t \tag{13}$$

The equation of motion at time $t+\Delta t$

$$M\ddot{U}_{t+\Delta t} + C\dot{U}_{t+\Delta t} + KU_{t+\Delta t} = f(t+\Delta t) \quad (14)$$

By replacing the equations (11), (12) and (13) in equation (14) and rearranging the terms we find:

$$(\frac{6}{\Delta t^2}M + \frac{3}{\Delta t}C + K)U_{t+\Delta t} = f(t+\Delta t) + (\frac{6M}{\Delta t^2} + \frac{3}{\Delta t}C)U_t + (2M + \frac{\Delta t}{2}C)\ddot{U}_t + (\frac{6M}{\Delta t} + 2C)\dot{U}_t$$

This last equation is of the form:

$\hat{K}U_{t+\Delta t} = \hat{f}$ With $\hat{K} = \frac{6}{\Delta t^2}M + \frac{3}{\Delta t}C + K$

And

$$\hat{f} = f(t+\Delta t) + (\frac{6M}{\Delta t^2} + \frac{3}{\Delta t}C)U_t + (2M + \frac{\Delta t}{2}C)\ddot{U}_t + (\frac{6M}{\Delta t} + 2C)\dot{U}_t$$

We notice that a dynamic problem has been transformed into a static problem.

-Stiffness K has been transformed by mass and damping.
-The force f (t +Dt) has been transformed by the preceding motion (initial condition) at time t.

If the increment of time is constant: $f_1(t)$ is the average value and if we suppose that the initial conditions are equal to zero: $t = 0 \Rightarrow U_t = \dot{U}_t = \ddot{U}_t = 0$ we obtain $U_{t+\Delta t} = \frac{f_1(t)}{\hat{K}}$

CENTRAL DIFFERENCE METHOD

If the equation of motion (6) is considered to be a set of differential equations with constant terms, any finite difference method can be used to assume the variation in acceleration and velocity

In the finite difference method, we assume that:

$$\ddot{U}(t) = \frac{1}{\Delta t^2}\left[U(t-\Delta t) - 2U(t) + U(t+\Delta t)\right] \quad (15)$$

$$\dot{U}(t) = \frac{1}{2\Delta t}\left[-U(t-\Delta t) + U(t+\Delta t)\right] \quad (16)$$

Let have the following motion equation

$$M\ddot{U}(t) + C\dot{U}(t) + KU(t) = f(t) \quad (17)$$

Figure 5. History of the response

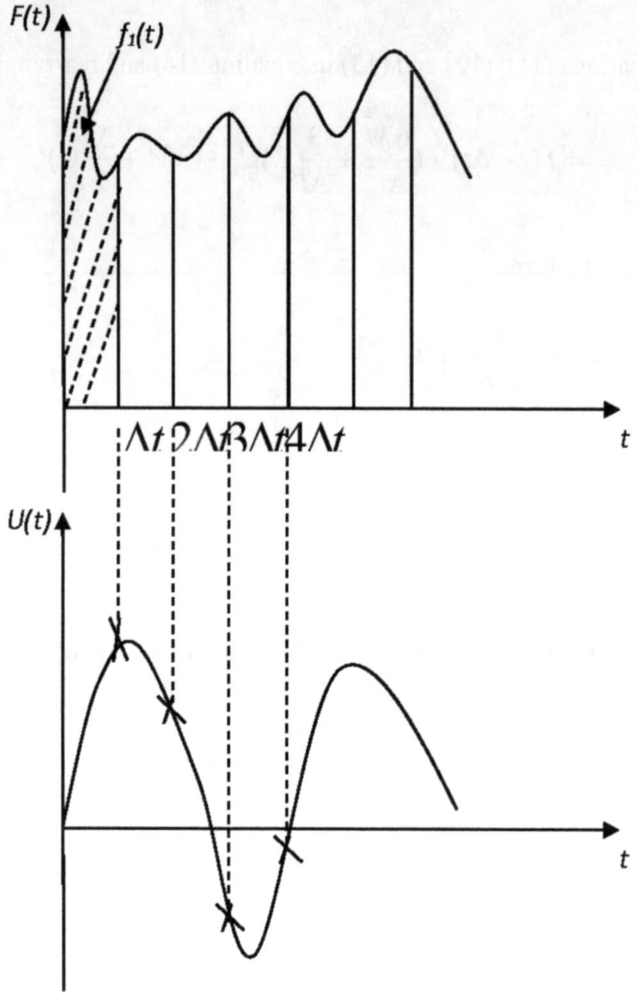

By replacing equations (15) and (16) in (17) we find

$$(\frac{1}{\Delta t^2}M + \frac{1}{2\Delta t}C)U(t+\Delta t) = \hat{f}(t) \tag{18}$$

With

$$\hat{f}(t) = f(t) - (K - \frac{2}{\Delta t^2}M)U(t) - (\frac{1}{\Delta t^2}M - \frac{1}{2\Delta t})U(t-\Delta t)$$

We can finally solve the equation (18) it is enough to know U (t + Dt) hence the need to know U (t) and U (t - Dt).

The problem will arise for t = 0 because in this case we must know U (t - Dt) therefore U (-Dt). In order to determine U (-Dt) we have to use equations (15) and (16) written at time t = 0, thus we will have these two equations

$$\ddot{U}(0) = \frac{1}{\Delta t^2}\left[U(-\Delta t) - 2U(0) + U(\Delta t)\right]$$

$$\dot{U}(0) = \frac{1}{2\Delta t}\left[-U(-\Delta t) + U(\Delta t)\right]$$

We take from the second equation the expression of U (Dt) that we will replace in the first equation, we determine U (-Dt), hence the following expression

$$U_i(-\Delta t) = U_i(0) - \Delta t\dot{U}_i(0) + \frac{\Delta t^2}{2}\ddot{U}_i(0)$$

STEP BY STEP SOLUTION USING CENTRAL DIFFERENCE METHOD

A. Initial Calculation

1- Form the matrix [M], [K]and [C]
2- Initialize initial conditions U_0 ,$\dot{U}_0$,$\ddot{U}_0$
3- Select the step $\Delta t \leq \Delta t_{cr}$ and calculate the following integration constants

$$a_0 = \frac{1}{\Delta t^2} \quad , a_1 = \frac{1}{2\Delta t} , a_2 = 2a_0 , a_3 = \frac{1}{a_2}$$

4- Calculation of $U_i(-\Delta t) = U_i(0) - \Delta t\dot{U}_i(0) + a_3\ddot{U}_i(0)$
5- Form the effective mass matrix $\hat{M} = a_0 M + a_1 C$
6- Triangularization of the matrix $\hat{M} = LDL^t$ (using Gauss method).

B. For Each Step

1- Calculation *of* $\hat{f}(t)$

$$\hat{f}(t) = f(t) - (K - a_2 M)U(t) - (a_0 M - a_1 C)U(t - \Delta t)$$

2- Solve the equation to find displacements at time t + Dt to do this we will use the equation $LDL^t U(t + \Delta t) = \hat{f}(t)$.

3- Calculate acceleration and velocities at time t, for this we have the equations

$$\ddot{U}(t) = a_0\left[U(t-\Delta t) - 2U(t) + U(t+\Delta t)\right]$$

$$\dot{U}(t) = a_1\left[-U(t-\Delta t) + U(t+\Delta t)\right]$$

WILSON'S q METHOD

It is an extension of linear acceleration method; Wilson introduced a coefficient q to ensure the stability of the algorithm. If q = 1 the method is reduced to that of the linear acceleration.

Figure 6. Linear variation of acceleration with Wilson extension

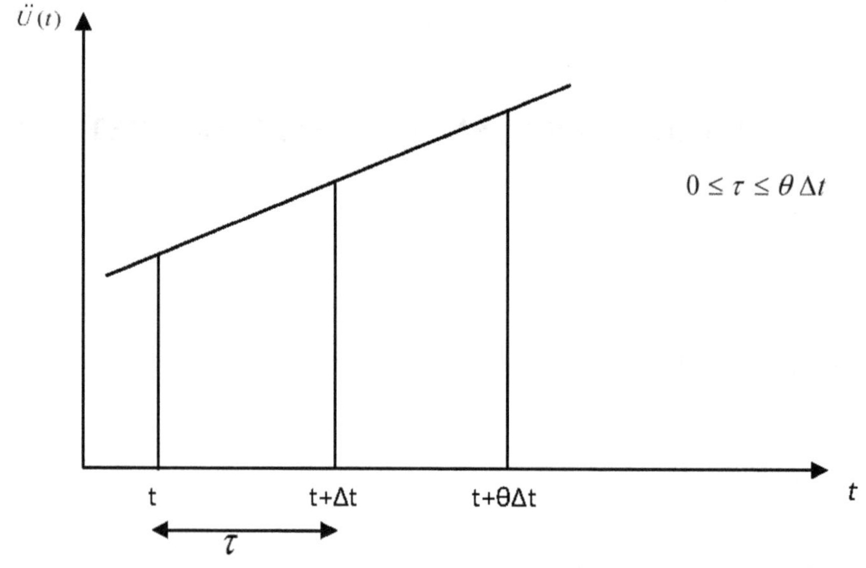

If we take q=1.40 whatever the time taken by the calculation, the method is unconditionally stable (θ³1.37 We take q = 1.40 by numerical experiment)

$$\ddot{U}(t+\tau) = \ddot{U}(t) + \frac{\tau}{\theta\,\Delta t}\left[\ddot{U}(t+\theta\,\Delta t) - \ddot{U}(t)\right] \tag{19}$$

There are two conditions to satisfy:

- The convergence of the algorithm
- The stability of the algorithm

By integrating the equation (19) with respect to τ we find:

$$\dot{U}(t+\tau)=\dot{U}(t)+\ddot{U}(t)\tau+\frac{\tau^2}{2\theta\,\Delta t}\left[\ddot{U}(t+\theta\,\Delta t)-\ddot{U}(t)\right] \tag{20}$$

By once again, integrating equation (20) we find

$$U(t+\tau)=U(t)+\dot{U}(t)\tau+\ddot{U}(t)\frac{\tau^2}{2}+\frac{\tau^3}{6\theta\,\Delta t}\left[\ddot{U}(t+\theta\,\Delta t)-\ddot{U}(t)\right] \tag{21}$$

Note

The condition on q is for the stability of the algorithm and the condition on Dt is for the convergence of the algorithm.

When $\tau=\theta\Delta t$ (at the end of the interval) equation (20) and (21) can be written:

$$\dot{U}(t+\theta\,\Delta t)=\dot{U}(t)+\frac{\theta\,\Delta t}{2}\left[\ddot{U}(t+\theta\,\Delta t)+\ddot{U}(t)\right] \tag{22}$$

$$U(t+\theta\,\Delta t)=U(t)+\dot{U}(t)\theta\,\Delta t+\frac{\theta^2\Delta t^2}{6}\left[\ddot{U}(t+\theta\,\Delta t)+2\ddot{U}(t)\right] \tag{23}$$

From equation (23), we take:

$$\ddot{U}(t+\theta\,\Delta t)=\frac{6}{\theta^2\Delta t^2}\left[U(t+\theta\,\Delta t)-U(t)\right]-\frac{6}{\theta\,\Delta t}\dot{U}(t)-2\ddot{U}(t) \tag{24}$$

By replacing equation (24) in equation (22), we find:

$$\dot{U}(t+\theta\,\Delta t)=\frac{3}{\theta\Delta t}\left[U(t+\theta\,\Delta t)-U(t)\right]-2\dot{U}(t)-\frac{\theta\,\Delta t}{2}\ddot{U}(t) \tag{25}$$

The equilibrium equation at time t + qDt is:

$$M\ddot{U}_{t+\theta\Delta t}+C\dot{U}_{t+\theta\Delta t}+KU_{t+\theta\Delta t}=f(t+\theta\Delta t) \tag{26}$$

The problem arises with $f(t+\theta\Delta t)$ because we do not have its value, we have only $f(t+\Delta t)$ and $f(t+2\Delta t)$, to solve the problem we must do the extension of the trapeze load

$$\hat{f}(t+\theta\,\Delta t)=f(t)+\theta\left[f(t+\Delta t)-f(t)\right]$$

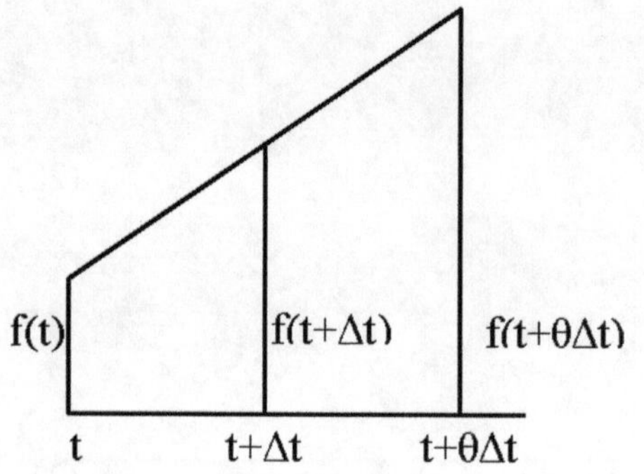

We have now $f(t+\theta\Delta t)$, $\ddot{U}(t+\theta\,\Delta t)$, and $\ddot{U}(t+\theta\,\Delta t)$ we replace in equation (26) to determine U(t) and then calculate again

$\ddot{U}(t+\theta\,\Delta t)$ and $\ddot{U}(t+\theta\,\Delta t)$.

WILSON'S q METHOD STEPS CALCULATION

Initial calculation

1- Form the matrix [M], [K]and [C]
2- *Initialize* U_0 ,$\dot{U}_0$,$\ddot{U}_0$ (can be given or calculated)
3- Select the step Dt, and calculate the integration coefficients with q=1.40

$$a_0 = \frac{6}{\theta^2 \Delta t^2} \quad , a_1 = \frac{3}{\theta \Delta t}, a_2 = 2a_1 , a_3 = \frac{\theta \Delta t}{2} \quad , a_4 = \frac{a_0}{\theta}$$

$$a_5 = \frac{-a_2}{\theta}, \quad a_6 = 1 - \frac{3}{\theta}, \quad a_7 = \frac{\Delta t}{2}, \quad a_8 = \frac{\Delta t^2}{6}$$

4- form the effective stiffness matrix

$$\hat{K} = K + a_0 M + a_1 C$$

5- Triangularization of the matrix $\hat{K} = LDL^t$

For Each Step

1- Calculate the effective loading in time t+qDt

$$\hat{f}(t+\theta\,\Delta t) = f(t) + \theta\left[f(t+\Delta t) - f(t)\right]$$
$$+M\left[a_0 U(t) + a_2 \dot{U}(t) + 2\ddot{U}(t)\right] + C\left[a_1 U(t) + 2\dot{U}(t) + a_3 \ddot{U}(t)\right]$$

We will have an equation $\hat{K}U(t+\theta\Delta t) = \hat{f}(t+\theta\Delta t)$

2-Solve for displacements at time t + qDt

$$LDL^{t}U(t+\theta\Delta t) = \hat{f}(t+\theta\Delta t)$$

3-Calculate displacement velocity and acceleration

$$\ddot{U}(t+\Delta t) = a_4\left[U(t+\theta\,\Delta t) - U(t)\right] + a_5\dot{U}(t) + a_6\ddot{U}(t)$$

$$\dot{U}(t+\Delta t) = \dot{U}(t) + a_7(\ddot{U}(t+\Delta t) + \ddot{U}(t))$$

$$U(t+\Delta t) = U(t) + \Delta t\dot{U}(t) + a_8\left[\ddot{U}(t+\Delta t) - 2\ddot{U}(t)\right]$$

NEWMARK'S METHOD

It is an extension of linear acceleration method.

Figure 7. Linear variation of acceleration with Newmark extension

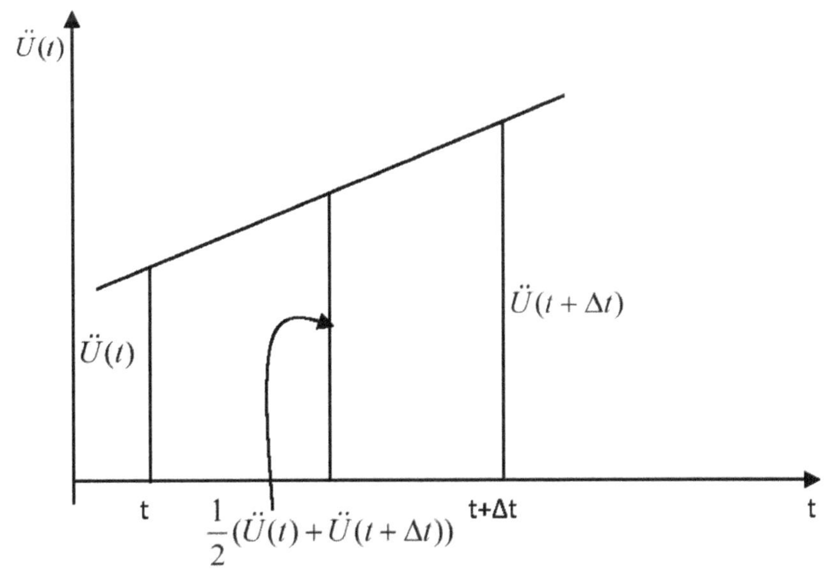

Newmark introduced two coefficients d and b to control the variation of the acceleration within the interval Dt. The coefficient d controls the damping and the coefficient b controls the variation of the acceleration within the interval Dt

$$\dot{U}(t+\Delta t)=\dot{U}(t)+\left[(1-\delta)\ddot{U}(t)+\delta\ddot{U}(t+\Delta t)\right]\Delta t \tag{27}$$

$$U(t+\Delta t)=U(t)+\dot{U}(t)\Delta t+\left[(\frac{1}{2}-\beta)\ddot{U}(t)+\beta\ddot{U}(t+\Delta t)\right]\Delta t^2 \tag{28}$$

From Equation (28):

$$\ddot{U}(t+\Delta t)=\frac{1}{\beta\Delta t^2}\left[U(t+\Delta t)-U(t)-\dot{U}(t)\Delta t-(\frac{1}{2}-\beta)\Delta t^2\ddot{U}(t)\right] \tag{29}$$

We replace the equation (29) in (27) we find:

$$\dot{U}(t+\Delta t)=\frac{\delta}{\beta\Delta t}(U(t+\Delta t)-U(t))+(1-\frac{\delta}{\beta})\dot{U}(t)+(1-\frac{\delta}{2\beta})\Delta t\ddot{U}(t) \tag{30}$$

The equation of motion at time t + Dt is written as follows:

$$M\ddot{U}_{t+\Delta t}+C\dot{U}_{t+\Delta t}+KU_{t+\Delta t}=f(t+\Delta t) \tag{31}$$

We replace equation (29) and (30) in equation (31), we obtain:

$$\left[K+\frac{1}{\beta\Delta t^2}M+\frac{\delta}{\beta\Delta t}C\right]U(t+\Delta t)=f(t+\Delta t)$$

$$+\left[\frac{1}{\beta\Delta t^2}M+\frac{\delta}{\beta\Delta t}C\right]U(t)+\left[\frac{1}{\beta\Delta t}M+(1-\frac{\delta}{\beta})C\right]\dot{U}(t)+\left[(\frac{1}{2\beta}-1)M-(1-\frac{\delta}{2\beta})\Delta tC\right]\ddot{U}(t)$$

This relationship is of the form $\hat{K}U(t+\Delta t)=\hat{f}$

We have just transformed a dynamic problem into a static problem, the stiffness matrix has been transformed by the mass and the damping, the force f has been transformed by the initial conditions at time Dt.

In order to have stable result, it is necessary to have $\Delta t\leq\frac{T}{20}$, d=1/2 and b=1/4. If d=1/2and b=1/6 Newmark's method becomes that of linear acceleration.

CALCULATION STEPS OF NEWMARK'S METHOD

Initial Calculation

1- Form the matrix [K],[M],[C]
2-Initialize $U_0 ,\dot{U}_0 ,\ddot{U}_0$
3- Select the time step Dt and the parameters δ³0.5, β³0.25(0.5+δ)2 i^n order to have the algorithm stability it is necessary to take d=1/2 and b=1/4 .Calculate integration constants

$$a_0 = \frac{1}{\beta \Delta t^2} \quad , a_1 = \frac{\delta}{\beta \Delta t}, \quad a_2 = \frac{1}{\beta \Delta t} \quad , a_3 = \frac{1}{2\beta} - 1$$

$$a_4 = \frac{\delta}{\beta} - 1, \quad a_5 = \frac{\Delta t}{2}(\frac{\delta}{\beta} - 2), \quad a_6 = \Delta t(1-\delta), \quad a_7 = \delta \Delta t$$

4-Form the effective stiffness matrix

$$\hat{K} = K + a_0 M + a_1 C$$

5- Triangularization of the matrix $\hat{K} = LDL^t$

For Each Step

1- Calculate the effective loading in time t+Dt

$$\hat{f}(t+\Delta t) = f(t+\Delta t) + \left[a_0 U(t) + a_2 \dot{U}(t) + a_3 \ddot{U}(t)\right] M + \left[(a_1 U(t) + a_4 \dot{U}(t) + a_5 \ddot{U}(t)\right] C$$

2- Solve for displacements at time t+Dt

$$LDL^t U(t+\Delta t) = \hat{f}(t+\Delta t)$$

3- Calculate displacement velocity and acceleration t+Dt

$$\ddot{U}(t+\Delta t) = a_0 (U(t+\Delta t) - U(t)) - a_2 \dot{U}(t) - a_3 \ddot{U}(t)$$

$$\dot{U}(t+\Delta t) = \dot{U}(t) + a_6 \ddot{U}(t) + a_7 \ddot{U}(t+\Delta t)$$

As with any numerical method, the accuracy depends on the time step (no integration) Dt; Three factors must be considered: the rate of change of the external load f(t), the rate of change of the nonlinear properties; the period of vibration of the structure (Bhatt, 2001).

STABILITY

We are interested almost exclusively in the dynamic analysis of the structures and the continuous media, to "one-step" algorithms in preference to "multi-step" algorithms, it is mainly because the systems to be solved are large and it is important to minimize the number of operations and value to be memorized, to the detriment, if necessary, of the precision (Hyghes,2000).

A "one-step" algorithm, like that of Newmark, makes it possible to calculate the solution at time t_{n+1}, starting from the solution at time t_n. In addition to the disadvantage of requiring the storage of several steps, the "multi-step" algorithms such as that of Houbolt described after, requires a startup procedure.

The algorithms of this family are:

Table 2. Algorithmic stability

Unconditional stability	2b £ g ³1/2
Conditional stability	g ³1/2 b< g/2 w^hDt £ W_{crit}

Note

1- The stability condition must be checked for all modes of the system. Therefore the maximum frequency, which is critical. However, $\omega_{\max}^h \le \max\left(\omega_i^{el}\right)$, the maximum frequency of the system $\omega_{\max}^h$ is limited by the maximum frequency of the individual elements ω_i^{el}.

2- It is always assumed that the damping is Rayleih type let us put C=aM+bK in this case $\xi = \dfrac{\left(\dfrac{a}{\omega}+\dfrac{b}{\omega}\right)}{2}$ is the rate of depreciation at frequency w. It is observed that for g = 1/2 are effect is to increase W_{crit} and therefore critical Dt, weakly. (For example, for g =0.6, b=0.2, Ω_{crit}=3.16 and Ω_{crit}=3.26 for ξ=0 and ξ=0.1 respectively.

CLASSIFICATION

The table below represents the family of Newmark algorithms to some well-known members.

Table 3. Summary of algorithm's types and accuracy

Method	Type	b	g	Stability	Precision
Linear acceleration	Implicit	1/6	1/2	$\Omega_{crit} = 2\sqrt{3}$	Dt^2
Average acceleration	Implicit	1/4	1/2	Unconditional	Dt^2
Central difference	Explicit	0	>1/2	Ω_{crit}=2	Dt

DESIRED CHARACTERISTIC OF A GOOD ALGORITHM

For the calculation of the structures we are mainly interested in the low frequencies and the system to be solved are often of big size. In this case, it is generally accepted that a high-performance algorithm must have the following characteristics:

1. Unconditional stability for linear problems
2. A single implicit equation set to solve at each step
3. Second order accuracy (Dt^2)
4. Algorithmic control of the dissipation of high frequencies, by dissipation in the low frequencies
5. Auto-start (no start procedure)

APPLICATION

Example (Commend & Zimmermann 2001)

The structure below represents a three-storey building with $k = 12\ EI / h^3$ and m are the stiffness of the two columns and mass of each storey respectively with h is the height of the storey: k_1=100N/m,

k_2 = 50N/m, k_3 = 25N/m, m = 100Kg.

1. Calculate frequencies and eigen-modes using Jacobi method
2. with $U_0 = 0\ ,\dot{U}_0 = \begin{bmatrix}1 & 2 & 2\end{bmatrix}^T$, calculate the response of the structure using Newmark's algorithm (n Dt = 10 T_1 knowing $T_1 = T_{max}$) for this two cases:
   a. g = 0.5, b = 0.25, Dt = ?
   b. g = 0.5, b = 0, Dt = ?

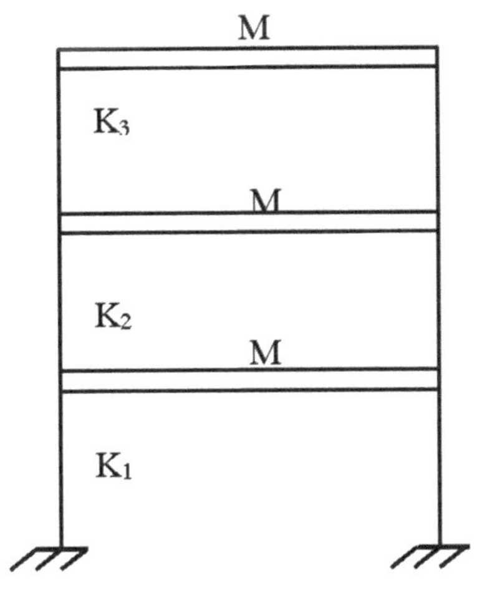

Solution

Extraction of frequencies and eigen-modes by Jacobi method.

$\omega_1 = 0.336$ ras/s	$T_1 = 8.681$ s	$f_1 = 0.054\ s^{-1}$	Mode 1 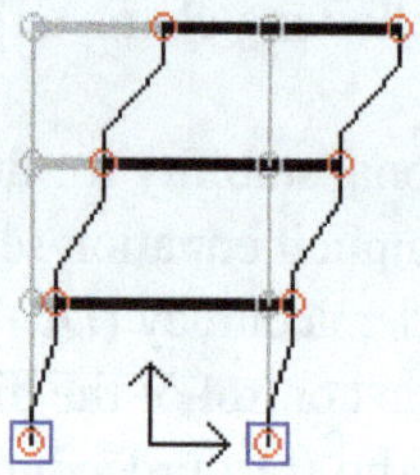
$\omega_2 = 0.793$ rad/s	$T_2 = 7.926$ s	$f_2 = 0.126\ s^{-1}$	Mode 2 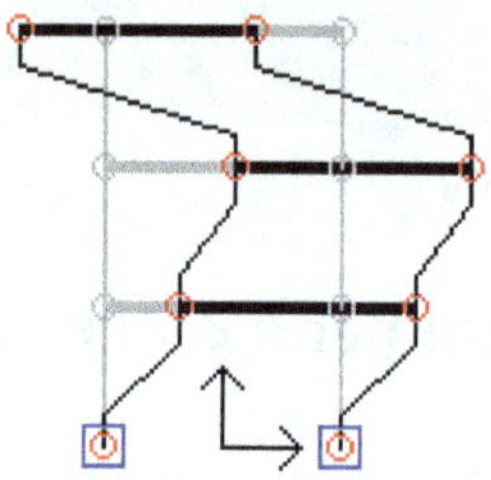
$\omega_3 = 1.326$ rad/s	$T_3 = 4.738$ s	$f_3 = 0.211\ s^{-1}$	Mode 3 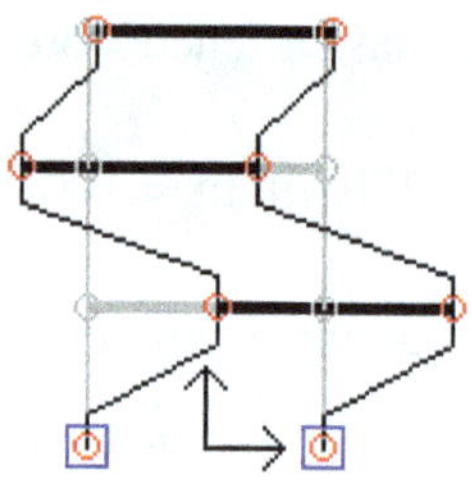

$$U_0 = 0, \dot{U}_0 = \begin{bmatrix} 1 & 2 & 2 \end{bmatrix}^T$$

Horizontal displacement representation of the second floor:

<table>
<tr><th></th><th>Newmark, $\gamma = 0.5$, $\beta = 0.25$</th><th>Newmark, $\gamma = 0.5$, $\beta = 0.0$</th></tr>
<tr><td>Δt = 0.15 n = 1250</td><td>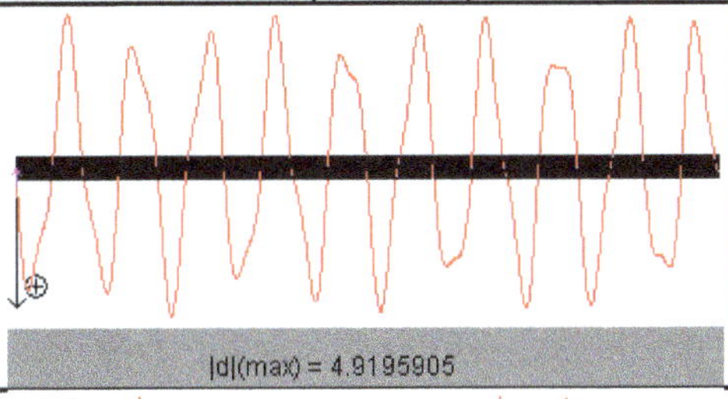
</td><td>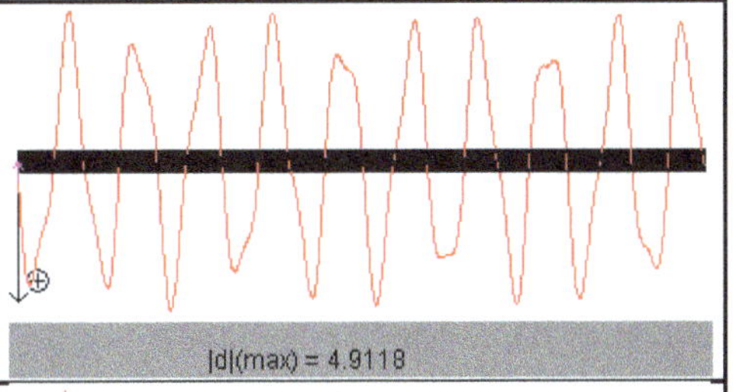
</td></tr>
<tr><td>Δt = 1.5 n = 125</td><td>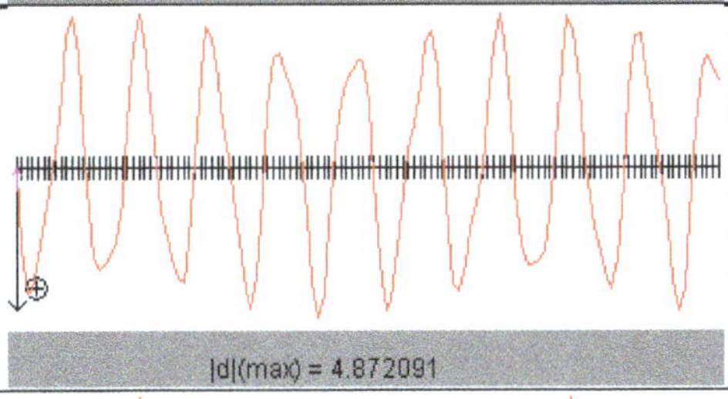
</td><td>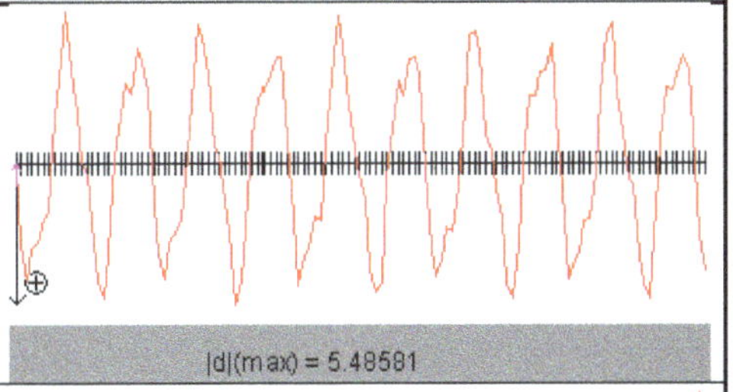
</td></tr>
<tr><td>Δt = 1.6 n = 118</td><td>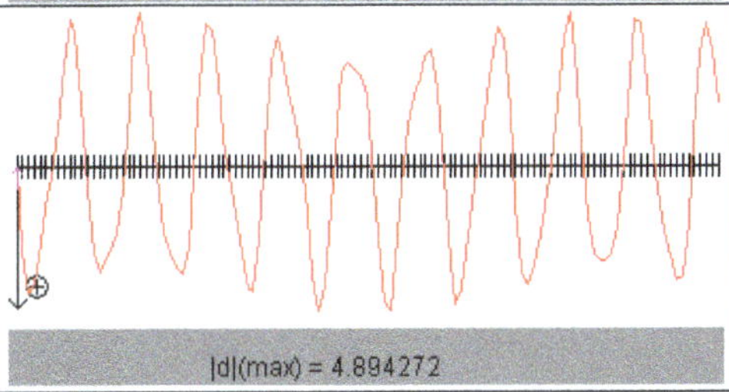
</td><td>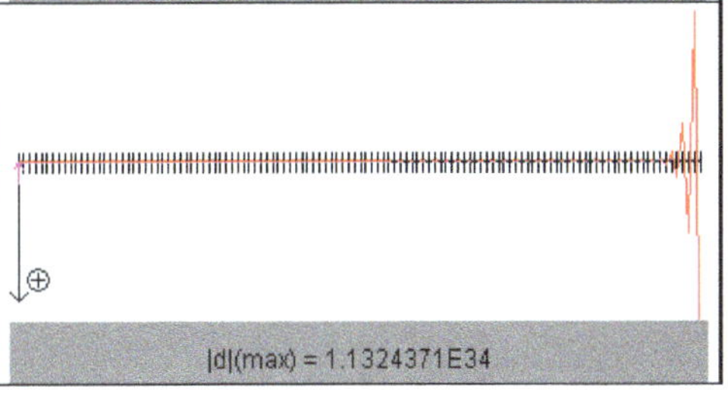
</td></tr>
</table>

Note

1) For Dt = 0.15, the two algorithms (b = 0.25 and b = 0) give an identical response
2) Newmark's algorithm with g = 0.5, b = 0.25 is Unconditional stability.

On the other hand, for the Newmark algorithm with g = 0.5, b = 0.0

$\omega m_{ax\Delta} t \leq \Omega c_{rit=} 2$

So here Δt_{crit}=1.508. The signal is stable for Dt = 1.5 but instable for Dt = 1.6.

REFERENCES

Bhatt, P. (2001). Programming the Dynamic Analysis of Structures. Academic Press.

Commend, S., & Zimmermann, T. (2001). *LSCDYN 1DOF and nDOF: two applications for computer-assisted teaching of structural dynamics. LSC Internal Report.* Swiss Federal Institute of Technology.

Hyghes, T. J. R. (2000). *The Finite Element Method Linear Static and dynamic Finite Element Analysis.* Dover Publications, Inc.

KEY TERMS AND DEFINITIONS

Algorithms: Is a process or set of rules to be followed in calculations or opter problem-solving operations, especially by a computer.

Algorithms Stability: A stable algorithm is when rounding errors does not cause large errors in the result.

Linear Acceleration: It is a method, in which we assume that the acceleration varies linearly in time.

Multi-Step Algorithm: It an algorithm which require the storage of several steps, the "multi-step" and, requires a startup procedure.

One-Step Algorithm: It an algorithm which allow us to calculate the solution at time t_{n+1}, starting from the solution at time t_n (we will divide the time into a time increment and we will consider equilibrium of the equation only in the interval Δt).

Wilson's q Method: It is an extension of linear acceleration method.

Section 5

Seismics Analysis and Modal Response Contribution of the Structure

Chapter 8
Seismic Analysis

ABSTRACT

An earthquake is caused by sudden motion of the earth's crust. Every year, tens of thousands of earthquakes of all sizes occur all over the world. Some cause tiny or major tremors, others occur in remote areas where no one lives. This chapter allow readers to find out more about the earth structure as well as earthquake nature. Therefore, to detail the definition and construction of a spectrum, a presentation of earthquake analysis is given. In order to become familiar with this analysis, two applications are presented at the end of the chapter with a detailed solution.

INTRODUCTION

The structural support motion represents another form of solicitation; in which the study constitutes the subject of this chapter, which allows us to shed light on the so-called seismic study. Before we get to the heart of the subject, we will first outline some points, namely the earth structure, the causes of an earthquake, the definition and construction of a spectrum.

The study of the response of structures to earthquakes is the most widely used application of structural dynamics in civil engineering. The main reason is that earthquakes generate large forces of inertia for the vast majority of buildings and civil engineering structures. Earthquakes remain the dominant dynamic excitation mode.

EARTH STRUCTURE

The earth is composed of four layers see figure 1, the crust, the mantle, the outer core and the inner core.

A brief definition of each layer is presented as follows:

Crust: the crust is the outer layer of hard rock; whose thickness varies from 4 to 60 kilometers.

Inner Core: The inner core with a radius of 1.37 km is considered as a Solid metal body.

DOI: 10.4018/978-1-7998-4399-3.ch008

Copyright © 2021, IGI Global. Copying or distributing in print or electronic forms without written permission of IGI Global is prohibited.

Figure 1. Earth structure (Clough & Penzien, 1995)

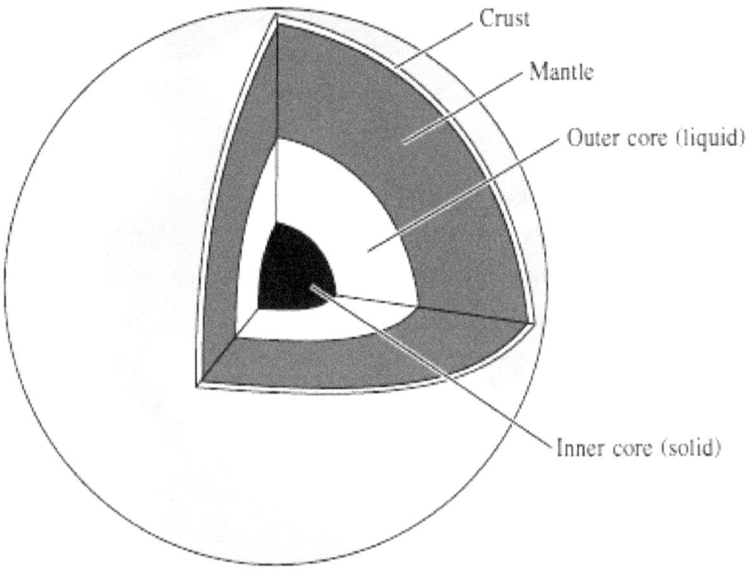

Coat: The mantle is the layer below the crust and is made of dense rock, a thickness of 2900 kilometers. The upper layer of the mantle is called the lithosphere; which is 100 kilometers thick and quite rigid. There is a soft layer below called asthenosphere that is 400 kilometers thick.

Outer core: The outer core is a 2000-kilometer thick layer that is considered a layer of liquid metal.

These different layers are distinguished from each other by variations in the wave propagation speed.

EARTHQUAKES CAUSES

Earthquakes can be caused by natural events or human activities. Here are some of the different causes of earthquakes:

Tectonic Plate's Motion

This is the most common cause of earthquakes. The earthquake occurs when the pressure that has accumulated in the tectonic plates causes a sudden rock breeze. This usually occurs within the boundaries of tectonic plates and along existing faults.

Volcanic Activity

The flow of magma can cause earthquakes; a volcanic eruption is often accompanied by small shakes just before, during and after the eruption. Deep below the volcano's orifice, the hot magma moves slowly through subterranean passages, exerting pressure, and trying to make its way to the surface of the earth. When this happens, the surrounding rock is pressurized by the magma thrust. This causes fractures of the rock so small earthquakes that occur.

Sometimes we thought the rupture of a fault causes an earthquake that shakes the magma stored deep under a volcano. This can cause an accumulation of magma's gases and vapors that cause the eruption of the volcano

Explosion

Humans have caused earthquakes due to explosions, for example, underground nuclear explosions have caused moderate-sized earthquakes that have been felt by people up to 50 kilometers. Blasting in quarries also, produce small seismic events that sometimes people mistakenly believe in earthquakes.

Mines Rupture

Small seismic events can occur when the roofs of underground caves or mines collapse.

FAULT'S RUPTURES MECHANISM

The Lithosphere behaves like an elasto-brittle material, the stress accumulates and then when it exceeds the threshold of resistance of the rocks there is rupture we can classify the faults according to the ruptures types.

- Normal fault.
- Reverse fault or overlap.
- Staggering fault.
- Mixed fault.

SEISMIC WAVES DEFINITION

When the rock filles, there is a sudden release of energy because of the pressure that has accumulated due to the tectonic plate's motion. Shock waves propagate through and around the earth in all directions, from the earthquake focus. When the train of waves crosses the surface of the earth, the ground vibrates.

P Wave

P waves move faster than other seismic waves. They are longitudinal waves, where the soil particles vibrate in the same direction of P waves that move like sound waves in the air. The P wave moves at a speed of 4 to 8 kilometers per second

P waves can move in rock and through the liquid. They can transmitte in the air as sound waves where they can heard as a low rolling sound.

S Wave

S waves propagate at a speed that is about half the speed of P-waves. These are transverse waves, where soil particles, vibrate from one side to the other, perpendicularly to the direction in which the wave is moving. S waves propagate at a speed of 2 to 5 kilometers per second. The waves can spread in the rock, but not in the air and the liquid.

Surface Wave

Surface waves are the slowest seismic waves. They are like waves on the sea that have a complicated rolling motion. Since these waves propagate, their motion is greater on the surface of the earth. The movement becomes smaller and smaller at greater depths below the surface. Surface waves propagate at a speed of 1.8 to 4.5 kilometers per second.

Surface waves take time and distance to develop. Their vibrations are small near the epicenter of the earthquake, while further away they may be the main cause of the felt shake.

The volumes waves that arrive at the surface of the earth produce surface waves affecting the ground only at an extremely low depth; we distinguish mainly two types of surface waves:

a) R-waves or Rayleigh waves: they are waves that propagate parallel to the free surface. They cause compressions (or action) as well as shears in the ground.

Q waves or love waves are waves for which the ground points move in the plane tangent to the surface, perpendicularly to the direction of propagation, thus generating shear stresses.

RESPONSE SPECTRUM CONSTRUCTION AND DEFINITION

We take a simple oscillator (mass spring system with one Dof) of known stiffness (frequency or pulsation ω is a fixed value), I apply to this oscillator an earthquake that remains unchanged during all the tests that will succeed; then I calculate the max displacement of the response of the oscillator. I do the same thing again by changing oscillator stiffness (period), to cover a wide band of period. See figures 2 and 3. The response spectrum is the graphical representation of the maximum displacements of a set of one degree of freedom systems as a function of the frequencies or periods of the system subjected to the same excitatory force.

Another graphical way to see the stages of spectrum construction is shown in Figure bellow

RESPONSE SPECTRUM AND DESIGN SPECTRUM

Let us consider the equation of the motion of a simple oscillator, solicited by an acceleration of the support, which is the most frequent form of seismic input:

$$\ddot{U}_r + 2\xi\omega\,\dot{U}_r + \omega^2 U_r = -\ddot{U}_g(t) \quad (1)$$

with $U_r = U_r(t, \omega, \xi)$, is a response, function of time, in terme of relative displacement, the frequency of the oscillator ω and the damping ratio ξ.

Figure 2. Different stages of spectrum construction (Lestuzzi 2007)

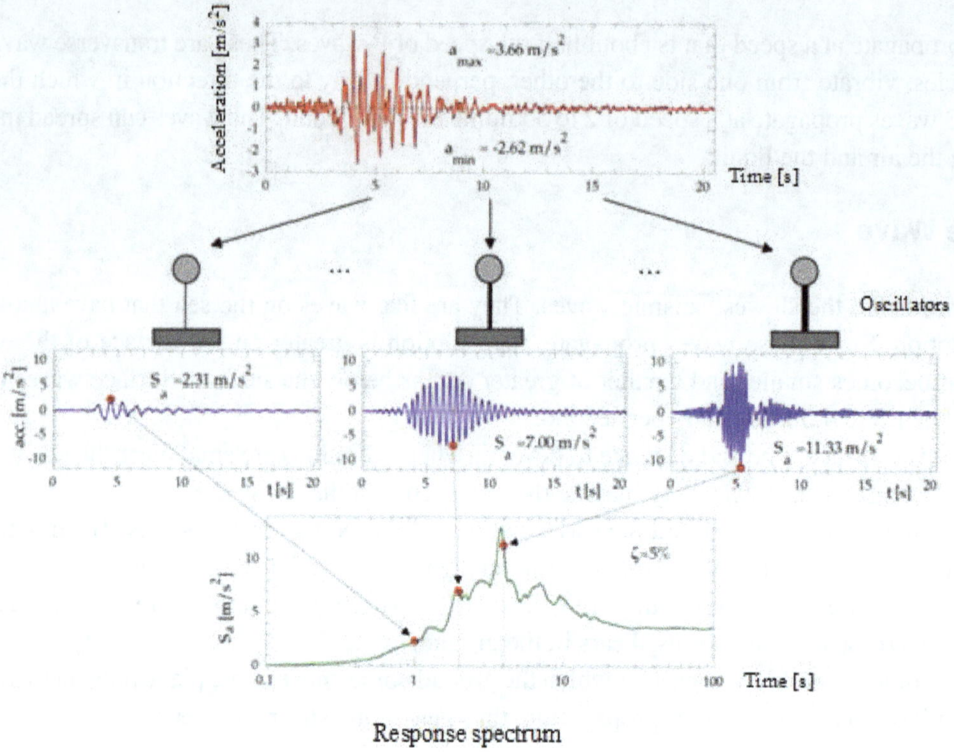

Figure 3. Spectrum construction

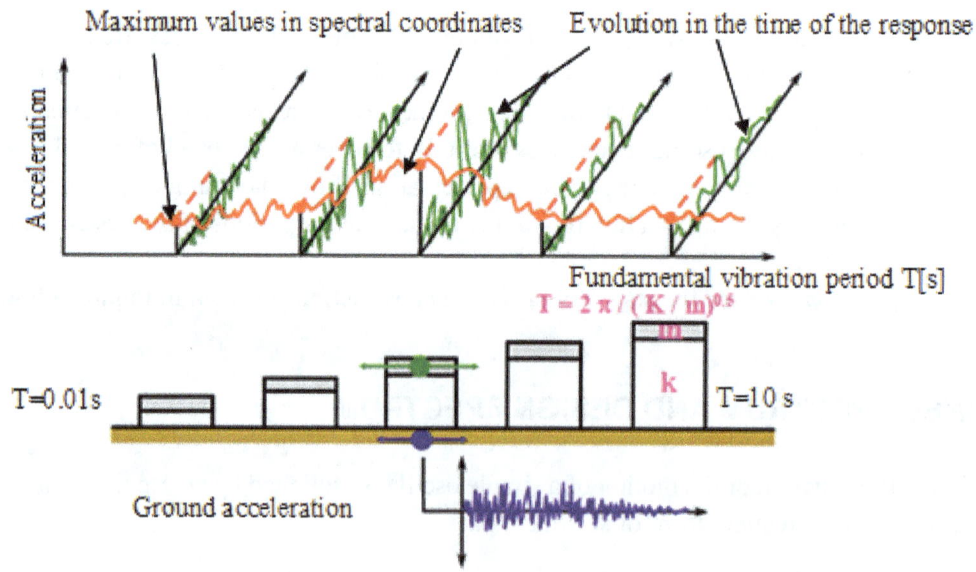

Figure 4. Simple oscillator solicited by an acceleration of the support

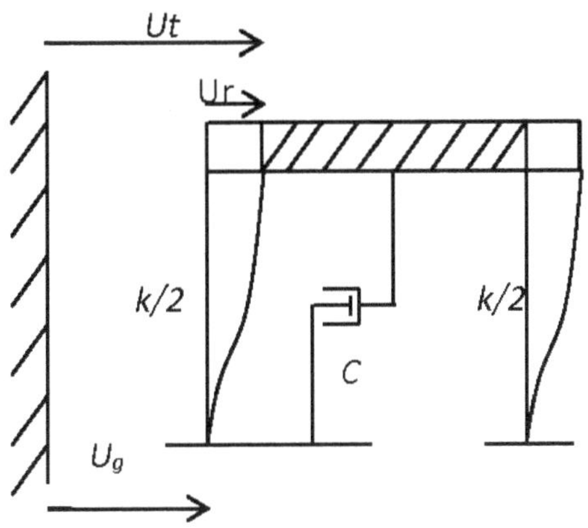

From the response $U_r(t)$, we define the equivalent static force by:

$$f_s(t) = KU_r(t) \tag{2}$$

We can write the equation 2 too

$$f_s(t) = M\omega_n^2 U_r(t) = M\,A(t) \tag{3}$$

With $\omega^2 = \frac{K}{M}$

A(t) has the dimension of an acceleration; A(t) is called pseudo-acceleration, obviously different from the total acceleration $\ddot{U}_t$.

The maximals of U(t) and A(t) are therefore deduced from $U_r(t)$ and are reported in the corresponding figures. For the simple shear-building oscillator, we can deduce statically the moments and shear forces as a function of time, directly from fs, see figure bellow.

Figure 5. Static equivalente force

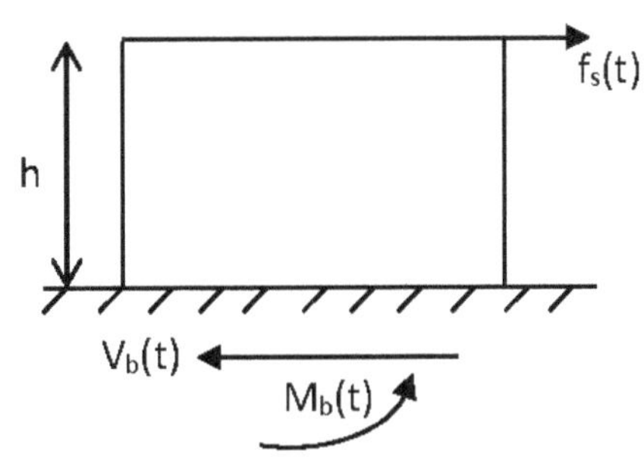

RESPONSE SPECTRUM

We call the response spectrum of a variable $U_r(t)$, A(t), etc., at a given earthquake, the graph of the maximum response (U_r) max, A_{max} as a function of ω (or f_n or T_n) for ratio damping ξ fixed.

Let suppose the equation bellow:

$$M\ddot{U}_a + C\dot{U}_r + kU_r = 0 \quad (4)$$

With $\ddot{U}_a$ is the absolute displacement and U_r is the relative displacement $U_a=U_r+U_g$

$$M(\ddot{U}_r + \ddot{U}_g) + C\dot{U}_r + kU_r = 0$$

$$M\ddot{U}_r + C\dot{U}_r + kU_r = -M\ddot{U}_g(t) \quad (5)$$

The resolution of this equation system is done by two methods; numerical or using Duhamel's integral

From equation (5), I can write:

$$\ddot{U}_r + 2\xi\omega\dot{U}_r + \omega^2 U_r = \ddot{U}_g \quad (6)$$

If we neglect damping (ξ = 0) we obtain:

$$\ddot{U}_r = -\omega^2 U_r$$

Knowing $\max(\ddot{U}_r) = S_a$ and $\max(U_r)=S_d$

Thus

$$S_a=\omega 2^S d \quad (7)$$

S_a or A(t) spectral acceleration

S_d or D(t) spectral displacement

Note: The negative sign - does not interest us when we study the earthquake because there is reversibility of the earthquake

We have

$$M\ddot{U}_a = F_e = kU_r$$

from where

$$\ddot{U}_a = \frac{k}{M}U_r = \omega^2 U_r$$

The spectral velocity is obtained by equating the max of the strain energy with the kinetic energy:

$$\frac{1}{2}mV^2 = \frac{1}{2}ku^2$$

Thus

$$\frac{1}{2}mS_V{}^2 = \frac{1}{2}kS_d{}^2$$

max(v)=S_V and max(U)=S_d

From where

$$\frac{1}{2}mS_v{}^2 = \frac{1}{2}kS_d{}^2$$

We will have

$$S_v^2 = \omega^2 S_d^2 \tag{8}$$

S_v: is spectral velocity

PSEUDO-VELOCITY RESPONSE SPECTRUM

Let us consider a quantity of eigenfrequency ω and fixed damping, and maximum response ($S_d = (U_r)_{max}$) associated to the one degree of freedom system. The quantity V has the dimension of a velocity. It is related to the maximum deformation energy Es_{max} stored by the system; indeed:

$$E_{s_{max}} = \frac{K(U_r)^2_{max}}{2} = \frac{kS_d{}^2}{2} = \frac{k(S_v/\omega)^2}{2} = \frac{MS_v{}^2}{2} \tag{9}$$

The response spectrum in terms of relative pseudo-velocity can be established from the displacement spectrum since; $S_v = \omega S_d = \frac{2\pi}{T}S_d$

PSEUDO-ACCELERATION RESPONSE SPECTRUM

We defined previously that S_a=ω2$^S d$, we take:

$$S_a = \omega^2 S_d = \left(\frac{2\pi}{T}\right)S_d$$

Note

The maximum shear V_{max} at the base corresponds to

$$V_{max} = f_{s\,max} = mA = \frac{A}{g}w$$

with $\frac{A}{g}$ corresponds to the coefficient of lateral force to be introduced for the earthquake (support motion) considred.

Figure 6. Design spectrum a) Displacement spectrum b) Velocity spectrum c) Acceleration spectrum (Capra & Davidovici 1982)

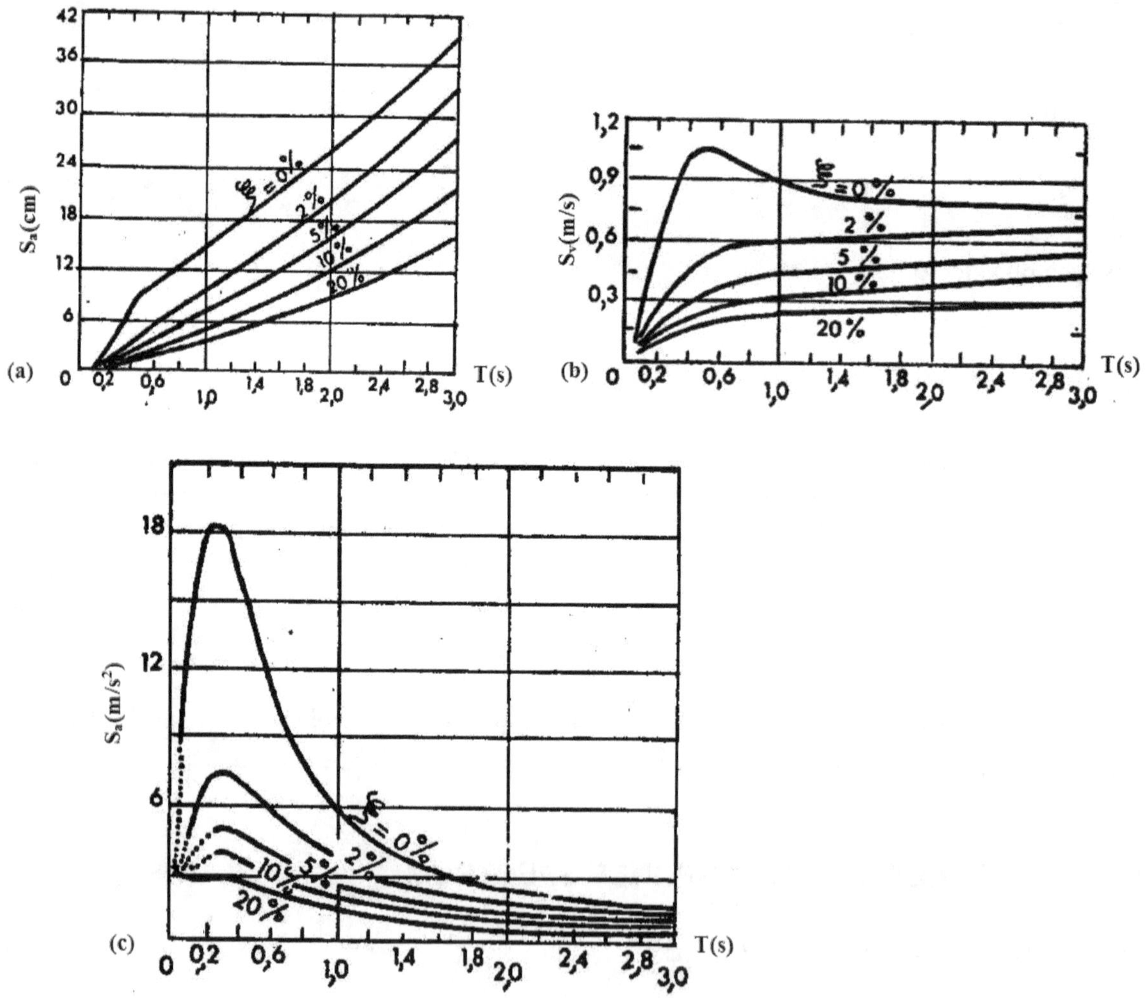

COMBINED DISPLACEMENT VELOCITY AND ACCELERATION SPECTRUM

The three-combined spectram actually contain the same information are related by: $\frac{S_a}{\omega} = S_v = \omega S_d$

However, each of them is directly associated with a physical information. S_d provides the maximum deformation of the system. The pseudo-velocity is related to the maximum deformation energy and the pseudo-acceleration is related to the equivalent static force or maximum shear. It is possible to represent the three quantities on a single graph, whose interest will appear in relation to the design spectrum.

We have seen that quantities are linked by: $\frac{S_a}{\omega} = S_v = \omega S_d$

PSEUDO-SPECTRAL RESPONSES TRIPARTITE REPRESENTATION

To understand the operation of the qaudrilogaritmic graph we consider the variation of $\log_{10}$T for constant values S_a or S_v (Rajasekaran 2009).

1) S_a = constant = C_1

If the pseudo acceleration spectrum is equal to a constant C_1, we can write

$$S_a = C_1 = \omega^2 S_d = \omega(\omega S_d) = \frac{2\pi}{T} S_v$$

Taking the $\log_{10}$ of this equation, we obtain:

$$\log_{10} C_1 = \log_{10} 2\pi - \log_{10} T + \log_{10} S_v$$

$$\log_{10} S_v = \log_{10} T + [\log_{10} C_1 - \log_{10} 2\pi]$$

Tthus

$$\frac{d(\log_{10} S_v)}{d(\log_{10} T)} = +1 \tag{10}$$

This results indicates that a 45 ° straight line on a tripartite graph represents a spectral acceleration, S_a, constant.

2) S_d = Constant = C_2

If the relative displacement spectrum is equal to a constant C_2, we can write

$$S_d = C_2 = \frac{S_v}{\omega} = \frac{T}{2\pi} S_v$$

Taking $\log_{10}$ of this equation, we obtain:

$$\log_{10} C_2 = \log_{10} T - \log_{10} 2\pi + \log_{10} S_v$$

$$\log_{10} S_v = -\log_{10} T + \left[\log_{10} C_1 + \log_{10} 2\pi\right]$$

$$\frac{d\left(\log_{10} S_v\right)}{d\left(\log_{10} T\right)} = -1 \quad (11)$$

This results indicates that a line at -45° on a tripartite graph represents a spectral shift, S_d, constant.

Figure 7. Tripartite diagram

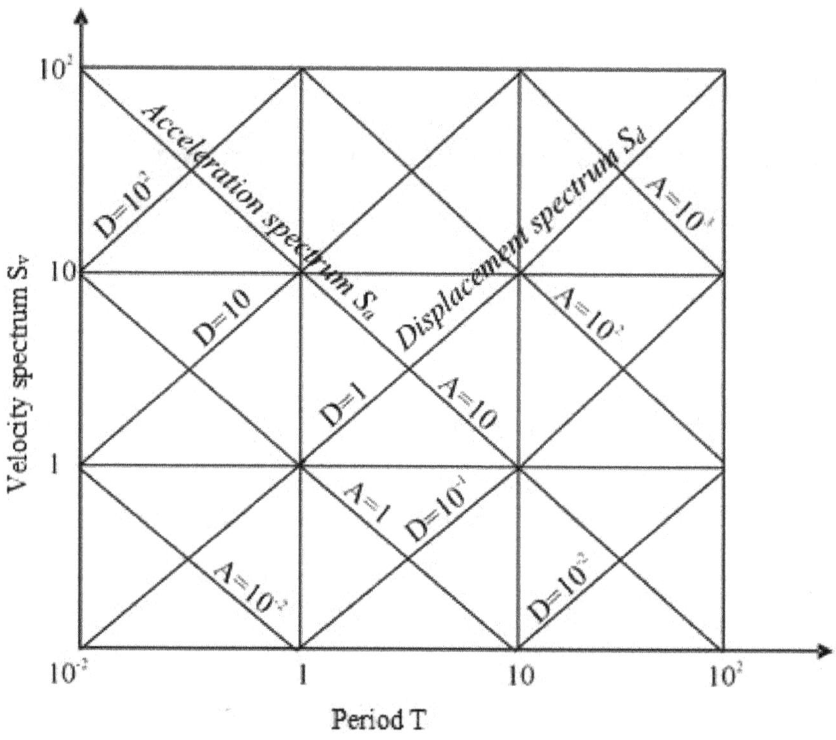

Since log D = *log V - log* ω_n

log A = log V + log ω_n

Similar representations are obviously possible by replacing ω by T or by changing the unit of accelerations to g. Figure bellow represents normalized spectrum response of El Centro earthquake

Response spectra of this type are usually established after significant earthquakes, and it is possible to deduce the maximum structural responses undergone by different structures comparable to a simple oscillator as follows:

Figure 8. Response spectrum of El Centro (Filiatrault.1996)

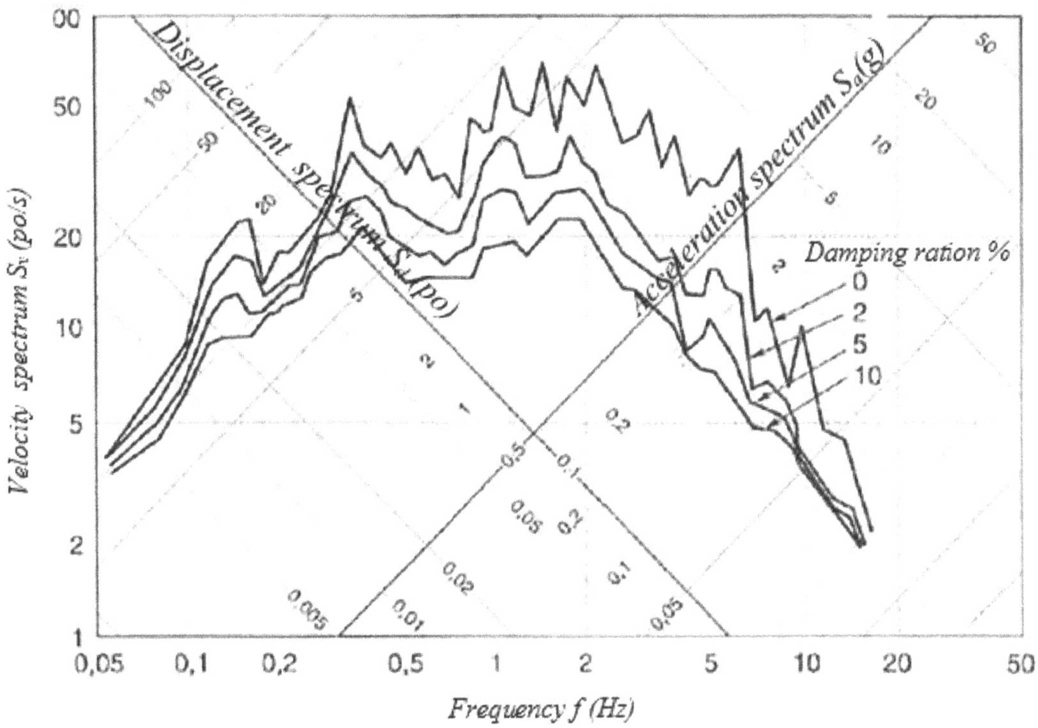

Maximum deformation:

$$(U_r)_{\max} = S_D = \frac{T}{2\pi} S_V = \left(\frac{T}{2\pi}\right)^2 S_A$$

Maximum equivalent static force:

$$f_{so} = K (U_r)_{\max} = m\ddot{U}_{\max}$$

From where V_{max} and M_{max} see figure 5.

SPECTRUM RESPONSE CARACTERISTICS

The response spectrum of structures is a tool for estimate the response of a building to the earthquake.

- In general it is an acceleration response, but there are displacement and velocity spectrum.
- The response spectrum is a "curve" on which the maximum values of the amplification of the ground motion.
- It characterizes the type of soil.

- It is evaluated for the peak of seismic motion.
- It is established for a given relative damping.
- From a calculation point of view, the values are given for a level (case of the simple oscillator) and are extrapolated for the multiple oscillator.

Figure 9. Normalized response spectrum of the El Centro earthquake (Filiatrault.1996)

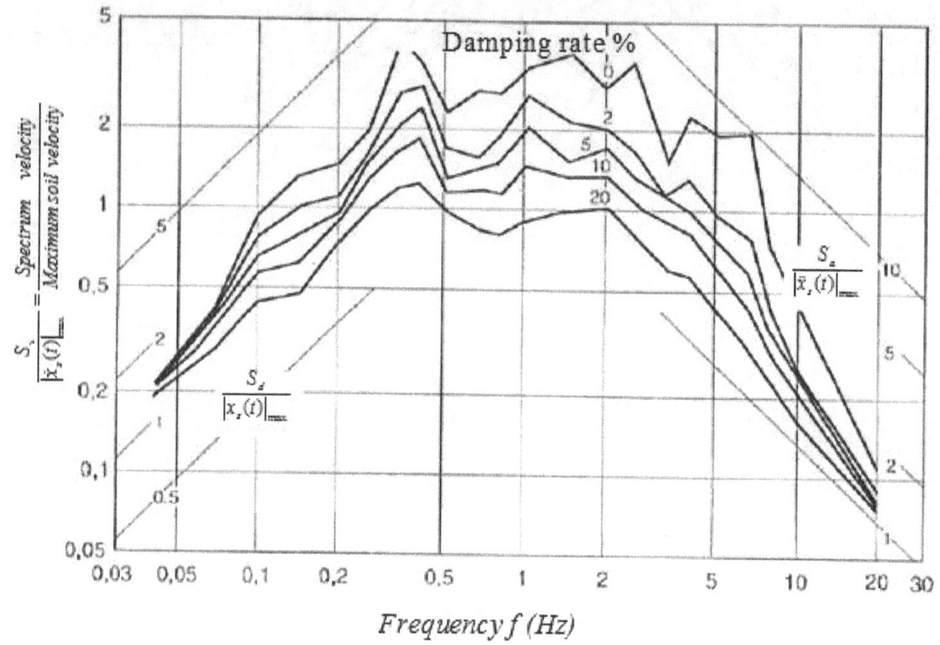

Based on these observations we deduce that we can divide the spectrum into three regions sensitive to either acceleration, velocity, or displacement of the ground.

1. In the region of low frequencies (long periods) the relative displacement is equal to maximu displacement of the soil while the soil acceleration tend to zero this region is called displacement-sensitive region because the response of the structure is connected directly to the displacement of the ground.
2. For high frequencies (short periods) region we notice that the maximum absolute acceleration is equal to the maximum soil acceleration while the relative displacement tends to zero.This region is called acceleration-sensitive region, since the response of the structure is directly related to the acceleration of the ground.
3. The region of the intermediate frequencies (periods,) there is a disparities of values in terms of accelerations displacements and velocities. We think, that the response of the structure appears to be better related (represented) by the velocity of the soil compared to the other earthquake parameters; thus this region is called the velocity-sensitive region

ELASTIC SPECTRUM DESIGN

The spectrum of design is used for either the design of new structures or the evaluation of existing structures. For this task the response spectrum corresponding to a single event is insufficient; information from several events (earthquakes) recorded at the same site will preferably be used.

DISTINCTIONS BETWEEN DESIGN AND RESPONSE SPECTRUM

The response spectrum represents the characteristic of a particular ground motion applied to different single oscillators over a chosen frequency range.

On the other hand, the spectrum of design represents a specification of the level of the forces, or deformations, seismic to be taken into account in the design, in a given site, according to the period of vibration of the structure to be constructed.

The spectrum of design at a given site can result from a combination of several design spectrums if the site to be built is likely to be hit by very different seismic events.

SEISMIC ANALYSIS

Static Equivalent Method

It is called the lateral force method; it replaces the inertial forces with horizontal (lateral) forces, which act on the mass. The basic concept of this method is that during a movement of the ground, the inertia of the building is replaced by a shear force at the base of the structure proportional to the weight of the building and the movement of the ground.

$$V_s = C_s Mg \tag{12}$$

V_s: Shear force at the base of the structure.

C_s: Seismic coefficient, which depends on the local conditions of the soil, the period of the structure and the ability of the structure to absorb dynamic energy by a ductile deformation (passage from an elastic phase to the elasto-plastic phase) this coefficient varies by country.

M: the mass of the building

g: gravity

$$C_s = ADBQ \tag{13}$$

A: site coefficient.

B: structure behavior factor.

D: response coefficient (dynamic amplification factor).

Q: quality factor.

The equivalent static method is applicable for structures having a regular shape, in other cases we must use other methods; which are:

- Modal superposition method
- Numerical integration method (see chapter 8)
- Spectral method.

Spectral Method

It is usable when we have a response spectrum; this method forces us to work with a one DDL system (because the spectrum is taken from a 1DDL system), more details of this part is developed in tne next chapter (see chapter 10).

$$m_i\ddot{U}_i + k_i(u_i - u_{i-1}) + k_{i+1}(u_i - u_{i+1}) = 0 \tag{14}$$

In terms of relative displacement, we can write:

$$m_i\ddot{U}_i(r) + k_i((u_i)_r - (u_{i-1})_r) + k_{i+1}((u_i)_r - (u_{i+1})_r) = -m_i\ddot{U}_g \tag{15}$$

To solve this system we must introduce the modal matrix to decouple the system into a set of equations with a single degree of freedom.

$$\ddot{y}_i + 2\xi_i\omega_i\dot{y}_i + \omega_i^2 y_i = \Gamma_i g(t) \tag{16}$$

y_i: Generalized (fictitious) coordinates

$$\Gamma_i = \frac{\{\varphi_i\}^t[M]\{I\}}{M_i^*} \tag{17}$$

Γ: Modal participation factor (this is the factor that will give us the importance of the mode)

We must calculate the modal response Yn (max), for that we have the frequency of our system we can therefore determine on the graph (spectrum) the value read directly from the spectrum.

$$y_i = \Gamma_i d_i \tag{18}$$

Note: We multiplie d_i (which is a one degree of freedom system displacement) by the modal participation factor to have a multi-degree of freedom displacement

The relative displacement of the mass is:

$$[(U_{in})_r]_{max} = \varphi_{in}(y_n)_{max} \tag{19}$$

It is the relative displacement of the mass i in the mode n.

The inter-storey displacement (for successive storey) is given by:

$$(\Delta_{in})_{\max} = \left[(U_{in})_r\right]_{\max} - \left[(U_{i-1,n})_r\right]_{\max} \tag{20}$$

The dynamic force of the structure can be obtained by calculating the total shear force at the base of the structure (see chapter 10).

MODAL COMBINATION RULES

If we take the response or the effort of each mode for example: For mode 1, we have the effort r_{1max} at time t_1, for mode 2, we have the effort r_{2max} at time t_2, for mode 3, we have the effort r_{3max} at time t_3, untill mode n, we have the effort r_{nmax} at time t_n.

Knowinga that, effect of each mode does not come separately it comes across in a combined form. Hence, the need to have a mathematical model allowing the taking into account of this combined effect.

The commonly used methods for obtaining the peak response quantity of interest for a MDOF system are as follows:

Absolute Sum Method

In absolute sum method(ABSSUM), the peak responses of all the modes are added algebraically, assuming that all modal peaks occur at same time. The maxi mum respo nse is given by:

$$r_{\max} = \sum_{i=1}^{n} |r_i|$$

The ABSSUM method provides a much conservative estimate of resulting response quantity and thus provides an upper bound to peak value of total response.

Square Root of Sum of Squares (SRSS) Method

In the SRSS method, the maxi mum response is obtained by square root of sum of square of response in each mode of vibration and is expressed by:

$$r_{\max} = \sqrt{\sum_{i=1}^{n} r_i^2}$$

The SRSS method of combining maximum modal responses is fundamentally sound where the modal frequencies are well separated. However, this method yield poor results where frequencies of major contributing modes are very close together.

Complete Quadratic Combination (CQC) Method

The alternative procedure is the Complete Quadratic Combination (CQC) method. The maximum response from all the modes is calculated as:

$$r = \sqrt{\sum_{i=1}^{n}\sum_{m=1}^{n} r_n \alpha_{nm} r_m}$$

With

$$\alpha_{nm} = \frac{8(\xi_n \xi_m)^{\frac{1}{2}}(\xi_n + \rho\xi_m)\rho^{\frac{3}{2}}}{(1-\rho^2)^2 + 4\xi_n\xi_m\rho(1+\rho^2) + 4(\xi_n^2 + \xi_m^2)\rho^2}$$

And

$$\rho = \frac{\omega_m}{\omega_n}$$

$\xi n_{,m}$ Is the damping coefficient of the mode n and m respectively.
$\alpha n_{,m}$ Is a correction coefficient, which allow us to take into account two approximate modes value.

APPLICATIONS

Example1 (Filiatrault.1996)

We submit the simple frame shown in the figure below to an earthquake identical to that of Elcentro.

We ask to calculate the maximum displacement of the frame as well as the maximum shear at the base assuming a critical damping of 2%.

E=200000MPA

I=0.34x10^{-8} m^4

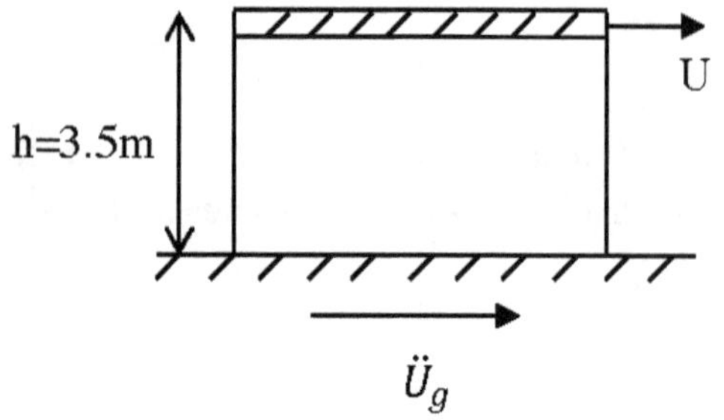

Figure 10. Response spectrum of El Centro (Filiatrault.1996)

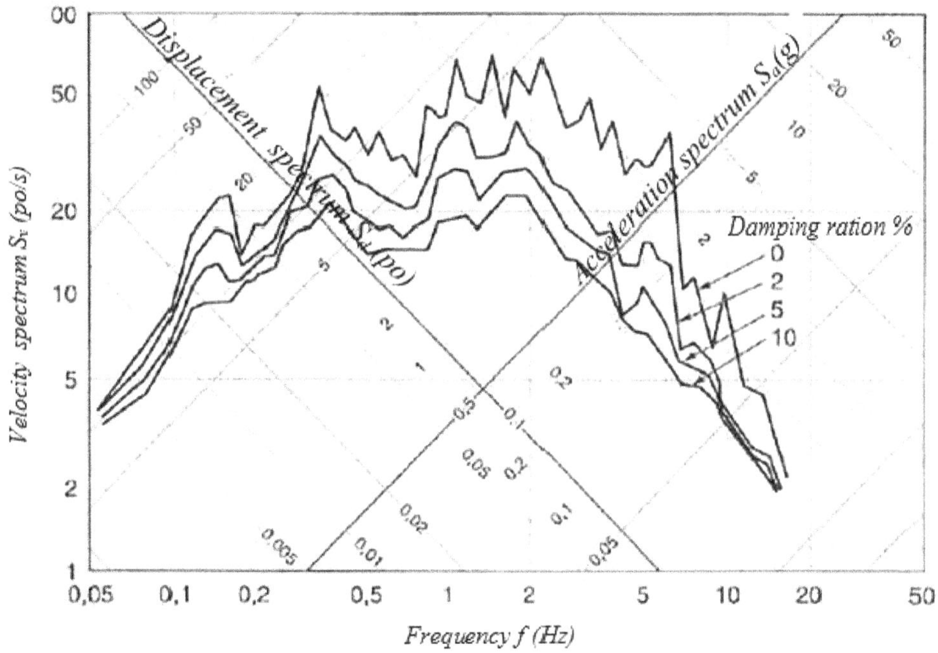

Solution

The coefficient of lateral stiffness is obtained by:

$$K = 2\left(\frac{3EI}{h^3}\right)$$

$$K = 2\left(\frac{3(200)x10^3 x10^6 x(0.34x10^{-8}}{(3.5)^3}\right)$$

$K = 9516.035$ *N/m*

W=Mg

Thus $\mathrm{M} = \frac{\mathrm{W}}{\mathrm{g}}$ as a value M=305.81*kg*

We have $\omega = \sqrt{\frac{K}{M}}$ as a value ω=5.85 r*ad/s*, thus $T = \frac{2\pi}{\omega} = 1.13s$

According to the response spectrum of figure 10 for a critical damping of 2%, one can read. Knowing that 1po =25.4mm

S_D=5 po=127mm=0.127m

S_V=26 po/s=660.4mm/s=0.6604m/s

S_A=0.4g =0.4(9.81)=3.924m/s^2

The maximum relative displacement on the top level is equal to

$U_{max}=S_D$=0.127m.

The base shear is equal to

$V_{base/max}=KU_{max}=K\ S_D$

$V_{base/max}$=9516.035 (0.127) N

$V_{base/max}$=1208.54 N

Example 2 (Commend & Zimmermann 2001)

We suppose a six-storey buiding represented bellow:

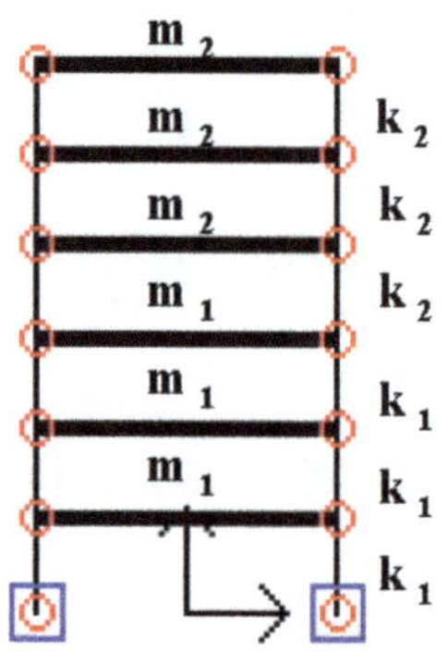

With $k_1 = 10^5$N/m $k_2 = 10^4$kg N/m $m_1 = 10^4$ kg and $m_2 = 10^2$ kg

1. Give the modes and frequencies of the structure.
2. It is assumed that this building has a Rayleigh damping characterized by ξ_a=10%.for $\omega_a = 0.3$ and ξ_b=10% for $\omega_b = 1.0$. What is the damping rate of the natural modes of the structure?
3. The damped building is subject to an El Centro type design earthquake, characterized by $u_s(t)$ displacement imposed for a duration of 60 seconds ($\Delta t = 0.1$).
    a. Represent the accelerogram at the 4th floor.
    b. Define from the damped accelerogram obtained under 3 a response spectrum ($\xi = 0, \xi = 0.05, \xi = 0.1$) for mechanical components to be placed on the 4th stage, whose basic pulsations are in the range $10^{-2} \rightarrow 10^2$. Use the software (Commend & Zimmermann 2001) with one degree of freedom and apply on several simple oscillators

Solution

The table below gives the eigen modes and frequencies of the structure calculated numerically of the structure.

Table 1. Eigenmodes and frequencies

$\omega_1 = 1.395$ rad/s	$T_1 = 4.504$ s	Mode 1 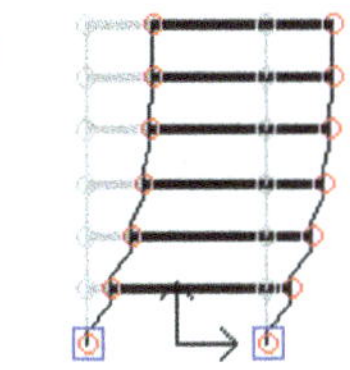
$\omega_2 = 3.871$ rad/s	$T_2 = 1.623$ s	Mode 2 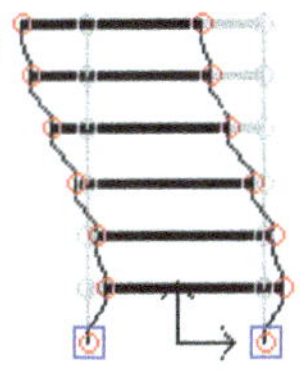
$\omega_3 = 4.558$ rad/s	$T_3 = 1.378$ s	Mode 3 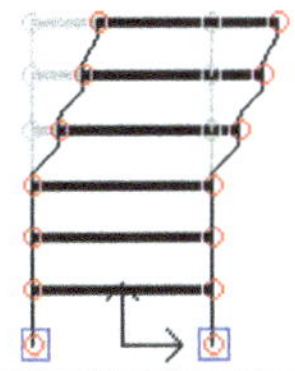
$\omega_4 = 5.711$ rad/s	$T_4 = 1.100$ s	Mode 4 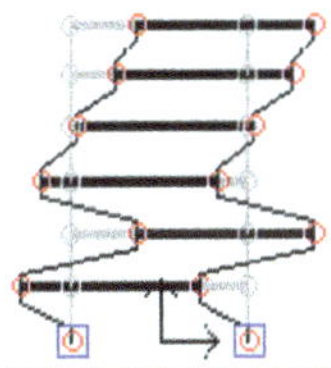
$\omega_5 = 12.485$ rad/s	$T_5 = 0.503$ s	Mode 5 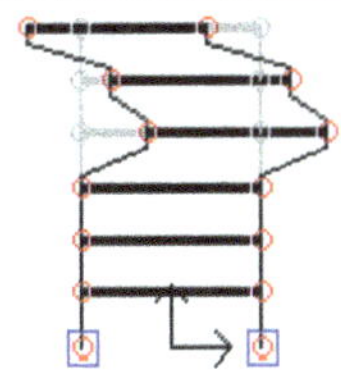
$\omega_6 = 18.022$ rad/s	$T_6 = 0.349$ s	Mode 6 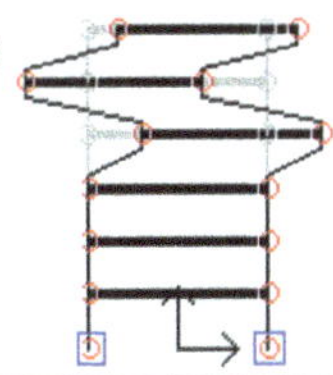

The coefficients α and β of Rayleigh damping are deduced as follows:

$$\mathbf{C} = \alpha\mathbf{M} + \beta\mathbf{K}$$
$$¦\ ^{T}\mathbf{C}¦\ = \alpha¦\ ^{T}\mathbf{M}¦\ + \beta¦\ ^{T}\mathbf{K}¦$$
$$¦\ ^{T}\mathbf{C}¦\ = \alpha\mathbf{I} + \beta›$$
$$2\omega_i\xi_i m_i = \alpha + \beta\lambda_i$$
$$2\omega_i\xi_i = \alpha + \beta\omega_i^2 \ \ (\mathrm{m_i} = 1) \Rightarrow \xi_i = \frac{1}{2}\left(\frac{1}{\omega_i}\alpha + \omega_i\beta\right)$$

we deduce couples ξ_a-ω_a and ξ_b-ω_b α = 0.056 and β = 0.044, then we obtain:

Box 1.

$\xi(\omega_1)$	$\xi(\omega_2)$	$\xi(\omega_3)$	$\xi(\omega_4)$	$\xi(\omega_5)$	$\xi(\omega_6)$
5.1%	9.2%	10.6%	13.1%	27.7%	39.8%

The damped building is subject to an El Centro earthquake applied in node 5 fouth floor, see figure above

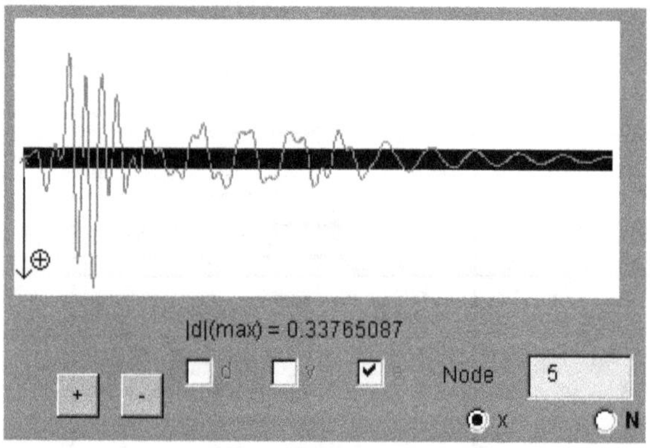

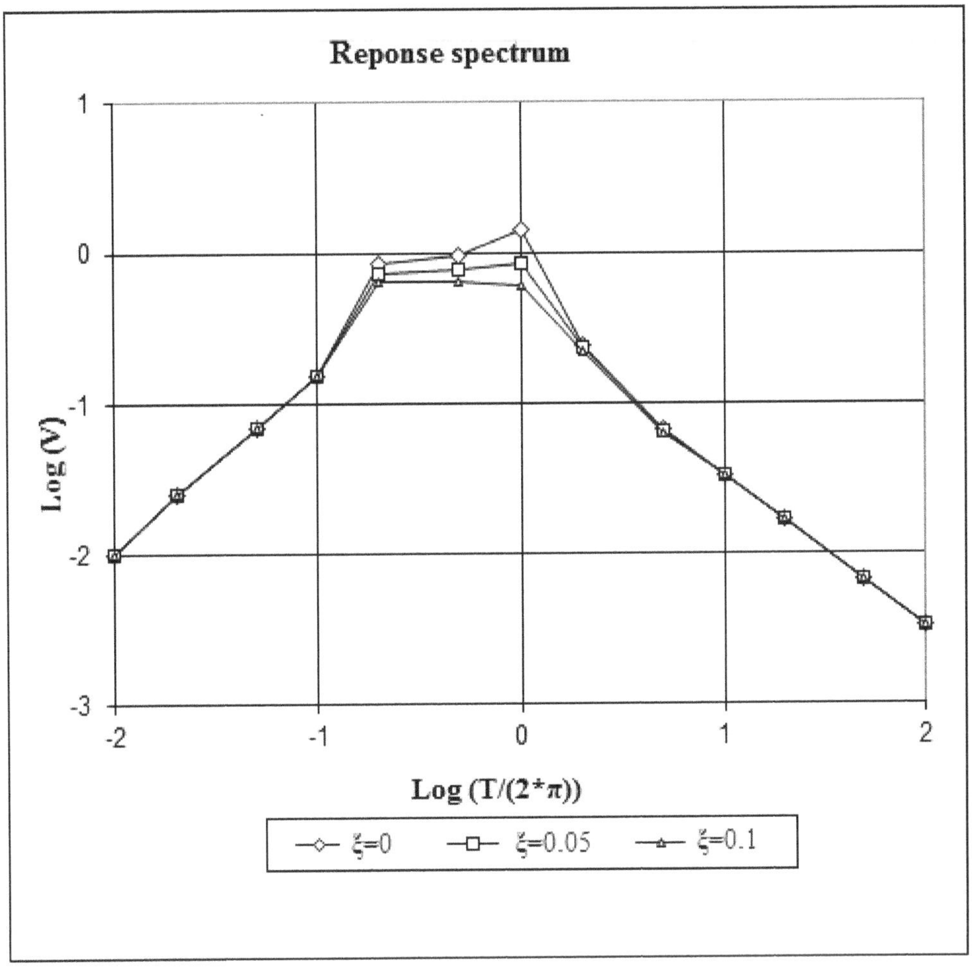

Box 2.

ω	1 /ω	$\lvert u_{r(max)} \rvert_{(\xi=0)}$	V	$\lvert u_{r(max)} \rvert_{(\xi=0.05)}$	V	$\lvert u_{r(max)} \rvert_{(\xi=0.1)}$	V
100.0000	0.0100	0.0001	0.0100	0.0001	0.0100	0.0001	0.0100
50.0000	0.0200	0.0005	0.0250	0.0005	0.0250	0.0005	0.0250
20.0000	0.0500	0.0034	0.0680	0.0034	0.0680	0.0034	0.0680
10.0000	0.1000	0.0154	0.1540	0.0153	0.1530	0.0153	0.1530
5.0000	0.2000	0.1674	0.8370	0.1438	0.7190	0.1276	0.6380
2.0000	0.5000	0.4816	0.9632	0.3913	0.7826	0.3233	0.6466
1.0000	1.0000	1.4363	1.4363	0.8492	0.8492	0.6078	0.6078
0.5000	2.0000	0.4885	0.2443	0.4642	0.2321	0.4423	0.2212
0.2000	5.0000	0.3304	0.0661	0.3241	0.0648	0.3180	0.0636
0.1000	10.0000	0.3357	0.0336	0.3322	0.0332	0.3288	0.0329
0.0500	20.0000	0.3371	0.0169	0.3354	0.0168	0.3336	0.0167
0.0200	50.0000	0.3375	0.0068	0.3368	0.0067	0.3361	0.0067
0.0100	100.0000	0.3376	0.0034	0.3372	0.0034	0.3369	0.0034

Box 3.

Log(1 /ω=T/(2π)) (ξ=0)	Log(V) (ξ=0)	Log(V) (ξ=0.05)	Log(V) (ξ=0.1)
-2	-2	-2	-2
-1.699	-1.602	-1.602	-1.602
-1.3	-1.167	-1.167	-1.167
-1	-0.812	-0.815	0.815
-0.699	-0.077	-0.143	-0.195
-0.3	-0.0163	-0.106	-0.189
0.0	0.157	-0.071	-0.216
0.3	-0.612	-0.634	-0.655
0.699	-1.179	-1.188	-1.196
1.0000	-1.4736	-1.4788	-1.4828
1.300	-1.772	-1.7747	-1.7773
1.699	-2.1675	-2.1739	-2.1739
2.000	-2.4685	-2.4685	-2.4685

REFERENCES

Capra, A & Davidovici, V. (1982). Calcul Dynamique des structures en zone sismiques. *Eyrolles.*

Clough, R. W. & Penzien, J. (1995). *Dynamic of structures.* Compute and Structure, Inc.

Commend, S., & Zimmermann, T. (2001). *LSCDYN 1DOF and nDOF: two applications for computer-assisted teaching of structural dynamics. LSC Internal Report.* Swiss Federal Institute of Technology.

Filiatrault, A. (1996). Element de génie parasismique et de calcul dynamique des structures. Ecole polytechnique de Monreal Canada.

Lestuzzi, P. (2007). *Génie parasismique semestre d'été.* Swiss Federal Institute of Technology.

Rajasekaran, S. (2009). *Structural dynamics of earthquake engineering Theory and application using mathematica and matlab.* Woodhead Publishing in Materials.

KEY TERMS AND DEFINITIONS

Acceleration Spectrum: Defined as the maximum acceleration values caused by the acceleration history.

Design Spectrum: Is used for either the design of new structures or the evaluation of existing structures based on response spectrum corresponding from several events (earthquakes) recorded at the same site.

Displacement Spectrum: Is the maximum displacement values caused by the acceleration history.

Earth Structure: Is the different major components of the earth.

Response Spectrum: Is a plot of the peak or steady state *response* (displacement, velocity or acceleration) of a series of oscillators of varying natural frequency that are forced into motion by the same base vibration or shock.

Seismic Waves: Is a wave that travels through the Earth, most often as the result of a tectonic earthquake.

Velocity Spectrum: Is the maximum velocity values caused by the acceleration history.

Chapter 9
Modal Response Contribution

ABSTRACT

In order to optimize structure calculation, it is inconceivable to miss the understanding of the modal response contribution and truncation error. This chapter enlightens the reader on the subject by dealing with certain points, namely the determination of the elastic forces modal contribution, modal participation factors and truncation error, and static correction procedure. At the end of the chapter, examples will be treated in order to bring clarity to the reader on the points cited before.

INTRODUCTION

This chapter is imperative in order to understand the optimum calculation to better approach the actual behavior of the structure hence the development of certain key points namely, determination of the elastic forces modal contribution, modal participation factors, truncation error, and static correction procedure.

In ordre to compute the response contribution of the nth mode it's imperative to perform static analysis of the structure, thus we can have modal static response. Then compte the pseudo-acceleration for the determination of the response due the nth mode. At this stage we can combine the modal contributions to determine the total response (Chopra.2001) and (Jerome. 2012).

DOI: 10.4018/978-1-7998-4399-3.ch009

Copyright © 2021, IGI Global. Copying or distributing in print or electronic forms without written permission of IGI Global is prohibited.

BRIEF REMINDER

Let us consider a system with several degrees of freedom.

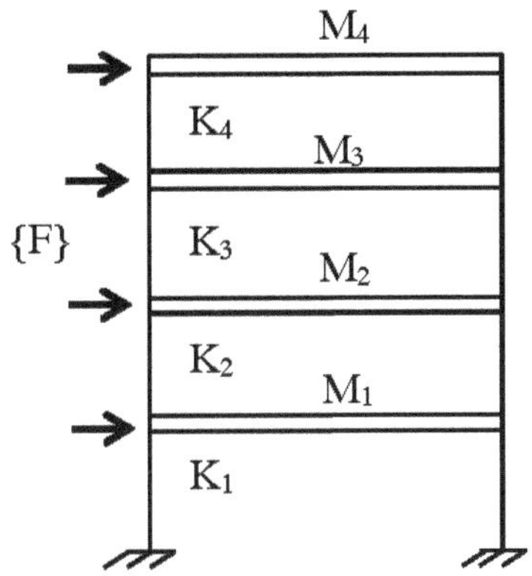

Assuming that $\{F(t)\}=\{S\}g(t)$

Where $\{S\}=\sum_{i=1}^{n}\{S_i\}=\sum_{i=1}^{n}\Gamma_i[M]\{\varphi_i\}$

Premultiplying by $\{\phi\}$t we deduce that $\Gamma_i=\frac{1}{m_i^*}\{\varphi_i\}^t\{S\}$ which represents the modal participation factor

{S}: is the static load pattern

$$[M]_{(nxn)}\{\ddot{U}\}_{(n)}+[C]_{(nxn)}\{\dot{U}\}_{(n)}[K]_{(nxn)}\{U\}_{(n)}=\{F(t)\}_{(n)}=\{S\}g(t)=\sum_{i=1}^{n}\{S\}_i\,g(t)$$

By substituting $\{U(t)\}=[\phi]\{y(t)\}$

In addition, premultiplying by $[\phi]$t we have

$$[\varphi]^t[M][\varphi]\{\ddot{y}\}+[\varphi]^t[C][\varphi]\{\dot{y}\}+[\varphi]^t[M][\varphi]\{y\}=[\varphi]^t\{F(t)\}=[\varphi]^t\{s\}g(t)=[\varphi]^t\sum_{i=1}^{n}\{s\}_i\,g(t)$$

$$[M^*]\{\ddot{y}\}+[C^*]\{\dot{y}\}+[K^*]\{y\}=[\varphi]^t\{F(t)\}=\sum_{i=1}^{n}\Gamma_i[\varphi]^t[M]\{\varphi_i\}g(t)$$

$$M_i^*\ddot{y}_i+C_i^*\dot{y}_i+K_i^*y_i=\Gamma_i M_i^* g(t)$$

$$\ddot{y}_i+2\xi_i\omega_i\dot{y}_i+\omega_i^2 y_i=\Gamma_i g(t) \tag{1}$$

The equation (1) describes i[th] modeWith $\Gamma_i=\frac{1}{M_i^*}\{\varphi_i\}^t\{S\}$ Modal participation factor (scales response of ith mode)

Dividing the equation (1) by Γ_i

$$\ddot{D}_i + 2\xi_i\omega_i\dot{D}_i + \omega_i^2 D_i = g(t) \quad (2)$$

With $D_i = \frac{y_i}{\Gamma_i}$ which is the Pseudo displacement

The final response of the system is:

$$\{U(t)\} = \sum_{i=1}^{n}\{U(t)\}_i$$

With $\{U(t)\}_i$ is the contribution from ith mode

$$\{U(t)\}_i = \{\varphi_i\}y_i(t) = \Gamma_i\{\varphi_i\}D_i(t) \quad (3)$$

DETERMINATION OF ELASTIC FORCES MODAL CONTRIBUTION OF THE Ith MODE

To determine the elastic forces modal contribution (Chopra. 2001) and (Jerome. 2012) we have:

$$\{f(t)\}_i = [K]\{U(t)\}_i = \omega_i^2[M]\{\varphi_i\}y_i(t)$$

$$\{f(t)\}_i = [K]\{U(t)\}_i = \omega_i^2[M]\{\varphi_i\}\Gamma_i D_i(t)$$

$$\{f(t)\}_i = [K]\{U(t)\}_i = \omega_i^2\{S\}_i D_i(t)$$

$$\{f(t)\}_i = [K]\{U(t)\}_i = \{S\}_i A_i(t) \quad (4)$$

With $A_i(t) = \omega_i^2 D_i(t)$ is Pseudo acceleration

$$\{f(t)\} = \sum_{i=1}^{n}\{f(t)\}_i = \sum_{i=1}^{n}\{S\}_i A_i(t) \quad (5)$$

Consider displacement response

$$\{U(t)\}_i = [K]^{-1}\{f(t)\}_i$$

$$\{U(t)\}_i = [K]^{-1}\{S\}_i A_i(t)$$

$$\{U(t)\}_i = \{U^{st}\}_i A_i(t) \tag{6}$$

With $\{U^{st}\}_i$ is static displacement due to time-invariant load pattern $\{S\}_i$

In general, any response r (floor shear, base shear, base moments, etc.), can be calculated as follows:

1- Calculate the static response r_i^{st}, due to the loading model S_i of ith mode
2- The modal response of ith mode

$$r_i(t) = r_i^{st}\omega_i^2 D_i(t) = r_i^{st} A_i(t) \tag{7}$$

With $D(t)$=max(U)

3- The total response

$$r(t) = \sum_{i=1}^{n} r_i^{st} A_i(t)$$

With r^{st} is the total static displacement

At this stage, we can define the modal contribution factor as follows:

$$\overline{r}_i = \frac{r_i^{st}}{r^{st}} \tag{8}$$

By definition

$$\sum_{i=1}^{n} \overline{r}_i = 1$$

If only the m first mode are considered in the analysis, in this case

$$\%Erreur = 1 - \sum_{i=1}^{m} \overline{r}_i = e_r \tag{9}$$

The number of modes to include depends on the spatial distribution of load

MODAL RESPONSE CONTRIBUTION AND EARTHQUAKES

For Multi degree of freedom system the contribution of the modal response can be summarized as follows:

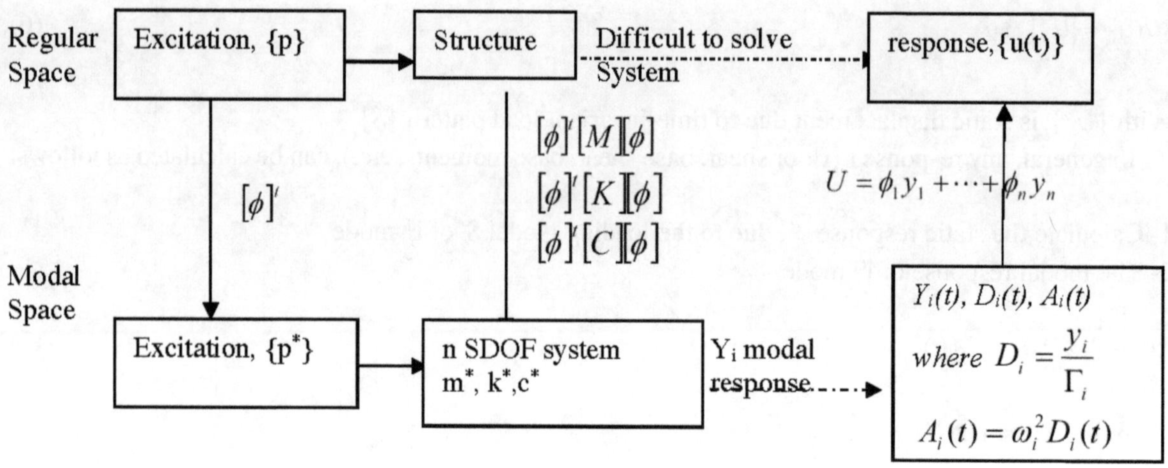

So we can write:

$$\{F(t)\} = \{S\}g(t) = \sum_{i=1}^{n}\{s\}_i g(t) = \sum_{i=1}^{n}\Gamma_i[M]\{\varphi_i\} \tag{10}$$

MODAL CONTRIBUTION

Modal contribution factor for mode i represents the ration between static response to load pattern $\{S_i\}$ and the static response to load pattern $\{S\}$

$$\bar{r}_i = \frac{r_i^{st}}{r^{st}}$$

r_i^{st} : Represents the static response to load pattern $\{S_i\}$
r^{st}: Represents the static response to load pattern $\{S\}$

Note: Before we could notice that elastic force $f_s = r$

SHEAR STRUCTURES GROUND MOTION (Chopra. 2001) and (Jerome.2012)

Let us consider a ground on shear as

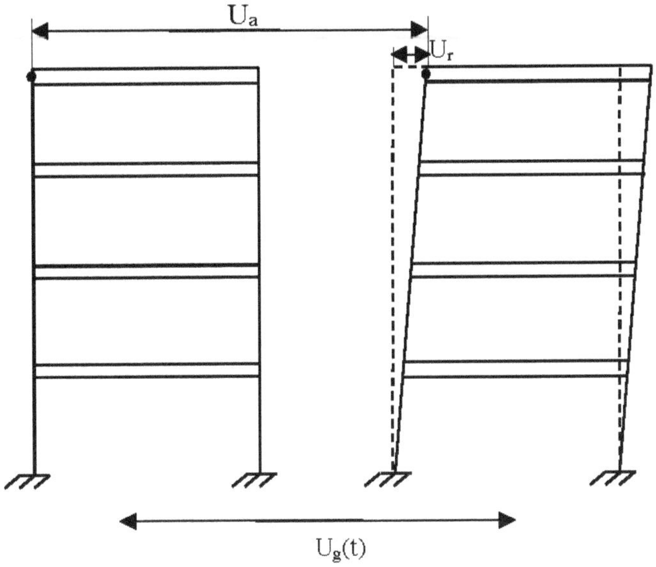

We know that:

$$U_a = U_r + U_g$$

$$\dot{U}_a = \dot{U}_r + \dot{U}_g$$

And $\ddot{U}_a = \ddot{U}_r + \ddot{U}_g$

$$[M]\{\ddot{U}_a\} + [C]\{\dot{U}_r\} + [K]\{U_r\} = \{0\}$$

$$[M](\{\ddot{U}_r\} + \{\ddot{U}_g\}) + [C]\{\dot{U}_r\} + [K]\{U_r\} = \{0\}$$

$$[M]\{\ddot{U}_r\} + [C]\{\dot{U}_r\} + [K]\{U_r\} = -[M]\{\ddot{U}_g\} \quad (11)$$

If matrix [M] is diagonal then the second member of equation (11) become $-[M]\{1\}\{\ddot{U}_g\}$
Now we can rewrite equation (11) as follows (Chopra. 2001) and (Jerome. 2012):

$$[M]\{\ddot{U}_r\} + [C]\{\dot{U}_r\} + [K]\{U_r\} = -[M]\begin{Bmatrix}1\\1\\1\\\vdots\\1\end{Bmatrix}\ddot{U}_g$$

$$[M]\{\ddot{U}_r\} + [C]\{\dot{U}_r\} + [K]\{U_r\} = -[M]\{\ell\}\ddot{U}_g(t) \quad (12)$$

$\{\ell\}$ Represent the influence vector

We assume

$$\{F_{eff}\} = -[M]\{\ell\}\ddot{U}_g(t)$$

$$\{F_{eff}\} = \{S\}\ddot{U}_g(t)$$

Where $\{S\}$ is the spatial loading vector $\{S\}= -[M]\{\ell\}$

Let us consider the equation of motion again

$$[M]_{(nxn)}\{\ddot{U}\}_{(n)} + [K]_{(nxn)}\{U\}_{(n)} = \{F_{eff}\}_{(n)} \quad (13)$$

By substituting $\{U(t)\}=[\phi]\{y(t)\}$

In addition, premultiplying by $[\phi]$t we have

$$[\varphi]^t[M][\varphi]\{\ddot{y}\} + [\varphi]^t[C][\varphi]\{\dot{y}\} + [\varphi]^t[M][\varphi]\{y\} = -[\varphi]^t[M]\{\ell\}\ddot{U}_g(t)$$

$$[M^*]\{\ddot{y}\} + [C^*]\{\dot{y}\} + [K^*]\{y\} = -[\varphi]^t[M]\{\ell\}\ddot{U}_g(t)$$

$$M_i^*\ddot{y}_i + C_i^*\dot{y}_i + K_i^*y_i = -\{\varphi\}^t\{S_i\}\ddot{U}_g(t)$$

$$\ddot{y}_i + 2\xi_i\omega_i\dot{y}_i + \omega_i^2 y_i = -\Gamma_i\ddot{U}_g(t)$$

$$\ddot{D}_i + 2\xi_i\omega_i\dot{D}_i + \omega_i^2 D_i = -\ddot{U}_g(t) \quad (14)$$

With

$$D_i = \frac{y_i}{\Gamma_i}$$

Where

$$\Gamma_i = \frac{1}{M_i^*}\{\varphi_i\}^t\{S\} = \frac{L_i}{M_i^*}$$

is the participation factor of mode i

$$L_i = \{\varphi_i\}^t\{S\} = \{\varphi_i\}^t[M]\{\ell\}$$

Is the modal earthquake excitation factor

$$\{S_i\} = \Gamma_i [M]\{\varphi_i\}$$

Spatial distribution loads.

MODAL PARTICIPATION FACTORS AND TRUNCATION ERROR

By limiting the number of modes considered, we introduce an error. Examining the mode i the corresponding equation is written

$$[M]_{(nxn)}\{\ddot{U}\}_{(n)} + [K]_{(nxn)}\{U\}_{(n)} = \{F(t)\}_{(n)}$$

By substituting $\{U(t)\}=[\phi]\{y(t)\}$
In addition, premultiplying by $[\phi]$t we have

$$[\varphi]^t[M][\varphi]\{\ddot{y}\} + [\varphi]^t[C][\varphi]\{\dot{y}\} + [\varphi]^t[M][\varphi]\{y\} = [\varphi]^t\{F(t)\}$$

$$M_i^*\ddot{y}_i + C_i^*\dot{y}_i + K_i^*y_i = \varphi_i^t F(t) \tag{15}$$

$$\ddot{y}_i(t) + 2\xi_i\omega_i\dot{y}_i(t) + \omega_i^2 y_i(t) = \frac{p_i(t)}{M_i^*}$$

$[\phi]$t With $P_i(t) = \varphi_i^t F(t)$
The modal participation factor

$$\Gamma_i = \frac{1}{M_i^*}\{\varphi_i\}^t\{S\}$$

with $M_1^* = \{\varphi_1\}^t[M]\{\varphi_1\} = 1$ if ϕi is normed with respect to the mass. Knowing that the amplitude of the response in one mode depends on the distribution of the load characterized by (Patrick 2011& Rajasekaran 2009).

The dynamic amplification factor can be evaluated for each mode in the same way as for a simple oscillator. We then obtain the figure above, on which the undamped case is represented.

ωt is the frequency of the mode considered and Ω is the frequency of the exciter force, F_{S_i} is the modal elastic resistance and Pi has been defined above.

For high frequencies with respect to Ω (β <<) one tends exactly to the behavior of the simple oscillator. The static case corresponds to β = 0.

For low frequencies with respect to Ω (β >>), we observe a reversal of the sign of the modal elastic resistance, which indicates a phase shift of 180 with respect to the load; at (β = -¥), the elastic resistance becomes insignificant. The symmetrical curves correspond to the modal inertial resistance FI; modal elastic and modal elastic inertial resistances combine to balance the load

$$\frac{F_{S_i}(\beta)}{P_i}+\frac{F_{I_i}(\beta)}{P_i}=1 \tag{16}$$

Figure 1. Response curve based on β

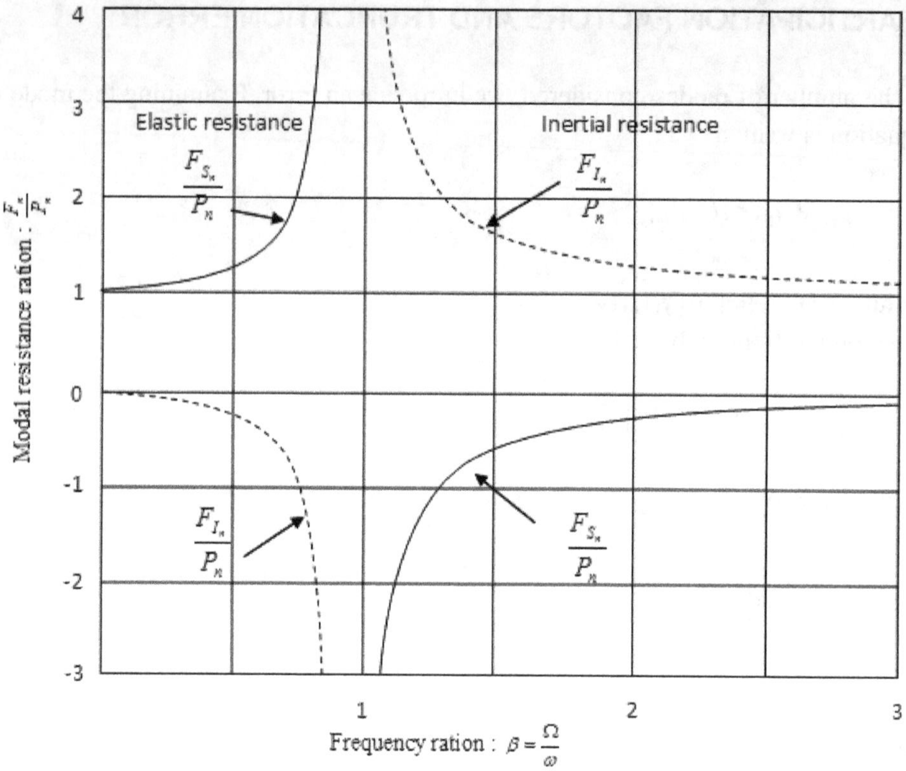

STATIC CORRECTION PROCEDURE

The fact that for the high frequencies (β <<) the response is essentially static makes it possible to define a correction procedure.

In the principle of modal superposition

$$U(t)=\sum_{i=1}^{n}\varphi_i y_i(t)$$

that can be divided into a first part (dynamic) and an essentially static part, let us write:

$$U(t)=U_d(t)+U_s(t)=\sum_{i=1}^{d}\varphi_i y_i(t)+\sum_{i=d+1}^{n}\varphi_i y_i(t) \tag{17}$$

The response of each modes contributing to the dynamic part will be evaluated individually, for example by the Duhamel integral or the direct integration. For the high modes, the static response will be obtained for each mode, dividing the modal load $P_i(t)$ by the modal stiffness from which.

$$y_{s_i} = \frac{P_i(t)}{K^*} = \frac{P_i(t)}{\left(\varphi_i^T K \varphi_i\right)} = \frac{\varphi_i^T F(t)}{\left(\varphi_i^T K \varphi_i\right)}$$

Hence a contribution to the global response of:

$$U_{s_i} = \varphi_i y_i = \varphi_i \frac{P_i(t)}{K^*} = \frac{\varphi_i P_i(t)}{\left(\varphi_i^T K \varphi_i\right)} = \frac{\varphi_i \varphi_i^T F(t)}{K^*} = f_i \; F(t)$$

Where f_i is the modal flexibility of the mode i, which defines the response of the mode to the load F(t), we deduce:

$$U_s(t) = \sum_{i=d+1}^{n} \varphi_i y_i(t) = \sum_{i=d+1}^{n} f_i F(t) = \sum_{i=d+1}^{n} f_i Sg(t)$$

This contribution can be calculated simply by calculating the total static response and then deducing the static response of the first "d" modes so:

$$U_s(t) = K^{-1} Sg(t) - \sum_{i=1}^{d} f_i Sg(t) \tag{18}$$

The global response is then finally:

$$U(t) = U_d(t) + U_s(t) = \sum_{i=1}^{d} \varphi_i y_i(t) + \left(K^{-1} - \sum_{i=1}^{d} f \right) Sg(t) \tag{19}$$

taking into account the higher modes by this approach, rather than ignoring them, makes it possible to reduce "d", the number of modes to be taken into account for the dynamic

APPLICATIONS

Example 1

Let us assume a two-storey structure shown below knowing that the damping rate of both first and second mode is $\xi 1_{=} \xi 1 =_{5} \%$.

We need to determine the overturn moment at the base as well as the modal contribution and consequently the number of mode to include. (Chopra. 2001) and (Jerome. 2012).

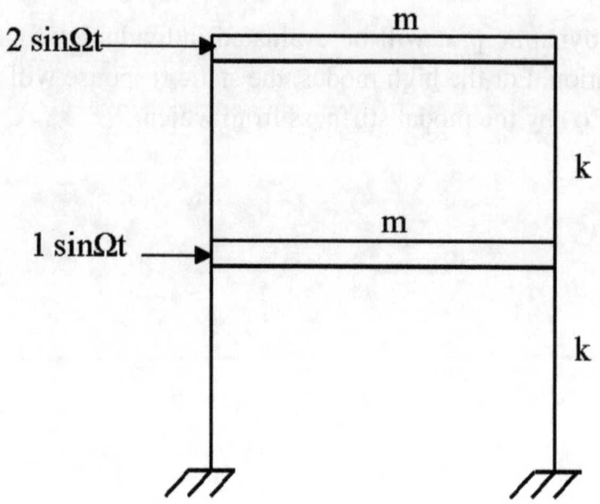

Solution

As the masse a lumped, the mass matrix is diagonal as follows:

$$[M]=\begin{bmatrix} m & 0 \\ 0 & m \end{bmatrix}=m\begin{bmatrix} 1 & 0 \\ 0 & 1 \end{bmatrix}$$

The stiffness matrix is:

$$[K]=\begin{bmatrix} 2k & -k \\ -k & k \end{bmatrix}=k\begin{bmatrix} 2 & -1 \\ -1 & 1 \end{bmatrix}$$

Determination of eigenvalues and eigenvectors

$$\omega_1^2 = 0.382\frac{k}{m} \quad \rightarrow \{\varphi_1\} = \begin{Bmatrix} 1 \\ 1.6182 \end{Bmatrix}$$
$$\omega_2^2 = 2.618\frac{k}{m} \quad \rightarrow \{\varphi_2\} = \begin{Bmatrix} -1 \\ 0.6180 \end{Bmatrix}$$

Applied load

$$\{F\} = \begin{Bmatrix} 2 \\ 1 \end{Bmatrix} \sin \Omega t$$

where

$$\{S\} = \begin{Bmatrix} 2 \\ 1 \end{Bmatrix} = \sum_{i=1}^{2} \Gamma_i [M]\{\varphi_i\} = \Gamma_1 \begin{Bmatrix} 1 \\ 1.6182 \end{Bmatrix} + \Gamma_2 \begin{Bmatrix} -1 \\ 0.6180 \end{Bmatrix}$$

$$\begin{Bmatrix} 2 \\ 1 \end{Bmatrix} = \begin{bmatrix} 1 & -1 \\ 1.6182 & 0.6180 \end{bmatrix} \begin{Bmatrix} \Gamma_1 \\ \Gamma_2 \end{Bmatrix}$$

We can deduce

$$\begin{Bmatrix} \Gamma_1 \\ \Gamma_2 \end{Bmatrix} = \begin{Bmatrix} 1.1707 \\ 0.1707 \end{Bmatrix} \Rightarrow \begin{cases} \{S_1\} = \Gamma_1 \{\varphi_1\} = \begin{Bmatrix} 1.1707 \\ 1.8944 \end{Bmatrix} \\ \{S_2\} = \Gamma_2 \{\varphi_2\} = \begin{Bmatrix} -0.1707 \\ 0.1055 \end{Bmatrix} \end{cases}$$

$$\{S_1\} = \Gamma_1 \{\varphi_1\} = \begin{Bmatrix} 1.1707 \\ 1.8944 \end{Bmatrix}$$

$$\{S_2\} = \Gamma_2 \{\varphi_2\} = \begin{Bmatrix} -0.1707 \\ 0.1055 \end{Bmatrix}$$

We consider the overturn moment

$$\{S_1\} = \Gamma_1 \{\phi_1\} = \begin{Bmatrix} 1.1707 \\ 1.8944 \end{Bmatrix}$$

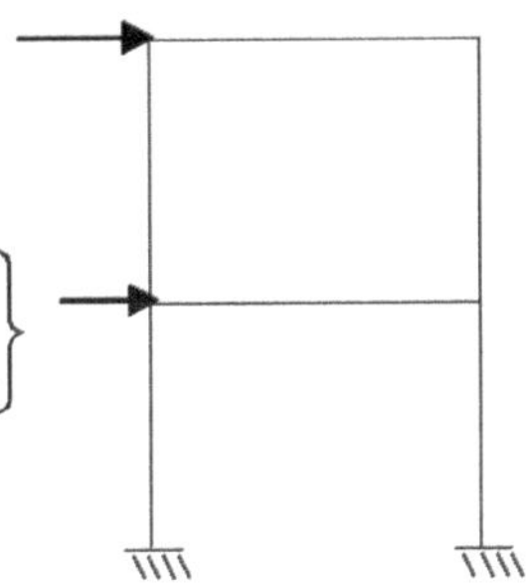

$$\{S_2\} = \Gamma_2 \{\phi_2\} = \begin{Bmatrix} -0.1707 \\ 0.1055 \end{Bmatrix}$$

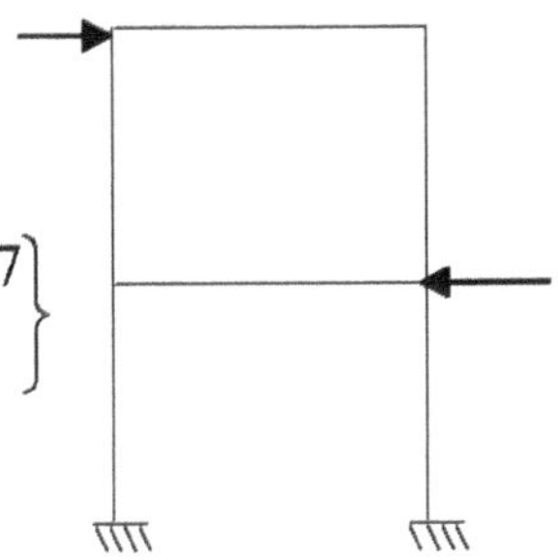

We consider the overturn moment

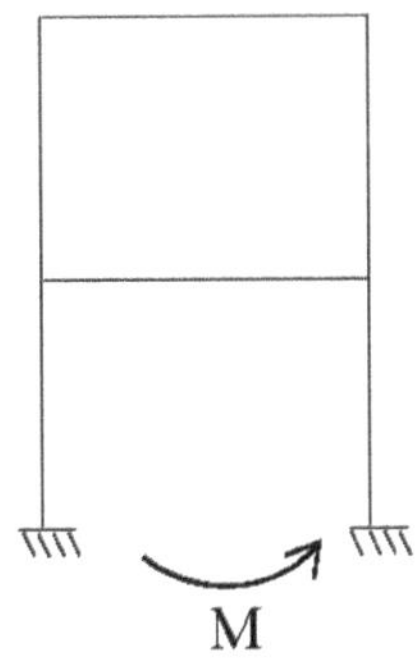

$$M_1^{st} = (1.1707)(h) + (1.8944)(2h)$$

$$M_1^{st} = 4.9595h$$

$$M_2^{st} = (-0.1707)(h) + (0.1055)(2h)$$

$$M_1^{st} = 0.0403h$$

$$M(t) = M_1(t) + M_1(t) = M_1^{st}\omega_1^2 D_1(t) + M_2^{st}\omega_2^2 D_2(t)$$

Modal contribution

$$\bar{r}_1 = \frac{4.9595}{4.9595 + 0.0403} = 0.992$$

$$\bar{r}_1 = \frac{0.0403}{4.9595 + 0.0403} = 0.008$$

In this case we consider only the first mode because the error e=0.8%.

Example 2 (Jerome. 2012)

Let us assume a two-story structure shown below, We need to determine the Equivalent static forces And the story Shears ?

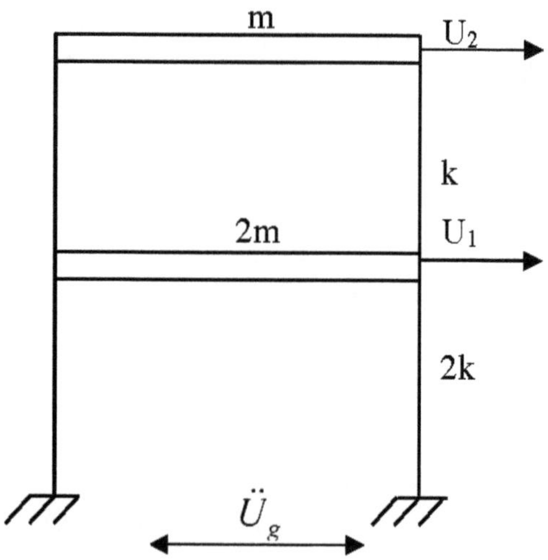

Solution

$$[M] = \begin{bmatrix} 2m & 0 \\ 0 & m \end{bmatrix} = m\begin{bmatrix} 2 & 0 \\ 0 & 1 \end{bmatrix}$$

$$[K] = \begin{bmatrix} 2k+k & -k \\ -k & k \end{bmatrix} = k\begin{bmatrix} 3 & -1 \\ -1 & 1 \end{bmatrix}$$

As the damping is neglected the equation of motion is as follows:

$$[M]\{\ddot{U}\}+[K]\{U\}=-m\begin{bmatrix}2 & 0\\ 0 & 1\end{bmatrix}\begin{Bmatrix}1\\1\end{Bmatrix}\ddot{U}_g(t)$$

With $\begin{Bmatrix}1\\1\end{Bmatrix}=\{\ell\}$

Determination of eigenvalues and eigenvectors

$$\omega_1=\sqrt{\frac{k}{2m}}\quad\rightarrow\{\varphi_1\}=\begin{Bmatrix}0.5\\1\end{Bmatrix}$$

$$\omega_2=\sqrt{\frac{2k}{m}}\quad\rightarrow\{\varphi_2\}=\begin{Bmatrix}-1\\1\end{Bmatrix}$$

Generalized mass and stiffness

$$m_1^*=\frac{3}{2}m\quad k_1^*=\frac{3}{4}k$$

$$m_2^*=3m\quad k_2^*=6k$$

Participation factor of mode i is

$$\Gamma_i=\frac{1}{m_i^*}\{\varphi_i\}^t\{S\}=\frac{L_i}{m_i^*}$$

With $\{S\}=[M]\{\ell\}$

$$\Gamma_i=\frac{1}{m_i^*}\{\varphi_i\}^t[M]\{\ell\}$$

$$\Gamma_1=\frac{1}{\frac{3}{2}m}\begin{Bmatrix}0,5\\1\end{Bmatrix}^t\begin{bmatrix}2m & 0\\ 0 & m\end{bmatrix}\begin{Bmatrix}1\\1\end{Bmatrix}=\frac{4}{3}$$

$$\Gamma_2=\frac{1}{3m}\begin{Bmatrix}-1\\1\end{Bmatrix}^t\begin{bmatrix}2m & 0\\ 0 & m\end{bmatrix}\begin{Bmatrix}1\\1\end{Bmatrix}=-\frac{1}{3}$$

Equivalent static forces are represented as follows:

$$\{S\}_1=\Gamma_1[M]\{\varphi\}_1=\frac{4}{3}m\begin{bmatrix}2 & 0\\ 0 & 1\end{bmatrix}\begin{Bmatrix}0.5\\1\end{Bmatrix}=\frac{4}{3}m\begin{Bmatrix}1\\1\end{Bmatrix}$$

$$\{S\}_2 = \Gamma_2[M]\{\varphi\}_2 = -\frac{1}{3}m\begin{bmatrix}2 & 0\\0 & 1\end{bmatrix}\begin{Bmatrix}-1\\1\end{Bmatrix} = -\frac{1}{3}m\begin{Bmatrix}-2\\1\end{Bmatrix}$$

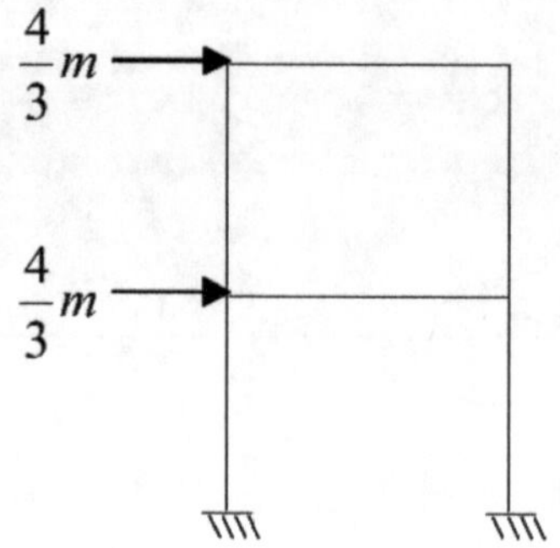

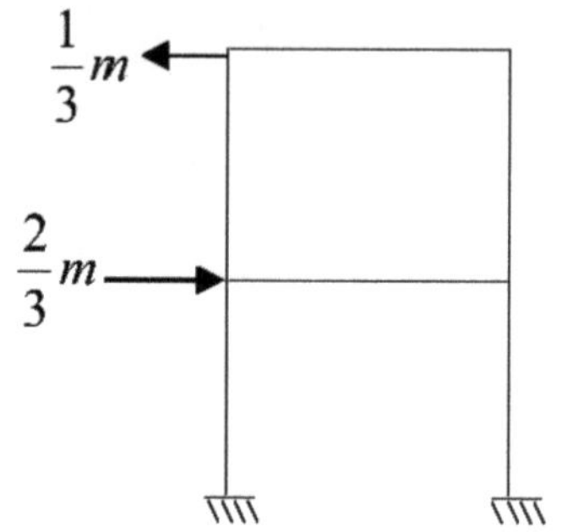

Thanks to the equation bellow we can have floor displacement

$$\{U(t)\} = \Gamma_1\{\varphi_1\}D_1 + \Gamma_2\{\varphi_2\}D_2$$

To determine the story shears we proceed as follows

$$\{V\} = \Gamma_1\{V_1^{st}\}A_1 + \Gamma_2\{V_2^{st}\}A_2$$

With V_1^{st} and V_2^{st} are the shear forces due to $\{S_1\}$ and $\{S_2\}$ respectively

$$\{V\} = \frac{4}{3}m\begin{Bmatrix}2\\1\end{Bmatrix}A_1 + -\frac{1}{3}m\begin{Bmatrix}-1\\1\end{Bmatrix}A_2$$

$$A_i(t) = \omega_i^2 D_i(t)$$

$$\{V\} = \frac{4}{3}m\begin{Bmatrix}2\\1\end{Bmatrix}\omega_1^2 D_1 + -\frac{1}{3}m\begin{Bmatrix}-1\\1\end{Bmatrix}\omega_2^2 D_2$$

Example3 (Filiatrault.1996)

We consider a 3-stage structure modeled in a 3 degree of freedom system, subjected to the El Centro earthquake (18/05/1940) whose response spectrum is represented in figure below.

It is asked to estimate the maximum probable shear at the base using Square root of sum of squares (SRSS) method combination. We assume a critical damping of 2% in each mode of vibration.

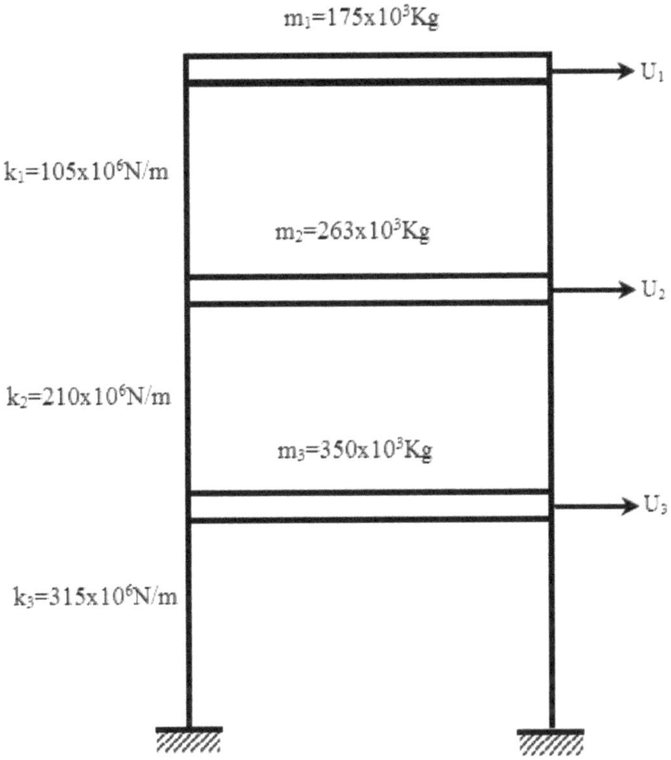

Solution

Step 1: We form the mass and stiffness matrix

$$[M]=\begin{bmatrix} M_1 & 0 & 0 \\ 0 & M_2 & 0 \\ 0 & 0 & M_3 \end{bmatrix}=\begin{bmatrix} 175 & 0 & 0 \\ 0 & 263 & 0 \\ 0 & 0 & 350 \end{bmatrix}10^3 kg$$

$$[K]=\begin{bmatrix} K_1 & -K_1 & 0 \\ -K_2 & K_1+K_2 & -K_2 \\ 0 & -K_2 & K_2+K_3 \end{bmatrix}=105\begin{bmatrix} 1 & -1 & 0 \\ -1 & 3 & -2 \\ 0 & -2 & 5 \end{bmatrix}10^6 N/m$$

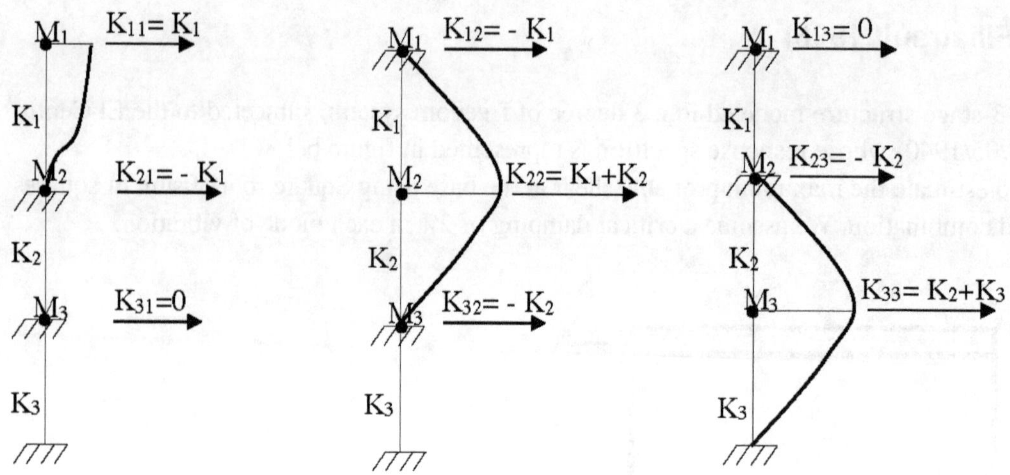

Figure 2. Response spectrum of El Centro (Filiatrault.1996)

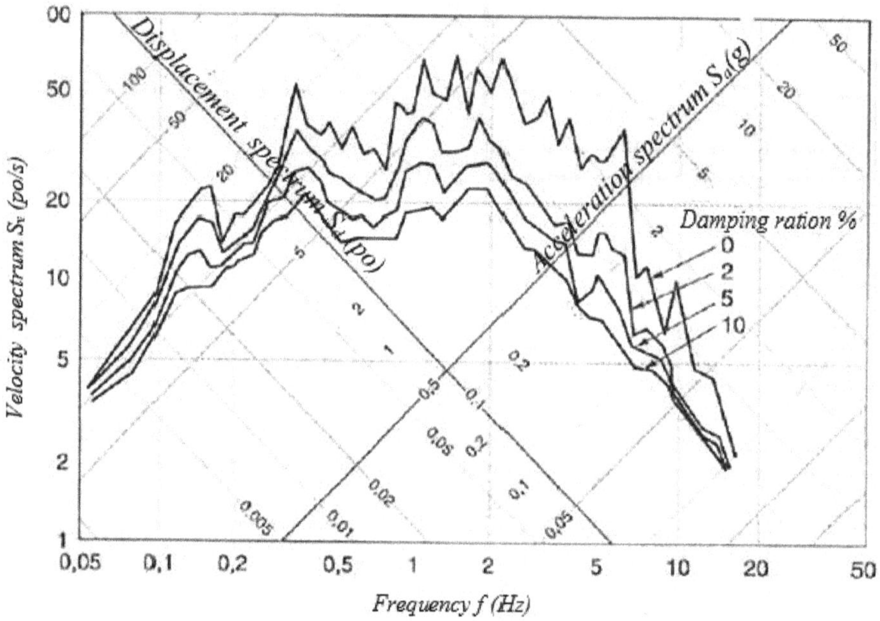

Step 2: Calculation of Natural Frequencies and Deformation Modes

$$\{\varphi_1\} = \begin{Bmatrix} 1 \\ 0.644 \\ 0.3 \end{Bmatrix} \; \{\varphi_2\} = \begin{Bmatrix} 1 \\ -0.644 \\ -0.676 \end{Bmatrix} \; \{\varphi_3\} = \begin{Bmatrix} 1 \\ -2.57 \\ 2.47 \end{Bmatrix}$$

Step 3: Calculation of the generalized masses

$$M_1^* = \{\varphi_1\}^t [M]\{\varphi_1\} = 0$$

$$M_1^* = \langle 1 \quad 0.664 \quad 0.3 \rangle \begin{bmatrix} 175 & 0 & 0 \\ 0 & 263 & 0 \\ 0 & 0 & 350 \end{bmatrix} 10^3 \begin{Bmatrix} 1 \\ 0.644 \\ 0.3 \end{Bmatrix} = 315.575x10^3 Kg$$

$$M_2^* = \{\varphi_2\}^t [M]\{\varphi_2\} = 0$$

$$M_2^* = \langle 1 \quad -0.601 \quad -0.676 \rangle \begin{bmatrix} 175 & 0 & 0 \\ 0 & 263 & 0 \\ 0 & 0 & 350 \end{bmatrix} 10^3 \begin{Bmatrix} 1 \\ -0.601 \\ -0.676 \end{Bmatrix} = 429.94x10^3 Kg$$

$$M_3^* = \{\varphi_3\}^t [M]\{\varphi_3\} = 0$$

$$M_3^* = \langle 1 \quad -2.57 \quad 2.47 \rangle \begin{bmatrix} 175 & 0 & 0 \\ 0 & 263 & 0 \\ 0 & 0 & 350 \end{bmatrix} 10^3 \begin{Bmatrix} 1 \\ -2.57 \\ 2.47 \end{Bmatrix} = 4047.404x10^3 Kg$$

Step 4: Calculation the modal participation factor

$$\Gamma_1 = \frac{\{\varphi_1\}^t [M]\{I\}}{M_i^*}$$

$$\Gamma_1 = \frac{\langle 1 \quad 0.664 \quad 0.3 \rangle \begin{bmatrix} 175 & 0 & 0 \\ 0 & 263 & 0 \\ 0 & 0 & 350 \end{bmatrix} 10^3 \begin{Bmatrix} 1 \\ 1 \\ 1 \end{Bmatrix}}{315.575x10^3} = 1.24$$

$$\Gamma_2 = \frac{\{\varphi_2\}^t [M]\{I\}}{M_2^*}$$

$$\Gamma_2 = \frac{\langle 1 \quad -0.601 \quad -0.676 \rangle \begin{bmatrix} 175 & 0 & 0 \\ 0 & 263 & 0 \\ 0 & 0 & 350 \end{bmatrix} 10^3 \begin{Bmatrix} 1 \\ 1 \\ 1 \end{Bmatrix}}{429.94x10^3} = -0.511$$

$$\Gamma_3 = \frac{\{\varphi_3\}^t [M]\{I\}}{M_3^*}$$

$$\Gamma_3 = \frac{\langle 1 \quad -2.57 \quad 2.47 \rangle \begin{bmatrix} 175 & 0 & 0 \\ 0 & 263 & 0 \\ 0 & 0 & 350 \end{bmatrix} 10^3 \begin{Bmatrix} 1 \\ 1 \\ 1 \end{Bmatrix}}{4047.404x10^3} = 0.09$$

Step 5: Using the response spectrum shown in the figure above, we can have the spectral response in each mode at 2% critical damping knowing that 1 inch = 25.4mm

- First mode $f_1 = \frac{1}{T_1} = \frac{\omega_1}{2\pi} = 2.308Hz$

From graph we take the spectral shift of the first mode Sd(1) = 2in = 50.8mm

- Second mode $f_2 = \frac{1}{T_2} = \frac{\omega_2}{2\pi} = 4.95Hz$

From graph we draw the spectral shift of the first mode Sd(2) = 0.46in = 11.684mm»11.7mm

- Third mode $f_3 = \frac{1}{T_3} = \frac{\omega_3}{2\pi} = 7.337Hz$

From graph we draw the spectral shift of the first mode Sd(3) = 0.2in = 5.08mm»5.1mm

Step 6: Calculate the maximum modal responses in each mode and for each degree of freedom.

- 1st mode:

$$U^{(1)}_{1\max} = \varphi_{11}\Gamma_1 S^{(1)}_d = (1)(1.424)(50.8) = 72.34mm$$

$$U^{(1)}_{2\max} = \varphi_{21}\Gamma_1 S^{(1)}_d = (0.644)(1.424)(50.8) = 46.6mm$$

$$U^{(1)}_{3\max} = \varphi_{31}\Gamma_1 S^{(1)}_d = (0.3)(1.424)(50.8) = 21.7mm$$

- 2nd mode:

$$U^{(2)}_{1\max} = \varphi_{12}\Gamma_2 S^{(2)}_d = (1)(-0.511)(11.7) = -5.98mm \approx -6mm$$

$$U^{(2)}_{2\max} = \varphi_{22}\Gamma_2 S^{(2)}_d = (-0.601)(-0.511)(11.7) = 3.6mm$$

$$U^{(2)}_{3\max} = \varphi_{32}\Gamma_2 S^{(2)}_d = (-0.676)(-0.511)(11.7) = 4.042mm$$

- 3th mode:

$$U^{(3)}_{1\max} = \varphi_{13}\Gamma_3 S^{(3)}_d = (1)(0.09)(5.1) = 0.46mm$$

$$U^{(3)}_{2\max} = \varphi_{23}\Gamma_3 S^{(3)}_d = (-2.57)(0.09)(5.1) = -1.18mm \approx -1.2mm$$

$$U^{(3)}_{3\max} = \varphi_{33}\Gamma_3 S_d^{(3)} = (2.47)(0.09)(5.1) = 1.13mm$$

Step7: Calculation of the maximum equivalent seismic forces for each mode by multiplying the maximum modal responses by stiffness matrix.

$$\{f\}_{i(\max)} = [K]\{U_i\}_{\max}$$

- 1st mode:

$$\{f\}_{1(\max)} = \begin{Bmatrix} f^{(1)}_{1\max} \\ f^{(1)}_{2\max} \\ f^{(1)}_{3\max} \end{Bmatrix} = 105\begin{bmatrix} 1 & -1 & 0 \\ -1 & 3 & -2 \\ 0 & -2 & 5 \end{bmatrix}\begin{Bmatrix} 72.4 \\ 46.6 \\ 21.7 \end{Bmatrix} = \begin{Bmatrix} 2709 \\ 2520 \\ 1606.5 \end{Bmatrix} KN$$

- 2nd mode:

$$\{f\}_{2(\max)} = \begin{Bmatrix} f^{(2)}_{1\max} \\ f^{(2)}_{2\max} \\ f^{(2)}_{3\max} \end{Bmatrix} = 105\begin{bmatrix} 1 & -1 & 0 \\ -1 & 3 & -2 \\ 0 & -2 & 5 \end{bmatrix}\begin{Bmatrix} -6 \\ 3.6 \\ 4.042 \end{Bmatrix} = \begin{Bmatrix} -1008 \\ 915.18 \\ 1366.05 \end{Bmatrix} KN$$

- 3th mode:

$$\{f\}_{3(\max)} = \begin{Bmatrix} f^{(3)}_{1\max} \\ f^{(3)}_{2\max} \\ f^{(3)}_{3\max} \end{Bmatrix} = 105\begin{bmatrix} 1 & -1 & 0 \\ -1 & 3 & -2 \\ 0 & -2 & 5 \end{bmatrix}\begin{Bmatrix} 0.46 \\ -1.2 \\ 1.13 \end{Bmatrix} = \begin{Bmatrix} 174.3 \\ -663.6 \\ 845.25 \end{Bmatrix} KN$$

These maximum equivalent seismic forces for each mode are illustrated in this of the 3-stage structure in the figure below.

Step 8: Obtain maximum base shear for each mode by adding the maximum seismic forces algebraically.

1st mode

$$V^{(1)}_{\max} = 2709 + 2520 + 1606.5 = 6835.5KN$$

2nd mode

$$V^{(2)}_{\max} = -1008 + 915.18 + 1366.05 = 1273.23KN$$

3th mode

$$V^{(3)}_{\max} = 174.3 - 663.6 + 845.25 = 355.95KN$$

Figure 3. Maximum equivalent seismic force for each mode of three-storey structure

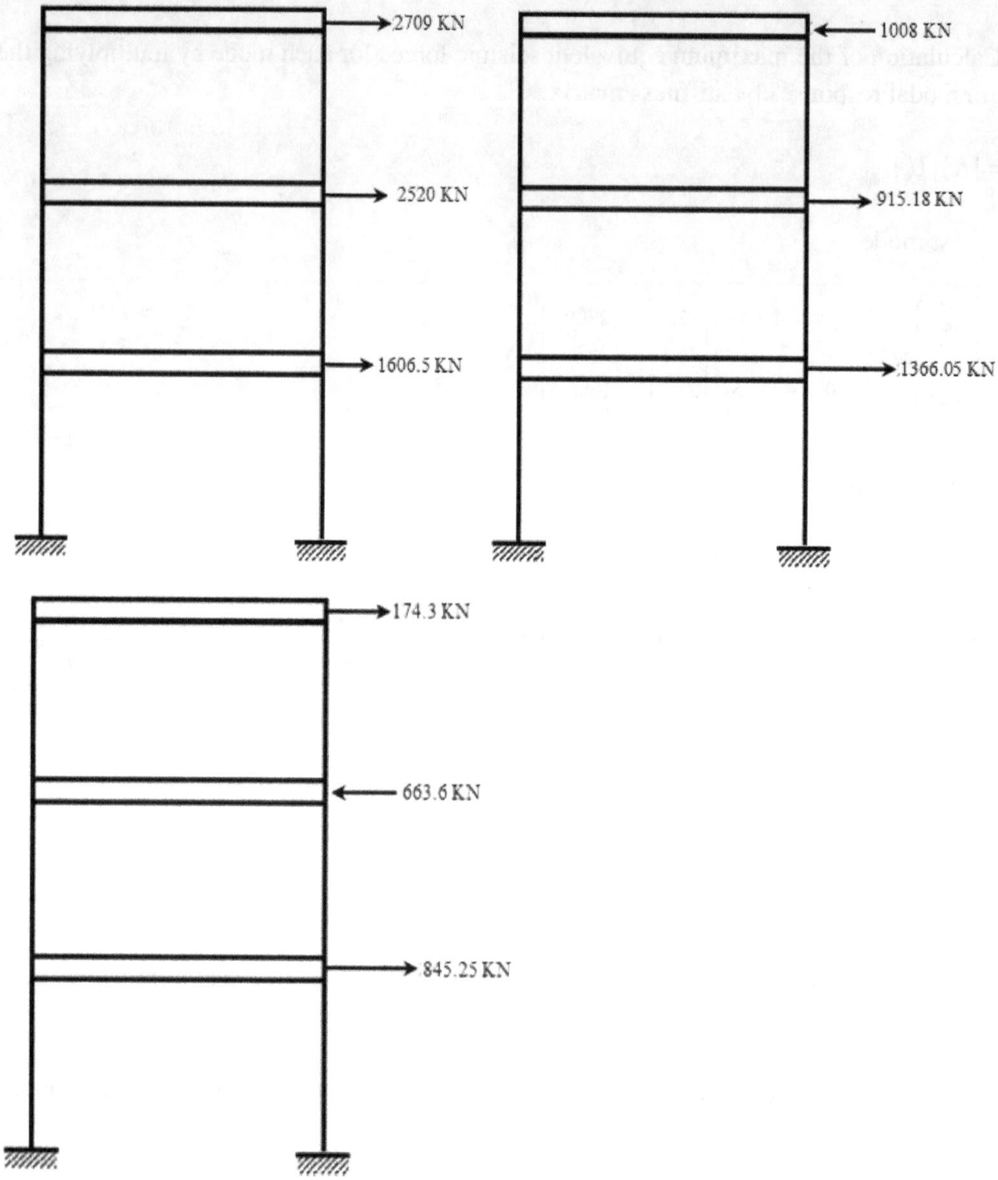

Step 9: Estimation of Probable Maximum Shear Base by Combining Modal Shears Using the SRSS Method

$$|V|_{\max} = \sqrt{|6835.5|^2 + |1273.23|^2 + |355.95|^2} = 6969.17KN$$

The probable maximum displacements of the three degrees of freedom, according to the rule of the average quadratic

Modal Response Contribution

$$\{U\}_{(\max)} = \begin{Bmatrix} U_{1\max} \\ U_{2\max} \\ U_{3\max} \end{Bmatrix} = \begin{Bmatrix} \sqrt{|72.34|^2 + |46.6|^2 + |21.7|^2} \\ \sqrt{|-6|^2 + |3.6|^2 + |4.042|^2} \\ \sqrt{|0.46|^2 + |-1.2|^2 + |1.13|^2} \end{Bmatrix} = \begin{Bmatrix} 88.74 \\ 8.08 \\ 1.71 \end{Bmatrix} mm$$

Modal contribution

$$\bar{r}_1 = \frac{88.74}{1.71 + 8.08 + 88.74} = 0.90 = 90\%$$

$$\bar{r}_2 = \frac{8.08}{1.71 + 8.08 + 88.74} = 0.082 = 8.2\%$$

$$\bar{r}_2 = \frac{8.08}{1.71 + 8.08 + 88.74} = 0.0173 = 1.73\%$$

In this problem the mode 1 contributes to the totality of the maximum response

$$\{f\}_{(\max)} = \begin{Bmatrix} f_{1\max} \\ f_{2\max} \\ f_{3\max} \end{Bmatrix} = \begin{Bmatrix} \sqrt{|2709|^2 + |1008|^2 + |174.3|^2} \\ \sqrt{|2520|^2 + |915.18|^2 + |-663.6|^2} \\ \sqrt{|1606.5|^2 + |1366.05|^2 + |845.25|^2} \end{Bmatrix} = \begin{Bmatrix} 2895.708 \\ 2761.941 \\ 2271.867 \end{Bmatrix} KN$$

Figure 4. Base maximum shear

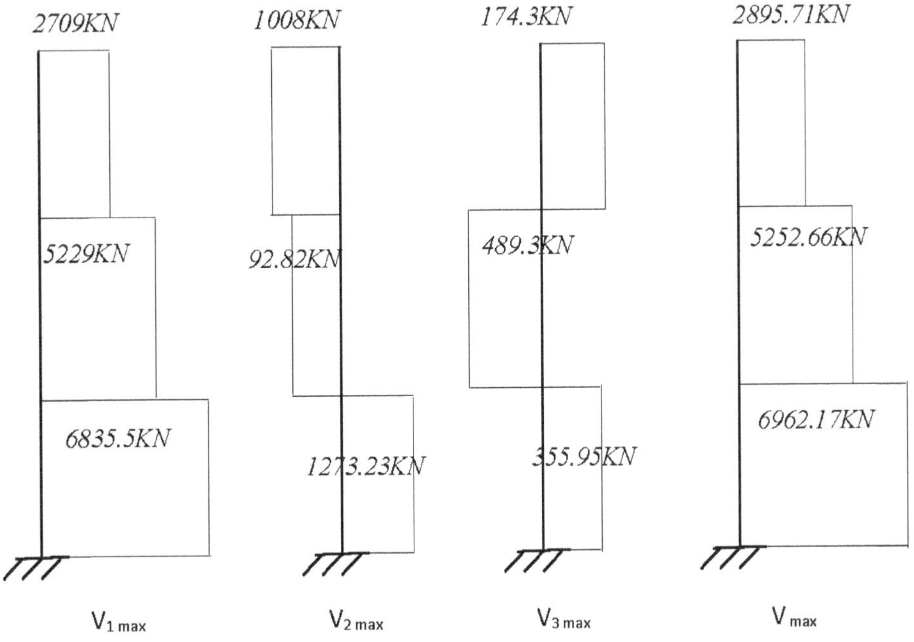

REFERENCES

Chopra, A. K. (2001). Dynamics of Structures: Theory and Applications to Earthquake Engineering (2nd ed.). Prentice Hall.

Filiatrault, A. (1996). Element de génie parasismique et de calcul dynamique des structures. Ecole polytechnique de Monreal Canada.

Jerome, P. L. (2012). *Structure dynamics course note*. Civil and Environmental Engineering University of Michigan.

Patrick, P (2011). *Dynamics of structures*. John Wiley & Sons, Inc.

Rajasekaran, S. (2009). *Structural dynamics of earthquake engineering Theory and application using mathematica and matlab*. Woodhead Publishing in Materials.

KEY TERMS AND DEFINITIONS

Influence Vector: Is a displacement transformation vector that expresses the displacement of each structure degree of freedom.

Inertial Resistance: Is the resistance of any physical object to any change in its state of motion.

Modal Contribution Factor for Mode i: Represents the ration between static response to load pattern $\{S_i\}$ and the static response to load pattern $\{S\}$.

Static Correction: Is an essential step in seismic processing and it has an effect on the later steps of seismic processing.

Truncation Error: Is the errors that result from using an approximation in place of an exact mathematical procedure.

Section 6

Nonlinear Transient and Static Nonlinear Analysis

Chapter 10
Nonlinear Transient Analysis

ABSTRACT

In this chapter, the author begin by presenting the main causes of non-linearity, which are geometric and material source, change in boundary condition. The last presented source is pretensions. Then they go to the physical understanding of non-linear behavior by presenting the different phases of hysteresis curve sequence of a reinforced concrete structure. In this chapter, readers pass over various numerical formulation, which allow them to deal with non-linearity, namely Lagrange and Euler formulation, total Lagrangian formulation, Piola-Kirchhoff 2, and corotational formulation. Some examples are exposed at the end of the chapter.

INTRODUCTION

The nonlinear mechanics of continuous, solid and structural media is an extremely vast and rich field of study and research. Since about 1970, the interest of researchers and practitioners for the "nonlinear" has increased considerably, because numerical methods coupled with electronic programs make it possible to effectively solve problems of amazing complexity, and this in all domain: structure, flow, diffusion, heat, fluid, dynamics, soils, plasticity, etc.

We begin by presenting the main causes of non-linearity (Belytschko, Kam, Moran. Elkhodary. 2014), there are usually 04 sources of non-linearity.

The first is the geometric (kinematic) source, which includes large displacements (second order, instabilities ...) and large deformations (James (2001)). The second is a material source that takes into account the non-linearity of the material used, plasticity (metals, concrete, soil ...), viscosity, creep (concrete, wood, plastics, soils ...), and the hyper-elasticity (rubber ...); in this same source we also take into account semi-rigid nodes, fissures, fractures, damage and finally temperature (fire ...), all these phenomena fall within the framework of material non-linearity.

The third source of non-linearity is that of changes in boundary conditions, namely: variable (essential and / or natural) limit conditions, "follower" forces (pressure, etc.), unilateral connections, contact, sliding, friction and construction phases. The fourth source is Haubans, cables; pretensions etc see (Belytschko, Kam, Moran and Elkhodary. 2014).The first two types will be covered in this chapter.

DOI: 10.4018/978-1-7998-4399-3.ch010

Copyright © 2021, IGI Global. Copying or distributing in print or electronic forms without written permission of IGI Global is prohibited.

LARGE DISPLACEMENTS (CINEMATIC)

Lagrange and Euler Formulation

In Euclidean space, a body can be considered to be composed of a set of particles. The position of each of them can be identified by three real numbers, for example its coordinates in an arbitrary system of axes. All of these positions define the configuration of the body. The evolution of this configuration in space and time is called the movement of the body.

Let W be a configuration of the body, in which the particles occupy positions well defined by the coordinates xi; if these quantities are taken as independent variables, W is called the reference configuration and the formulation (or description) of the movement is called Lagrangian (or material).

Let W 'be a subsequent configuration of the body; it is called the current configuration and the particles called xi in W occupy new positions characterized by the coordinates xi '. These coordinates are respectively called material (xi) and spatial (xi '). For simplicity, they will all be measured in a single straight Cartesian system figure 1. (Frey 2005)

We call initial configuration W0 a conventional configuration generally corresponding to the resting, unloaded state of the body. Any other configuration is called deformed configuration (W, W '...).

Figure 1. Movement and successive configurations

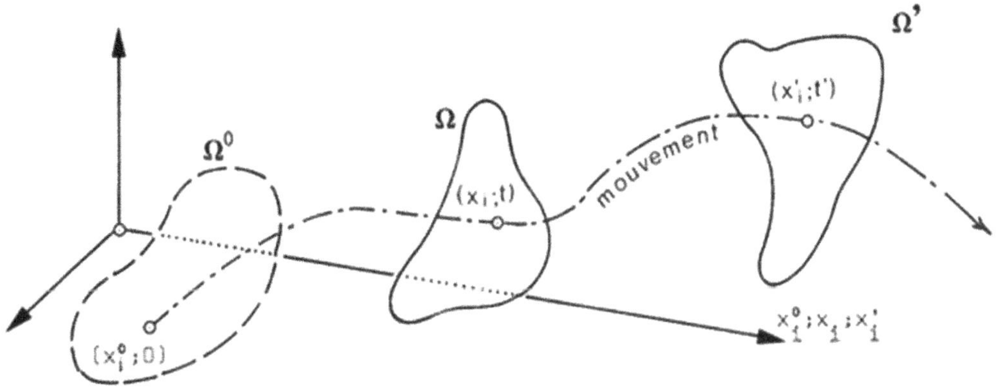

MEASUREMENT OF THE STRAIN (GREEN TENSOR)

Movement, Jacobian, Deformation Gradient

Let W (xi) be a reference configuration, usually initial, of the body and W '(xi ') a deformed configuration (fig. 2.2). The movement is described by the continuous and differentiable application (mapping) (Frey 2005)

$$x_i' = x_i' (x_1, x_2, x_3) \quad (1)$$

$$J_{ij} = \frac{\partial x'_i}{\partial x_j} \tag{2}$$

From the exact differential, form the Jacobian matrix J (3 × 3) whose determinant or Jacobian

$$J = \det \mathbf{J} = \frac{dV'}{dV} \tag{3}$$

Must be positive, not zero and finite, because it can be interpreted as the ratio of two corresponding volume elements dV and dV 'of W and W'

$$¥ > J > 0 \tag{4}$$

J (or Jij) is often called the strain gradient and denoted F (or F_{ij}); the exact differential

dx¢ = F dx (5)

But it cannot be used for measurement of deformation, because it is not zero in a rigid body movement. In such a case indeed.

F ° R

Where R is the (large) rigid rotation matrix

Green's Displacements and Tensor

The components of the displacement field are

$$u_i(x_1, x_2, x_3) = x'_i - x_i \tag{6}$$

We take x'_i from equation (6) and then we introduce it in equation (2) we have

$$J_{ij} = \frac{\partial u_i}{\partial x_j} + \delta_{ij} \tag{7}$$

Where $\frac{\partial u_i}{\partial x_j}$ are the gradients of the displacements and d_{ij} the symbol of Kronecker.

The deformation is measured by the variation of the square of the distance between two neighboring points; either dl and dl' two corresponding line elements; we have

$$d\ell^2 = dx_k dx_k \qquad d\ell'^2 = dx'_k dx'_k$$

The invariant $d\ell'^2 - d\ell^2$ is a function only of the deformation, not of the rigid displacement, and makes appear the symmetric Green E_{ij} strain tensor.

$$d\ell'^2 - d\ell^2 = 2E_{ij}\, dx_i\, dx_j \tag{8}$$

Depending on the displacements, the latter is written

$$E_{ij} = \frac{1}{2}\left(\frac{\partial u_i}{\partial x_j} + \frac{\partial u_j}{\partial x_i} + \frac{\partial u_k}{\partial x_i}\frac{\partial u_k}{\partial x_j} \right) \tag{9}$$

$$E_{xx} = \frac{\partial u}{\partial x} + \frac{1}{2}\left[(\frac{\partial u}{\partial x})^2 + (\frac{\partial v}{\partial x})^2 + (\frac{\partial w}{\partial x})^2 \right]$$

$$E_{yy} = \frac{\partial v}{\partial y} + \frac{1}{2}\left[(\frac{\partial u}{\partial y})^2 + (\frac{\partial v}{\partial y})^2 + (\frac{\partial w}{\partial y})^2 \right]$$

$$E_{zz} = \frac{\partial w}{\partial z} + \frac{1}{2}\left[(\frac{\partial u}{\partial z})^2 + (\frac{\partial v}{\partial z})^2 + (\frac{\partial w}{\partial z})^2 \right]$$

$$\begin{aligned} E_{xy} = E_{yx} &= \frac{1}{2}(\frac{\partial u}{\partial y} + \frac{\partial v}{\partial x} + \frac{\partial u}{\partial x}\frac{\partial u}{\partial y} + \frac{\partial v}{\partial x}\frac{\partial v}{\partial y} + \frac{\partial w}{\partial x}\frac{\partial w}{\partial y}) \\ E_{yz} = E_{zy} &= \frac{1}{2}(\frac{\partial v}{\partial z} + \frac{\partial w}{\partial y} + \frac{\partial u}{\partial y}\frac{\partial u}{\partial z} + \frac{\partial v}{\partial y}\frac{\partial v}{\partial z} + \frac{\partial w}{\partial y}\frac{\partial w}{\partial z}) \\ E_{xz} = E_{zx} &= \frac{1}{2}(\frac{\partial w}{\partial x} + \frac{\partial u}{\partial z} + \frac{\partial u}{\partial x}\frac{\partial u}{\partial z} + \frac{\partial v}{\partial x}\frac{\partial v}{\partial z} + \frac{\partial w}{\partial x}\frac{\partial w}{\partial z}) \end{aligned} \tag{10}$$

LAGRANGIAN FORMULATIONS

Depending on the choice of the reference configuration, various versions of the Lagrange formulation are obtained.

TOTAL LAGRANGIAN FORMULATION

Integral Equilibrium

Here, the initial configuration W_0 is constantly used as a reference. The forces, stresses, deformations, displacements and the equilibrium of the current configuration W are expressed in W_0 see figure 2. This procedure requires, in particular, the definition of special constraints, images of true constraints s_{ij}

of W, defined in W_0; they therefore lose their physical meaning. One ordinarily uses the constraints of Piola-Kirchhoff n ° 2 (abbreviated PK2), noted S_{ij}.

With these stresses, the volume equilibrium (differential or strong form) is written

$$\frac{\partial}{\partial x_i^0}\left(S_{im}J_{jm}\right)+b_j^0=0 \qquad S_{ij}=S_{ji} \tag{11}$$

Where $J_{jm}=\dfrac{\partial x_j}{\partial x_m^0}$ is the Jacobian of the application $W^0 \rightarrow W$ $x_i = x_i(x_j^0)$. The PK2 stresses are symmetrical and linked to the true stresses by

$$\sigma_{ij}=\frac{1}{J}J_{im}S_{mn}J_{jn} \tag{12}$$

The integral balance (weak form), interpreted in principle of virtual displacements, is

$$\int_{V^0} S_{ij}\,\delta E_{ij}\,dV^0=\int_{V^0} b_i^0\,\delta u_i^0\,dV^0+\int_{A_t^0} t_i^0\,\delta u_i^0\,dA^0 \tag{13}$$

In addition, shows Green's deformations (equation 2) as kinematics associated with the constraints of PK2. The symbol d should be considered as a variation bearing on the field of large displacements u_i^0; indeed, we have.

$$\delta E_{ij}=\frac{1}{2}\left(\frac{\partial\delta u_i^0}{\partial x_j^0}+\frac{\partial\delta u_j^0}{\partial x_i^0}+\frac{\partial u_k^0}{\partial x_i^0}\frac{\partial\delta u_k^0}{\partial x_j^0}+\frac{\partial u_k^0}{\partial x_j^0}\frac{\partial\delta u_k^0}{\partial x_i^0}\right) \tag{14}$$

PIOLA- KIRCHHOFF 2(PK2) STRESSES

Let dA and dA_0 be two corresponding facets in W and W_0 see figure 4. On dA act the true stresses vector f_s and the force $dQ = f_s\,dA$. To define the stresses vector $\mathbf{f}_S^0$ and the force $dQ_0 = dA_0$, of type PK2 and acting on the facet dA0, one applies to the element of force dQ the same geometrical transformation (5) as with the element of line (namely, with the notations of this chapter, dx = J dx0)

$d\,Q^0 = J^{-1}\,d\,Q$

We can then decompose $\mathbf{f}_S^0$ in a Cauchy tetrahedron constructed in W_0 to define the S_{ij} components of the constraints of PK2.

Figure 2. a) Total Lagrangian formulation b) PK2 stresses

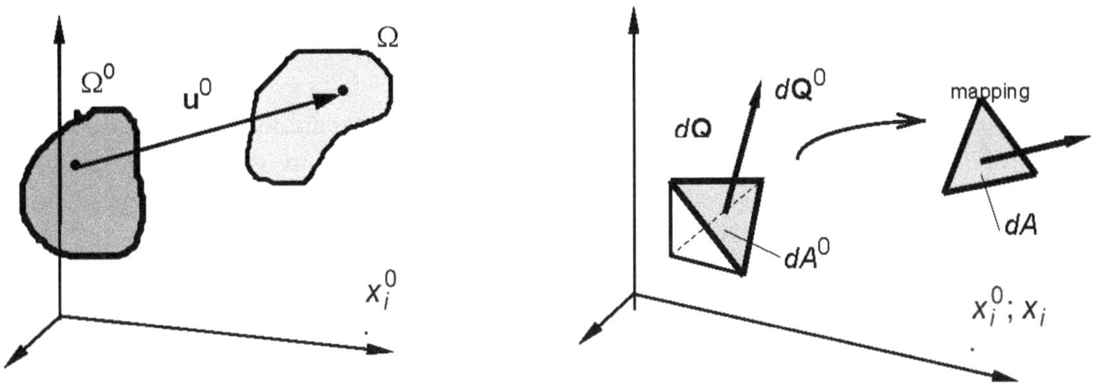

UPDATED AND CURRENT LAGRANGIAN FORMULATIONS

Now the current configuration W serves as a reference see figure 3. There are two 'variants': we can either use W to calculate a subsequent configuration W^D (up-dated, W known), or express the direct equilibrium of W^D to find its own position in relation to W (current). We work with real forces, constraints and deformation. The expression of equilibrium therefore has the same structure as in linear theory; for example, the integral equilibrium is

$$\int_V \sigma_{ij}\, \delta\varepsilon_{ij}\, dV = \int_V b_i\, \delta u_i\, dV + \int_{A_t} t_i\, \delta u_i\, dA \tag{15}$$

Figure 3. a) Lagrangian formulation updated b) Corotational Lagrangian formulation

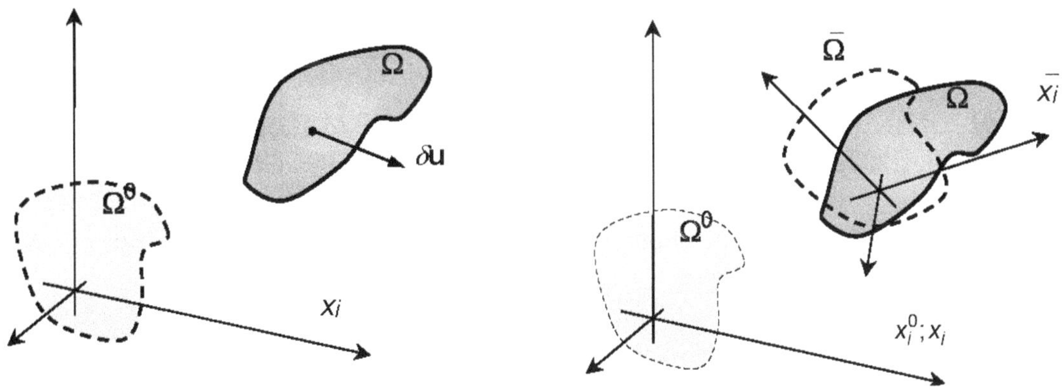

COROTATIONAL FORMULATION IN STRUCTURES

Assumptions and Advantages

It in the case of large displacements, but small deformations, of structures that the corotational formulation is effective. The assumption of small deformation is hardly restrictive: in the fields of civil engineering, aeronautics, vehicle construction, naval architecture and, often, mechanics, many structures have large displacements accompanied by small deformations.

Let us consider in space a small fragment of a structure, Which undergoes large displacements from a known initial configuration W^0 see figure 4. The remaining deformations being small, the shape of the fragment in any deformed configuration W remains very close to that of W^0 (and all the more that the fragment is small). Then rigidly move W^0 to a new chosen position as close to $\bar{\Omega}$ as possible. The axes are also driven in this large rigid rotation (at an angle in the drawing plane of Figure 3cb) to the new position ($\bar{x}_i$).

Figure 4. a) Large displacements and small deformation b)Corotational formulation

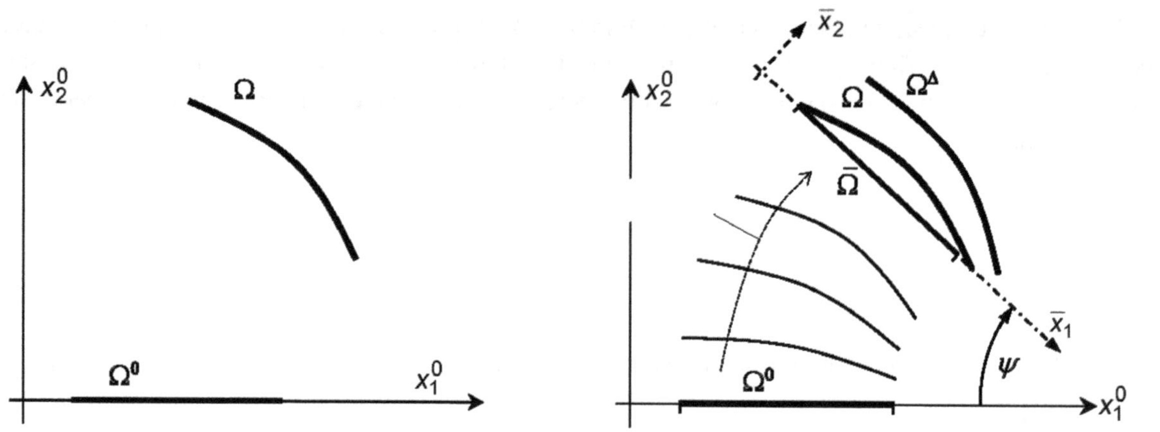

PHYSICAL UNDERSTANDING OF NON-LINEAR BEHAVIOR (HYSTERESIS CURVE SEQUENCE, REINFORCED CONCRETE STRUCTURE)

In this section we try to schematize in phase form the different sequences of hysteresis curve of the reinforced concrete structure see figure 5

- Phase 1 is an elastic loading phase, where the latter is reversible.
- Phase 2 is the one where we reach the maximum resistance from where the lamination of certain sections.
- Phase 3 is the phase of the discharge and this is where there are permanent deformations due to the plastification and just after the intersection of the hysteresis curve with the axis of the deformations, there is a stiffness loss, which appears (low rigidity).

- Phase 4 starts from the closing of the cracks, just after there is an increase of rigidity which will arise because there is the closing of the cracks; hence the recharge in the other direction.

Figure 5. Hysteresis curve of the reinforced concrete structure (Lestuzzi 2007)

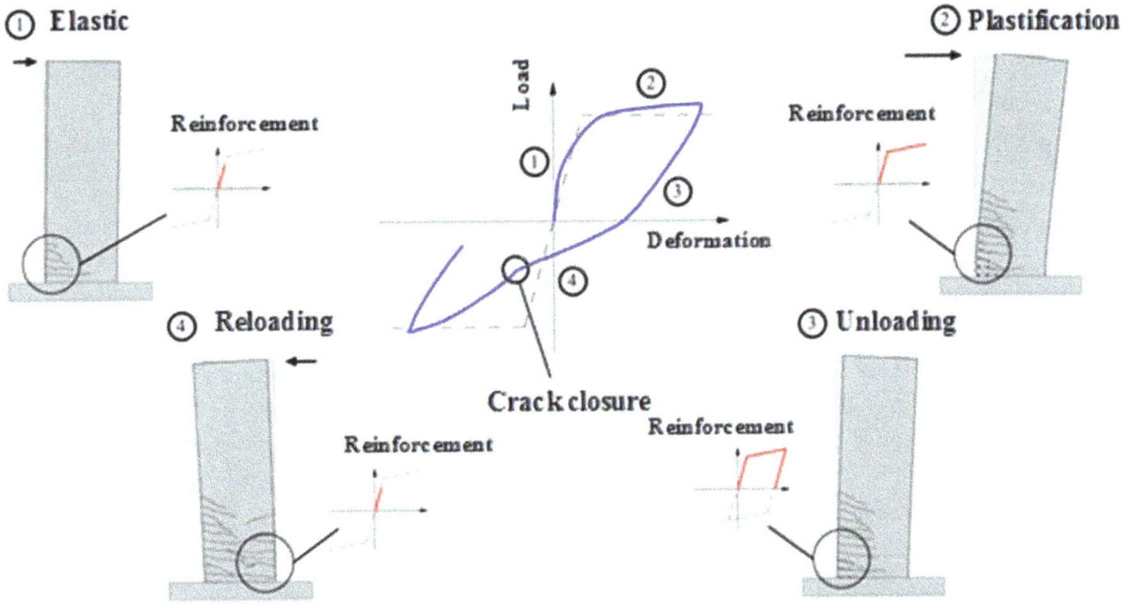

Let us see the necessary adjustments to the algorithms presented previously to solve equation (16). By generalizing and examining the following equation:

$$M\ddot{U} + N(U,\dot{U}) = F \tag{16}$$

This equation includes the different equations.

$N(U,\dot{U})$ Means that we are dealing with a nonlinear function N as a function of U and $\dot{U}$. Rewriting equation (16) in the form:

$$M\ddot{U} + F^{\text{int}} = F^{ext} \tag{17}$$

It is clear that this form is generic; this says that:

$F^{\text{int}} = C\dot{U} + KU$, corresponds to the case of the linear dynamics previously seen.
$F^{\text{int}} = N(U,\dot{U})$, corresponds to the case of the non-linear dynamics that will be approached.
$F^{\text{int}} = C\dot{U} + N(U,\dot{U})$, corresponds to the case of linear diffusion (thermal, Stokes, Darcy, etc.)
$F^{\text{int}} = C\dot{U} + N(\dot{U})$, corresponds to the Navier-Stokes equation (Penaly type incompressible formulation).

FORMULATION OF AN EXPLICIT NEWMARK ALGORITHM, PREDICTOR-CORRECTOR TYPE FOR THE NON-LINEAR CASE

Let the following non-linear differential equation system:

$$M\ddot{U} + N(U,\dot{U}) = F \tag{18}$$

Who should be satisfied whatever : $t_{n+1} = (n+1)\Delta t$
Let's put the following explicit form

$$M\ddot{U}_{n+1} + N(\tilde{U}_{n+1},\tilde{\dot{U}}_{n+1}) = F_{n+1} \tag{19}$$

If Dt tends to 0 the equation (19) tends to equation (18) and combined with the usual predictors and correctors of the Newmark algorithm, this form is explicit if the matrix M is diagonal; $(\tilde{U}_{n+1},\tilde{\dot{U}}_{n+1})$ are predictors. The following explicit algorithm results:

Table 1. Explicit algorithm of newmark for the non-lineair case

1-Initialization
2-For each Dt
2.1- Predictors calculation
$\tilde{U}_{n+1} = U_n + \Delta t\,\dot{U}_n + \frac{\Delta t^2}{2}(1-2\beta)\ddot{U}_n$
$\tilde{\dot{U}}_{n+1} = \dot{U}_n + \Delta t(1-\gamma)\ddot{U}_n$
2.2- Solve
$F^*_{n+1} = F_{n+1} - N\left(\tilde{U}_{n+1},\tilde{\dot{U}}_{n+1}\right)$
$\ddot{U}_{n+1} = \left(M^{-1}\right)\left(F^*_{n+1}\right)$
2.3Correcteors
$U_{n+1} = \tilde{U}_{n+1} + \Delta t^2\beta\,\ddot{U}_{n+1}$
$\dot{U}_{n+1} = \tilde{\dot{U}}_{n+1} + \Delta t\gamma\,\ddot{U}_{n+1}$
Next step
n=n+1 si $t<T_{max}$

Note
1. If, the mass matrix is diagonal no inversion of matrix is necessary
2. The algorithm is conditionally stable.

FORMULATION OF AN IMPLICIT NEWMARK ALGORITHM FOR THE NON-LINEAR CASE

We have to solve the equation of motion at time t n + 1:

$$M\ddot{U}_{n+1} + N(U_{n+1}, \dot{U}_{n+1}) = F_{n+1} \tag{20}$$

With

$$U_{n+1} = \tilde{U}_{n+1} + \Delta t^2 \beta \ddot{U}_{n+1} \tag{21}$$

$$\dot{U}_{n+1} = \tilde{\dot{U}}_{n+1} + \Delta t \gamma \ddot{U}_{n+1} \tag{22}$$

And

$$\tilde{U}_{n+1} = U_n + \Delta t\, \dot{U}_n + \frac{\Delta t^2}{2}(1-2\beta)\ddot{U}_n \tag{23}$$

$$\tilde{\dot{U}}_{n+1} = \dot{U}_n + \Delta t(1-\gamma)\ddot{U}_n \tag{24}$$

$$\Rightarrow \ddot{U}_{n+1} = \frac{1}{\Delta t^2 \beta} U_{n+1} - \frac{1}{\Delta t^2 \beta} \tilde{U}_{n+1} = \ddot{U}^d_{n+1} + \tilde{\ddot{U}}_{n+1} \tag{25}$$

By replacing equation (25) in equation (22) we find

$$\dot{U}_{n+1} = \frac{\gamma}{\Delta t \beta} U_{n+1} - \frac{\gamma}{\Delta t^2 \beta} \tilde{U}_{n+1} + \tilde{\dot{U}}_{n+1} = \dot{U}^d_{n+1} + \tilde{\tilde{\dot{U}}}_{n+1} \tag{26}$$

Both equations (21) and (22) show that

$\ddot{U}_{n+1}$ and $\dot{U}_{n+1}$ can be expressed based;
$\ddot{U}^d_{n+1}$ and $\dot{U}^d_{n+1}$ (function of Un+1 only) and of predictors;
$(\tilde{\ddot{U}}_{n+1})$ and $\tilde{\tilde{\dot{U}}}_{n+1}$ known at the time tn+1.

Replacing equations (25) and (26) in equation (20) yields an expression that is only a function of Un +1, is:

$$N^*(U_{n+1}) = F_{n+1}$$

With;

$$N^*(U_{n+1}) = M\ddot{U}_{n+1} + N(U_{n+1}, \dot{U}_{n+1})$$

Taylor's development, limited to the linear term then gives:

$$N^*(U_{n+1}^{i+1}) = N^*(U_{n+1}^{i}) + \left(\frac{dN^*}{dU}\right)^i \Delta U$$

And

$$\left(\frac{dN^*}{dU}\right)^i = \frac{M}{\Delta t^2 \beta} + \frac{\gamma}{\Delta t \beta} C_T^i + K_T^i = K_T^{*i}$$

Hence the system to solve

$$K_T^{*i}\, \Delta U = \Delta F$$

$$\Delta F = F_{n+1} - N(d_{n+1}^i)$$

$$d_{n+1}^{i+1} = d_{n+1}^i + \Delta U$$

FORMULATION OF AN ITERATIVE ALGORITHM FOR STATIC NONLINEAR CASE: NEWTON-RAPHSON METHOD

For simplicity, we first look at the static problem for a single degree of freedom system. Knowing that in the last chapter, we will deal in detail with the non-linear static case using a very elegant and fast graphic method, which is the pushover method. To solve the following scalar equation:

$$N(U) = F$$

In the vicinity of U_n, we have (by Taylor's development)

$$N(U_{n+1}) \cong N(U_n) + \frac{dN}{dU}\Delta U = N(U_n) + K_T \Delta U \tag{27}$$

Figure 6. Newton-Raphson step

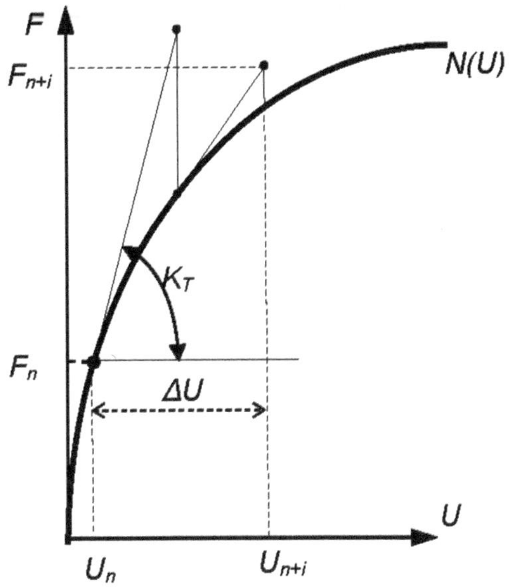

And in the vicinity of U_{n+1}

$$N\left(U_{n+1}^{\ i+1}\right) \cong N\left(U_n^{\ i}\right) + \frac{dN}{dU}\Delta U = N\left(U_n^{\ i}\right) + K_T \Delta U \tag{28}$$

From where

$$\begin{cases} K_T \Delta U = F_{n+1} - N\left(U_{n+1}^{\ i}\right) \qquad avec \quad \left(U_n = U_{n+1}^{\ 0}\right) \\ U_{n+1}^{\ i+1} = U_{n+1}^{\ i} + \Delta U \end{cases} \tag{29}$$

The algorithm is iterative we have convergence when:

$$\left\| F_{n+1} - N\left(U_{n+1}^{\ i+1}\right) \right\| < TOLERANCE$$

Note: Various algorithmic choices are possible:

1. Update KT at each iteration / Newton-Raphson algorithm
2. Opportunistic update: Modified Newton-Raphson Algorithm
3. $K_T = K_0$ (initial, "elastic") / algorithm with constant stiffness

This choice is irrelevant in the case of single degree of freedom above, but of course important in the case of multi degree of freedom where K is a matrix that must be inverted or not, depending on the case

Figure 7. Applied step load

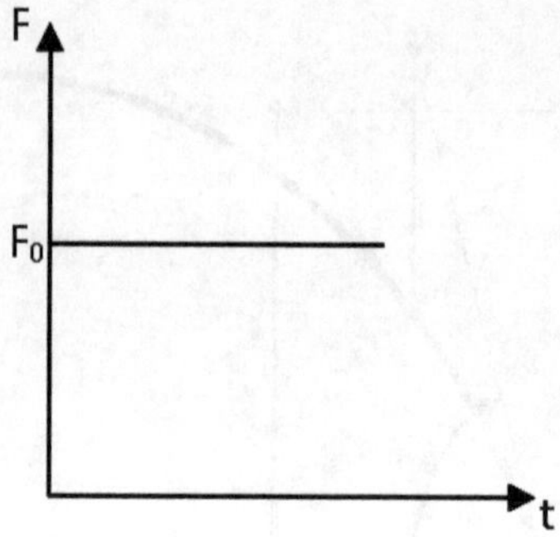

APPLICATIONS

Example 1 (Belgasmia & Moussaoui. 2015)

We consider a step load ($F = 0$ for $t < 0$ and $F = F_0$ for $t^3 0$) applied to SDOF nonlinear oscillator (figure 7). the oscillator, modeled as a bar, has the following characteristics: stiffness K =1 N/m (modulus of elasticity E =1 N/m², section A=1 m², length L = 1 m), mass M =1 kg, force F_0 = 1.

Nonliniarity is characterized by a yield stress $\sigma y_{ield=}$1.5N/m2· Elastic-perfectly plastic behavior is assumed.

Solution

The equation of motion is

$$M\ddot{u} + F(u) = F_0$$

There are there are 03 cases to consider in the response

case 1) $u \leq u_{yield}$

Since F_{yield}=1.5N, u_{yield}=1.5m. The time corresponding to yield can be computed by setting:

$$u_{yield} = \frac{\sigma_{yield}}{E} l = 1.5 = u_{st}(1 - \cos \omega t_{yield})$$

with $u_{si}=\omega=1$. We obtain $ty_{ield=}$arc cos(-0.5)=2.0944s

case 2) $u_{yield} \leq u \leq u_m$

After yielding the internal force is σy_{ieldA}, thus the equation of motion is

$$M\ddot{u} + F_{yield} = F_0 \Rightarrow M\ddot{u} = F_0 - F_{yield} = -0.5$$

$$M\ddot{u} + F_{yield} = F_0 \Rightarrow M\ddot{u} = F_0 - F_{yield} = -0.5$$

$$M\ddot{u} = F_0 - F_{yield} \Rightarrow M\dot{u} = (F_0 - F_{yield})t' + c_1$$

with $t' = t - t_{yield}$

$$Mu = (\frac{F_0 - F_{yield}}{2})t'^2 + c_1 t' + c_2$$

with $t' = t - t_{yield}$

The initial conditions are used to determine c1 and c2.

$$\begin{cases} u(t_{yield}) = u_{st}(1 - \cos\omega t) = 1.5 \\ \dot{u}(t_{yield}) = \omega\sin(\omega t_{yield}) = 0.866 \end{cases} \Rightarrow \quad c_1 = 0.866 \quad \text{and} \quad c_2 = 1.5$$

this case is true till $\dot{u} = 0$. Based on the previous equations:

$$\dot{u} = F_0 - F_{yield}(t - t_{yield}) + 0.866 = 0$$
$$\Rightarrow \quad t'_m = t - t_{yield} = 1.732$$
$$\Rightarrow \quad t_m = 3.8265s$$

thus

$$u_m = (\frac{F_0 - F_{yield}}{2})t_m'^{\,2} + c_1 t'_m + c_2 = -\frac{0.5}{2}(1.732)^2 + 0.866(1.732) + 1.5$$
$$u_m = 2.25m$$

case 3) t>3.8265 s

The displacement is

$$u = (u_m - \frac{F_{yield} - F_0}{K}) + \frac{F_{yield} - F_0}{K}\cos\omega t''$$

with $t'' = t - t_m$

$$u = (2.25 - 0.5) + 0.5\cos(t - 3.8265)$$

for $t = 7s \Rightarrow t'' = 3.1735s$ and u=1.25m

In this last case the amplitude of vibration will remain below u_m.

All values of the nonlinear three cases are checked in figure 8 and figure 9. We can see that the results are identical for both displacement- and force-based (flexibility based) see (Belgasmia. 2017) element and agree to the theory. A single element was used in both cases (by using Newmark algorithm γ=0.5, β=0.25)).

Example 2 (Commend & Zimmermann. 2001)

Let the structure below, loaded as shown.

1. Determine the motion equation and the Eigen frequencies?
2. Propose an explicit resolution algorithm and the stability condition?
3. Then propose an implicit algorithm in Da (take g = 0.5, b = 0.25)
4. Determine K_T and N(d) Give the appearance of the solution by comparing the explicit and implicit solutions.

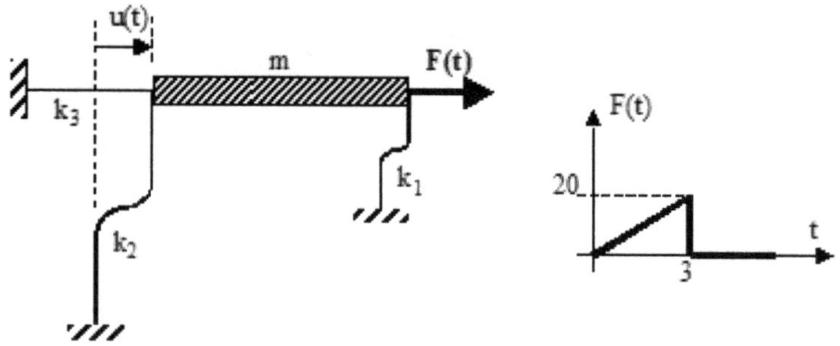

$k_1 = 12\ EI / h_1^3 = 32$N/m
$k_2 = 12\ EI / h_2^3 = 4$N/m
$k_3 = 12\ EA / L = 44$N/m
m = 20 kg

Solution

The motion equation is as follows:

$$m\ddot{U} + N(U) = F$$

With $N(U)=(k_1+k_2+k_3)U=80U$ if $U>0$

$N(U)=(k_1+k_2)U=36U$ if $U\leq 0$

Own frequencies are:

$$\omega_1 = \sqrt{\frac{k_1 + k_2}{m}} = 1.34\, rad/s,\ T_1 = 4.68s,$$

$$\omega_2 = \sqrt{\frac{k_1 + k_2 + k_3}{m}} = 2\, rad/s,\ T_2 = 3.14s$$

Algorithm of resolution used is that of Newmark explicit nonlinear, type predictors-correctors:

1. Calculation of predictors $\tilde{U}_{n+1}$ and $\tilde{\dot{U}}_{n+1}$
2. Solve:

$$\Delta F_{n+1} = F_{n+1} - N\left(\tilde{U}_{n+1}\right)$$

$$\ddot{U}_{n+1} = M^{-1} F^{*}_{n+1}$$

1. Calculation of correctors for U_{n+1} and $\dot{U}_{n+1}$
   Stability condition

$\omega m_{ax\Delta} t \leq \Omega = 2$, so $\Delta tc_{rit=}1$ for b = 0.

Implicit resolution algorithm in $\Delta\ddot{U}$

1. Calculation of predictors $\tilde{U}_{n+1}$ and $\tilde{\dot{U}}_{n+1}$
2. Solve:

$$\Delta F^{i}_{n+1} = F_{n+1} - N^{*}\left(U^{i}_{n+1}\right) = \Delta F^{i}$$

$$M^{*}_{T} = M + \Delta t \gamma C_T + \Delta t^2 \beta K_T$$

$$M^{*}_{T} \Delta \ddot{U} = \Delta F^{i}$$

3. Correctors:

$$\ddot{U}_{n+1}^{i+1} = \ddot{U}_{n+1}^{i} + \Delta\ddot{U}$$

$$\dot{U}_{n+1}^{i+1} = \tilde{\dot{U}}_{n+1} + \Delta t\gamma\ddot{U}_{n+1}^{i+1}$$

$$U_{n+1}^{i+1} = \tilde{U}_{n+1} + \Delta t^2\beta\ddot{U}_{n+1}^{i+1}$$

Note: the particular form of N (U) makes a single iteration sufficient.

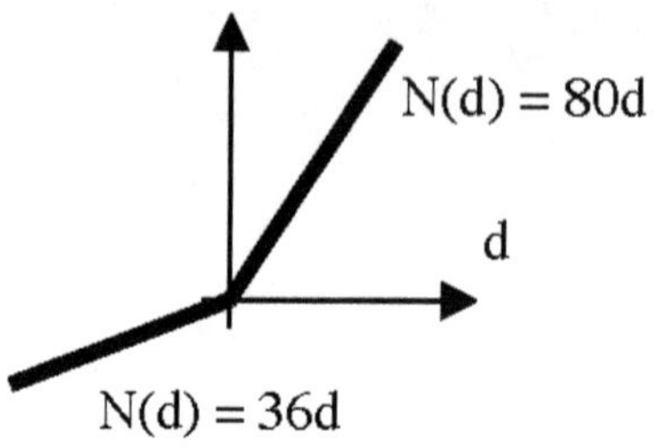

$$C_T = 0$$

$$K_T = \frac{\partial N(U)}{\partial U} = \begin{cases} 80 \text{ } if \text{ } U > 0 \\ 36 \text{ } if \text{ } U \leq 0 \end{cases}$$

Stability condition
For the implicit case, $\Delta t_{crit} = \infty$

Solution Comparison Obtained by the Two Algorithms with Different Dt:

Δt	Solution explicite, avec β = 0	Solution implicite, avec β = 0.25
0.10	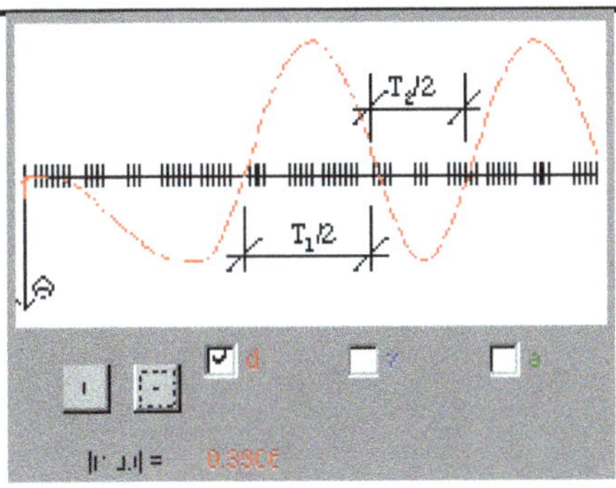	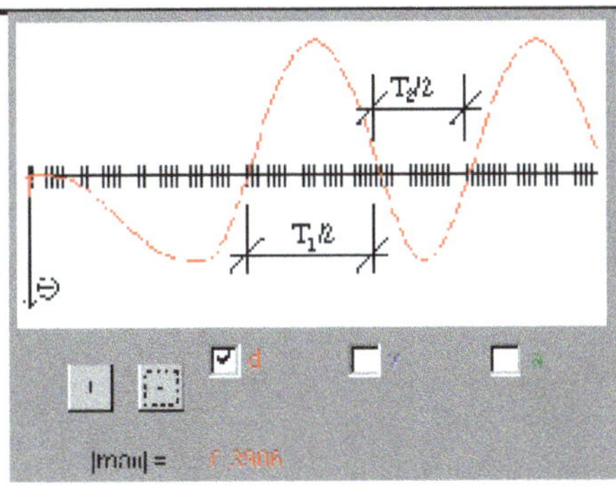
0.95	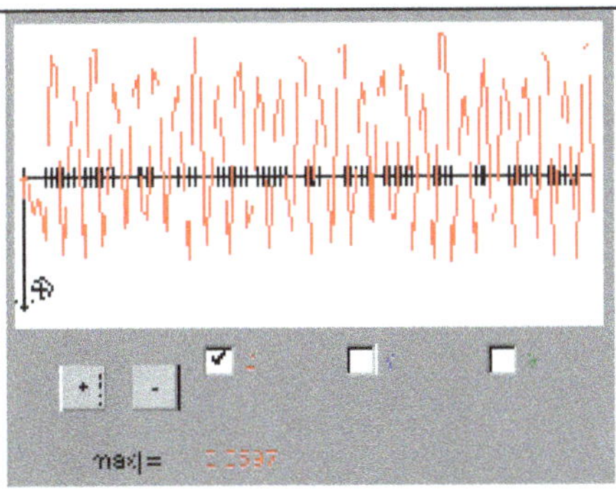	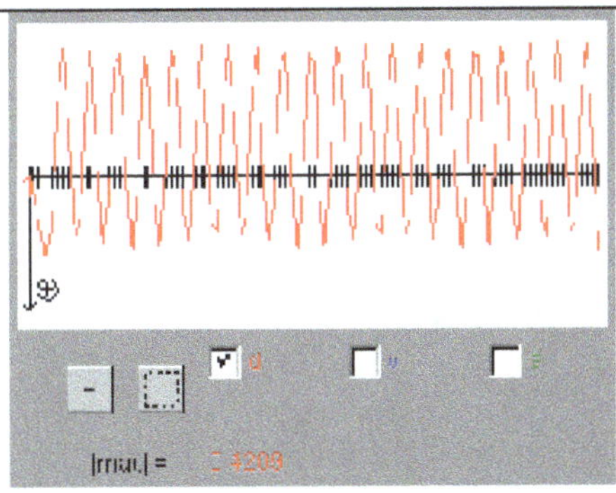
1.05	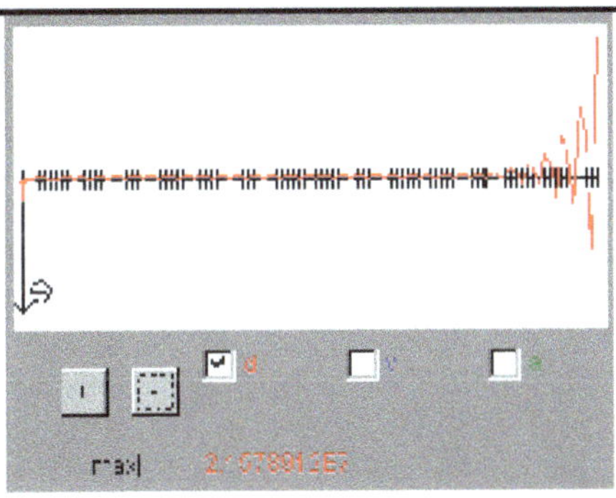 **unstable !!!**	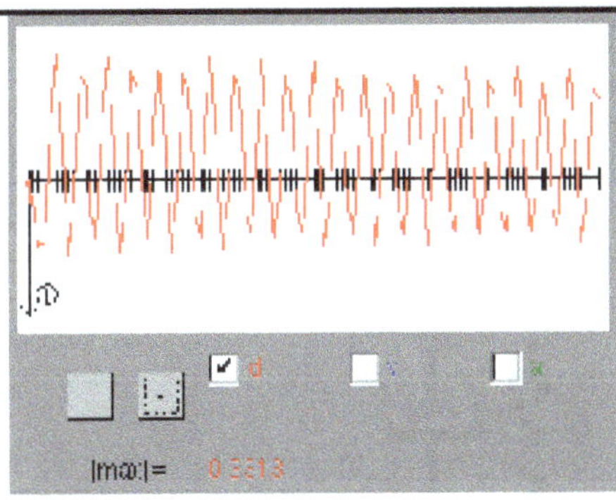

Note:

1. In this program d v and a represents the displacement the velocity and the acceleration respectively
2. After a transient phase during charging, there is a permanent oscillatory regime whose positive half-period is $T_2/2$ and the half-negative period $T_1/2$.

Example 3 (Commend & Zimmermann 2001)

We have the model below, we ask for:

1. Determine the equations of motion?
2. Propose a Resolution Algorithm?
3. Give the appearance of the solution for a) $k = 10^4$ *N/m and b)* $k = 10^6$ *N/m*

Solution

Motion equation is as follows:

$$m\ddot{U} + N(U,\dot{U}) = mg$$

with $N(U,\dot{U}) = c\dot{U} + kU$ is $U>0$

$N(U,\dot{U}) = 0$ is $U \leq 0$

$c = 2\xi m\omega$

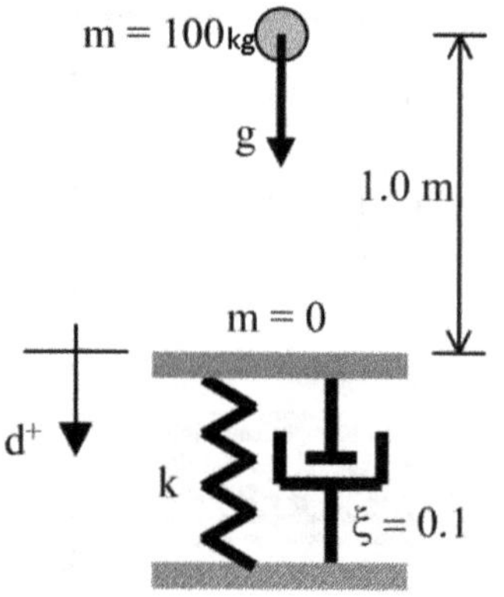

Explicit Algorithm (b = 0):

1. Calculation of predictors $\tilde{U}_{n+1}$ with $\tilde{\dot{U}}_{n+1}$
2. Solve:

$$m\ddot{U}_{n+1} = mg - N_1(\tilde{U}_{n+1}) - N_2(\tilde{U}_{n+1})$$

With $U_0 = -1$, $\dot{U}_0 = 0$.

3. Calculation of correctors for U_{n+1} and $\dot{U}_{n+1}$

Stability condition

$$\omega_{\max} = \sqrt{\frac{k}{m}},$$

$$\omega m_{ax\Delta} t \leq \Omega = 2$$

(for Newmark, b = 0):

case a) $\Delta t_{crit}=0.2$,
case b) $\Delta t_{crit}=0.02$

k = 1e4, Δt = 0.01, n = 200	**k = 1e6**, Δt = 0.01, n = 200
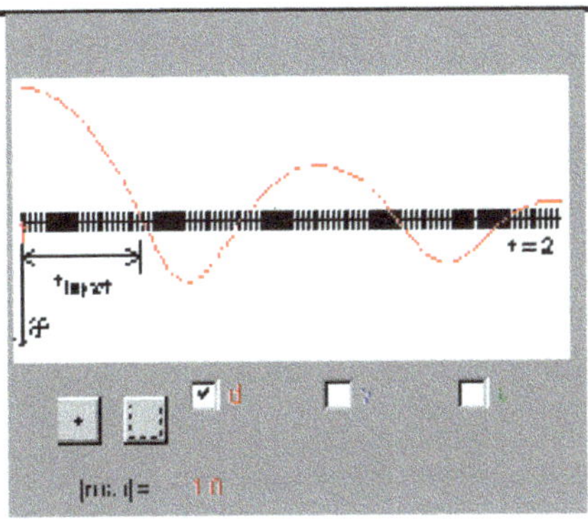	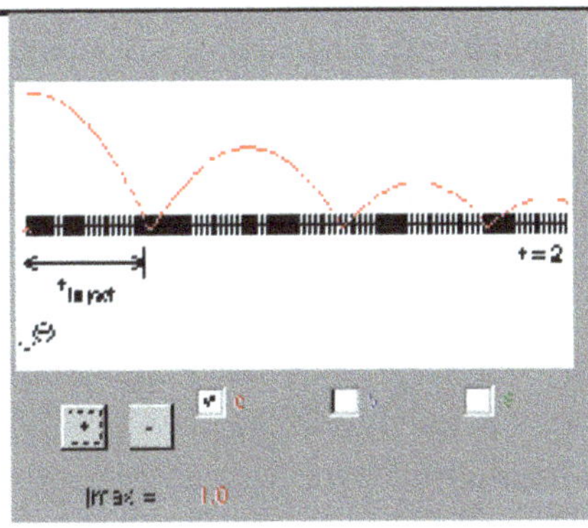

Remarks:

1. The rebound is even faster than the spring is rigid
2. The plate being without mass, it follows the ball only if the spring is compressed
3. t_{impact} is such that: $g\frac{t^2}{2} = -u_0$, so $t_{impact} = \sqrt{\frac{2}{g}} \cong 0.45$

REFERENCES

Belgasmia, M., & Moussaoui, S. (2015). Comparison Between Static Nonlinear and Time History Analysis using Flexibility-Based Model for an Existing Structure and Effect of Taking into Account Soil Using Domain Reduction Method for a Single Media. *KSCE Journal of Civil Engineering*, *19*(3), 651–663. doi:10.100712205-015-0351-y

Belgasmia, M. (2017). Definition of Static Nonlinear Procedure and Flexibility-Based model with Application on 2D Model for an Existing Structure and Comparing Results with Time History Analysis. In *Modeling and Simulation Techniques in Structural Engineering*. IGI Global.

Belytschko, T., Kam, W. L., Moran, B., & Elkhodary, K. L. (2014). Nonlinear Finite Elements For Continua and structutres. John Wiley & Sons, Ltd.

Commend, S., & Zimmermann, T. (2001). *LSCDYN 1DOF and nDOF: two applications for computer-assisted teaching of structural dynamics. LSC Internal Report.* Swiss Federal Institute of Technology.

Frey, F., & Rebora, B. (n.d.). *Analyse nonlineaire des structures LSC Internal Report.* Swiss Federal Institute of Technology.

James, F. D. (2001). *Nonlinear analysis of thin-walled structure.* Springer.

Lestuzzi, P. (2007). *Génie parasismique semestre d'été.* Swiss Federal Institute of Technology.

KEY TERMS AND DEFINITIONS

Explicit Algorithm: When a direct computation of the dependent variables can be made in terms of known quantities, the computation is said to be explicit.

Implicit Algorithm: When the dependent variables are defined by coupled sets of equations, and either a matrix or iterative technique is needed to obtain the solution, the numerical method is said to be implicit.

Iterative Algorithm: An iterative algorithm is a mathematical procedure that uses an initial guess to generate a sequence of improving approximate solutions for a class of problems, in which the n-th approximation is derived from the previous ones.

Large Deformation: Deals with deformations in which both rotations and strains are arbitrarily large. In this case, the undeformed and deformed configurations of the continuum are significantly different and a clear distinction has to be made between them.

Piola-Kirchhoff Stresses: Is a stress measure.

Predictor-Corrector: The idea behind the predictor-corrector methods is to use a suitable combination of an explicit and an implicit technique to obtain a method with better convergence characteristics.

Chapter 11
Nonlinear Static Analysis Method

ABSTRACT

This chapter presents the nonlinear static methods of analyses for seismic design of structures considered by Eurocode 8. The first method is the nonlinear pushover procedure, which is based on the N2 method. The second method is the classical nonlinear time history analysis. The first method is studied in more detail, because the second method is a well-established procedure whose only drawback is the time necessary for the analyses. Nonlinear solvers and procedure in program Z_Soil are described. After a simple nonlinear SDOF application, a test-bed application consisting of an existing two-story reinforced concrete building in Bonefro, Italy is used to compare the two nonlinear procedures. The selected building is representative of typical residential building construction in Italy in the 1970s and 1980s. The aim of this chapter section is to compare 2D and 3D procedures implemented in Z_Soil software. The second example is a 14-story reinforced concrete building designed according to the Algerian code using Sap2000 software.

INTRODUCTION

This chapter allows enlightening the reader, on an elegant method by its graphic power and the speed of diagnosing the existing or newly constructed structure called non-linear static method. By detailing all different steps, which allows the correct accomplishment of this method; and compare these results to those of time history analysis on a real structure example.

Modern seismic design codes allow engineers to use either linear or nonlinear analyses to compute design forces and design displacements. In particular, Eurocode 8 contains four methods of analysis: linear simplified static analysis, linear modal analysis, nonlinear pushover analysis and nonlinear time-history analysis, which can be summarized in the following table. These methods refer to the design and analysis of framed structures, mainly buildings and bridges. The two nonlinear methods require advanced models and advanced nonlinear procedures in order to be fully applicable by design engineers. This

DOI: 10.4018/978-1-7998-4399-3.ch011

Copyright © 2021, IGI Global. Copying or distributing in print or electronic forms without written permission of IGI Global is prohibited.

chapter gives an overview of the steps followed to implement state-of-the-art nonlinear models in Z_Soil and presents a test-bed application to a reinforced concrete building.

Displacement-based and force-base elements are used in this study. The first is a classical two-node, displacement-based, Euler-Bernoulli frame element. The second is a two-node, force-based, Euler Bernoulli frame element. The main advantage of the second element is that it is "exact" within the relevant frame element theory. This implies that one element per frame member (beam or column) is used in preparing the frame mesh, thus leading to a reduction of the global number of degrees of freedom. The complete theory for the force-based element can be found in (Aydinoglu. 2004).

The nonlinear response of 2D and 3D models of an existing building are presented as a first full-size application. The building is a residential two-storey reinforced concrete building in Bonefro, Italy(Belgasmia 2006), using Z_Soil software(Zimmermann 2006). It is representative of typical residential building construction in Italy in the 1970's and 1980's. The design spectrum for the building was obtained from Eurocode 8 (2003) using the local soil properties and the peak ground acceleration given by the new Italian seismic map.

Table 1. Summary of four methods analysis

Action Structure	*Static*	*Dynamic*
Elastic	*Remplacement Force*	*Response Spectrum*
Nonlinear	*Static nonlinear*	*Nonlinear Time- history*

ON-LINEAR FRAME ANALYSIS METHODS IN EUROCODE 8

In this section, the nonlinear pushover procedures given by Eurocode 8 are presented. The first procedure is the Nonlinear Pushover Method of Analysis; the second is the Nonlinear Time History Method of Analysis.

The nonlinear static analysis (or pushover analysis) consists in applying to the building the gravitational loads (which remain fixed). Then a system of horizontal forces that keeping unchanged the relative relationships between the same forces. This forces are all scaled to make the horizontal movement of a control point of the structure (for example a point at the top of the building) grow monotonic, until reaching the limit state of interest limit (usually the ultimate limit state, in some cases the collapse limit state). There are several pushover methodologies, some already present in today's seismic regulations, others under study and development. These methodologies are similar in the general philosophy of the method, but different in details and verifications. In this chapter, we will refer to the pushover methods of Eurocode 8 (2003).

According to EC8, pushover analysis may be used to verify the structural performance of newly designed buildings and of existing buildings. In particular, pushover analysis may be used for the following purposes:

1. To verify or revise the over-strength ratio values $\alpha u/\alpha 1$. The definition of the over-strength ratio is recalled with the aid of figure 1. If the structure is pushed with a lateral load distribution of con-

stant shape and increasing intensity, $\alpha u/\alpha_1$ is the ratio between the base shear αuVb c$_{o}$r$_{r}$esponding to the formation of a mechanism and the base shear αuVb co$_{r}$r$_{e}$sponding to the formation of the first plastic hinge. The over-strength ratio is used in linear analysis to compute the behavior factor q which allows to obtain nonlinear design spectrum for elastic analysis starting from the linear design spectrum

Figure 1. Over-strength ratio $\alpha u/\alpha 1$

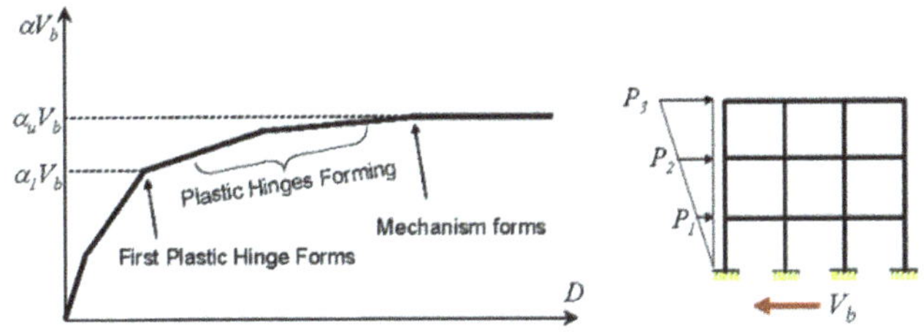

2. To estimate the expected plastic mechanisms and the damage distribution; more precisely to Verify the actual distribution of the inelastic demand of the buildings, in particular to estimate the sequence of plastic hinges, the request for local ductility and the distribution of the damage;
3. To assess the structural performance of existing or retrofitted buildings for the purposes of EC8 Part 3. As a design method for buildings of new construction replacing linear analysis methods. In this case, the so-called target displacement (defined below) becomes the basis of the structure design. This is the basis of displacement-based design;
4. As an alternative to the design based on linear-elastic analysis which uses the behavior factor q. In this case, the target displacement found from the pushover analysis should be used as the basis for the design, for assessing the capacity of existing buildings before and after seismic improvement or adaptation.

The pushover analysis is schematically illustrated in figure2. On the response curve (or capacity curve), the achievement of three limit states is indicated. The ratio $\alpha u/\alpha 1$ $_{o}$f the previous point represents the ratio between the shear force at the base to the ultimate limit state and formation time of the first plastic hinge (which is similar to that at the damage limit state).

GENERAL PROCEDURES FOR PUSHOVER METHODS

In general, the pushover method involves the following phases:

1. Construction of structure nonlinear model.
2. Application of constant gravitational loads.
3. Application of the horizontal pushover forces.
4. Checks at the limit state we are interested in.

Figure 2. Schematic illustration of the pushover method

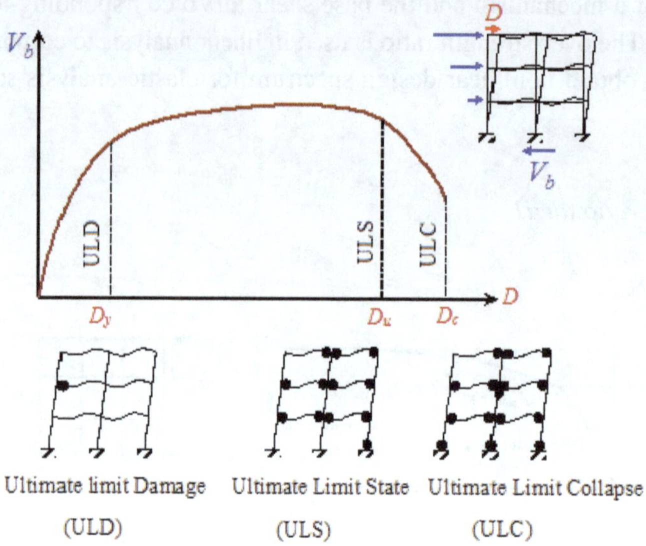

The pushover analysis consists in applying the constant gravitational load to the non-linear model of the structure under examination, followed by a distribution of horizontal forces that is gradually increased. During this monotonous load path, the displacement of a control node and the total base shear are monitored and represented in the so-called Capacity Curve. The seismic request on a given building is obtained by calculating from the design spectrum the so-called target displacement, the maximum displacement expected for the reference limit state. This shift is identified along the Capacity Curve. The forces, displacements and deformations read on the nonlinear model at the target displacement are compared with the respective capacities of the structure.

DETAILS ON THE PUSHOVER PROCEDURE AND CHECKS

The Pushover procedure according to Eurocode 8 (2003) is divided into the following steps:

1. Construction of the nonlinear model of the structure;
2. Application of gravitational loads;
3. Application of horizontal actions (pushover analysis) to obtain the Capacity Curve;
4. Reduction of the Capacity Curve to a system equivalent to 1GDL;
5. Bi-linearization of the Capacity Curve;
6. Evaluation of the maximum displacement for the limit state we are interested in (target displacement);
7. Conversion into maximum displacement of the control point;
8. Verifications at the limit state.

NON-LINEAR MODEL OF THE STRUCTURE

The non-linear model of the structure can be constructed with three-dimensional, two-dimensional or frame finite elements, or with mixed modeling, typically with frame-like elements for framework building and with bi-or three-dimensional finite elements for other elements (such as floors, the wall beams or other elements not easily assimilable to rigid elements or beam elements). We refer here for simplicity to a model with only frame type elements but the approach is general.

An important choice to make in the structure-modeling phase is that between three-dimensional or two-dimensional model. The Eurocode 8 (2003) in this regard sets out the following. In the case of a regular structure in plan, two orthogonal plane models can be used, loaded and verified separately (Figure 3). In the general case of an irregular structure in plan, a three-dimensional model of the structure must be considered, performing two separate analyzes with the earthquake acting in the two main directions (Figure 4). Eurocode 8 (2003) provides details on the definition of regularity in plan.

Figure 3. Load diagram for regular building in plan

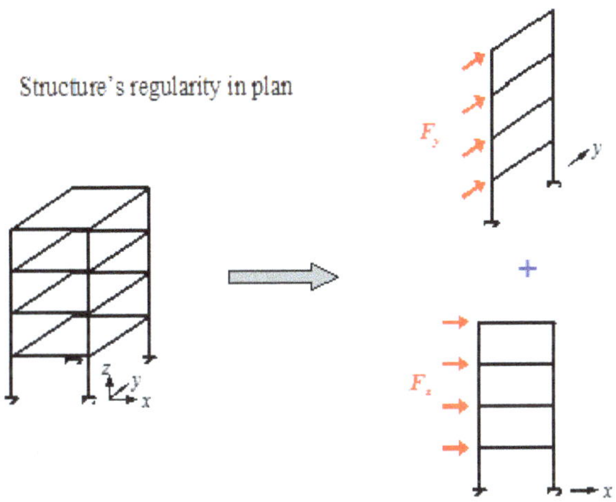

EC8 adds that buildings not conforming to the regularity criteria of Eurocode 8 (2003) shall be analyzed using a spatial (3D) structural model. Two independent analyses with lateral loads applied in one direction only may be performed. No indications are given in Eurocode 8 (2003), or in the published literature on how to perform a pushover analysis with two loads distributions applied simultaneously in two orthogonal directions. Therefore, it is assumed here that in a pushover analysis the structure is pushed with loads applied in one horizontal direction at a time (the vertical seismic is typically neglected in buildings). Indications are given on how to combine the effects of the actions applied separately in two horizontal directions. For buildings conforming to the regularity criteria of Eurocode 8 (2003) the analysis may be performed using two planar models, one for each principal direction horizontal direction.

Nonlinear time-history analysis is performed through direct numerical integration of the equations of motion of the structure. Accelerograms represent the design ground motion.

Figure 4. Load diagram for irregular building in plan

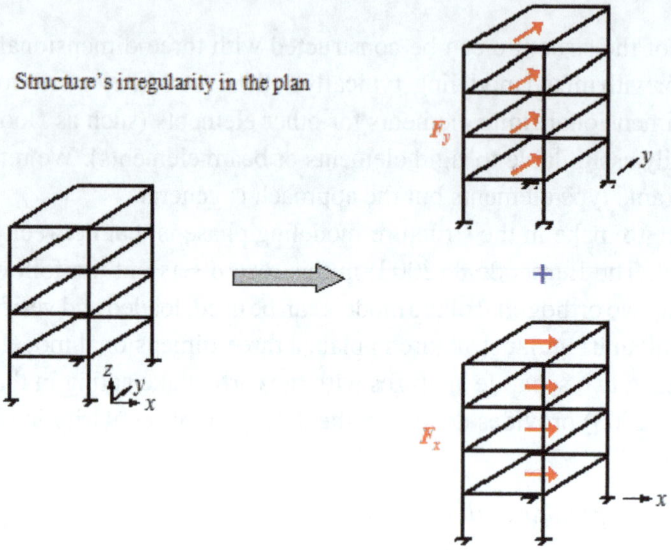

The structural element models and the resulting structural model of the overall building is very similar for pushover and nonlinear time-history analysis. The only difference lies in the need to have cyclic models for the time-history analysis.

GRAVITATIONAL LOADS APPLICATION

The structure is loaded first with the vertical gravitational loads foreseen by the norms for the seismic load condition. The final state resulting from this analysis is used as the initial state for the pushover analysis, performing the two analyzes in series. It is emphasized that the application of gravitational loads in a nonlinear model, has important effects, on the response to horizontal forces; and how these effects depend mainly on the type of nonlinearity used. In particular, if the model used involves the M N interaction between bending behavior and normal stress response (as is the case with fiber models) the interaction is automatic and the response of the columns varies according to the level of axial effort due both to gravitational loads and to pushover forces. If, on the other hand, the non-linear model provides independent constitutive laws for bending and normal effort, the application of vertical loads has a reduced effect on the response of the frame, with obvious limits on the validity of the results.

The initial steps of both nonlinear procedures are identical: construction of the nonlinear frame model and application of the gravity loads. The gravity loads remain constant during the nonlinear analysis (both static and dynamic). The application of the gravity loads is schematically shown in figure 5. The value of the constant gravity loads $G_k + P_k + \sum_i (\psi_{2i} Q_{ki})$ is given by Eurocode 8 (2003). This initial step is quite important because it may change the initial state of the structure. In a reinforced concrete building, for example, the gravity loads typically induce cracking in beams and apply high axial forces on columns.

Figure 5. Application of constant gravity loads

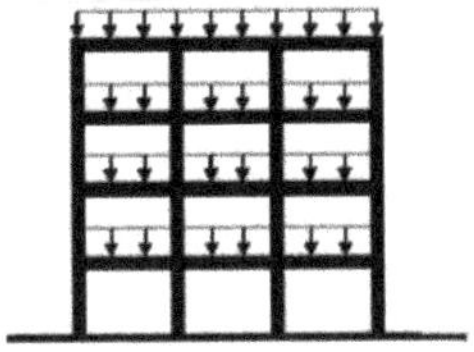

Finally, Eurocode 8 (2003) states that the seismic action in nonlinear methods shall be applied in both positive and negative directions (depending on the symmetry of the structure).

HORIZONTAL ACTIONS (PUSHOVER) AND CAPACITY CURVE

A form of load is applied to the structure provided by the reference legislation. The Eurocode 8 (2003) require that at least two distinct distributions of horizontal forces be considered, applying them to the mass centers of the masses on each floor:

1. A uniform distribution of forces proportional to the masses of the planes;
2. A modal distribution of forces proportional to the shape of the first oscillation mode.

NONLINEAR STATIC PUSHOVER PROCEDURE IN EUROCODE 8

The Nonlinear Static Pushover Procedure in Eurocode 8 (2003) follows the N2 method developed by Fajfar (1999) & Fajfar (2002). The method consists of applying constant load shapes to the building model. The load shapes represent the lateral loads applied by the ground motion. The load intensity is increased in a pseudo-static manner. The structure model can be planar (2D) or spatial (3D), depending on the regularity characteristics of the building. The load pattern, on the other hand, is always applied in one direction only. For analyses with input ground motion in more than one direction, for example input ground motion in the x and y directions, combination rules are given by Eurocode 8 (2003).

The nonlinear pushover analysis consists of applying monotonically increasing constant shape lateral load distributions to the structure under consideration. The structure model can be either 2D or 3D. In particular, Eurocode 8 (2003) states that for buildings with plan regularity, 2D analysis of single plane frames can be performed, while for buildings with plan irregularity a complete 3D model is necessary. Given that the nonlinear methods are particularly interesting for existing buildings, which are rarely regular, a 3D model is required in most cases.

The nonlinear pushover procedure in Eurocode 8 (2003) follows the N2 method developed by professor Fajfar, of the University of Ljubljana, Slovenia (Fajfar (1999) & Fajfar (2002) Fajfar, Kilar, Marusic & Perus(2005)). The N2 method was developed using a shear-building model, i.e. a frame model with floors rigid in their planes. Furthermore, vertical displacement are typically neglected in the method and only the two horizontal ground motion components, x and y, are considered. Extension to the general case of a fully deformable frame is straightforward.

The N2 method consists of applying two load distributions to the frame:

1. a "modal" (or "triangular") pattern, that is a load shape proportional to the mass matrix multiplied by the first elastic mode shape, $P^1 = M\phi 1_{;}$
2. a "uniform" pattern, that is a mass proportional load shape, $P^2 = MR$.

Where M is the mass matrix, f_1 is the first mode shape and R is a vector of 1s corresponding to the degrees of freedom parallel to the application of the ground motion and 0 for all other dofs. In the N2 method, f_1 is normalized so that the top floor displacement is 1, i.e. $f_{i,1}=1$. The two load distributions are schematically shown in figure 6. The applied lateral load distributions are increased and the response is plotted in terms of base shear V_b vs. top floor displacement D (for example center of mass of the top floor). This is the so-called pushover curve or capacity curve (also shown schematically in figure 6).

Figure 6. Load distribution for pushover analysis according to EC8 and pushover response curves

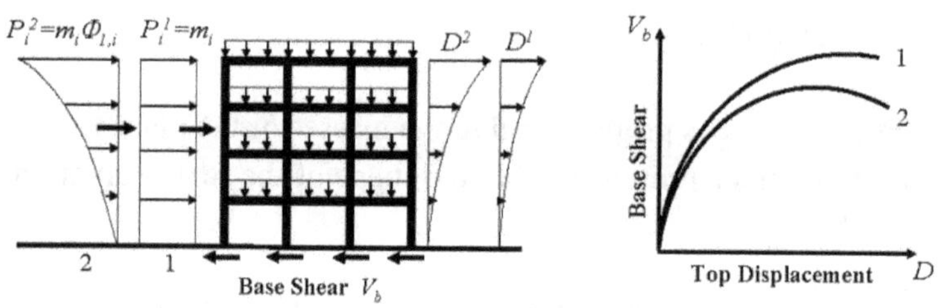

The N2 procedure transforms the response of the MDOF system into the response of an equivalent SDOF system. This is necessary in order to compare the building capacity curve of Figure 6 with the demand, expressed in the design codes by the design spectra, which refer to SDOF systems. The transformation MDOF – SDOF, as developed by Fajfar (1999) & Fajfar (2002), is treated in more detail in the following section.

In general, pushover analysis must be applied both in the positive and in the negative load direction, as the sections are not generally symmetrical. This leads to a minimum of four Capacity Curves, as illustrated in Figure 7.

EQUIVALENT SDOF MODEL AND CAPACITY DIAGRAM IN EUROCODE 8

As previously stated, Fajfar (1999) & Fajfar (2002) assumes that the building is a shear frame, i.e. the floors are rigid in their own plane. If the vertical displacements of the building are neglected, the floor displacements are given by the three degrees of freedom shown in figure 8. The degrees of freedom are typically taken at the center of mass. Note that the beams can deform outside the floor plane that is the nodes have out-of-floor plane rotational degrees of freedom.

Figure 7. General scheme of the pushover analysis on flat frame and Capacity Curves

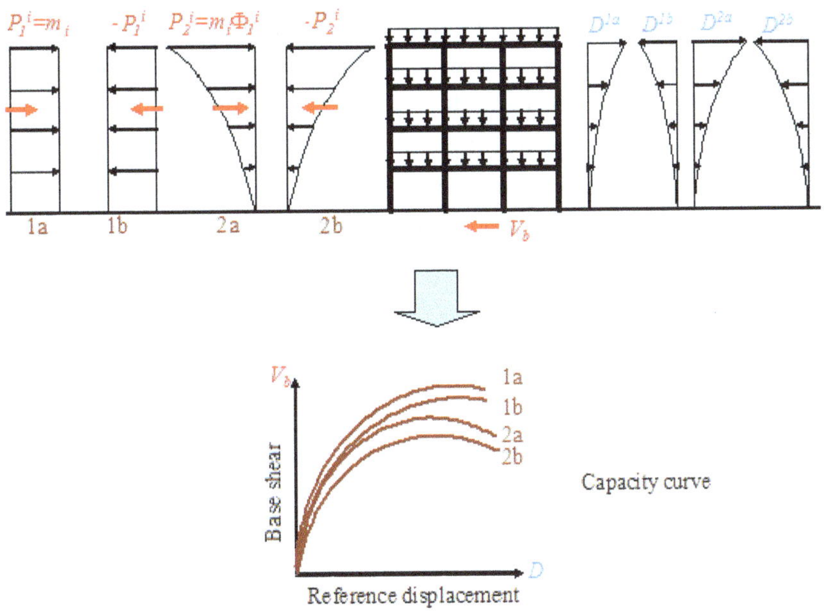

Figure 8. Rigid slab degrees of freedom in 3D shear frame (for simplicity, column axial deformability is neglected)

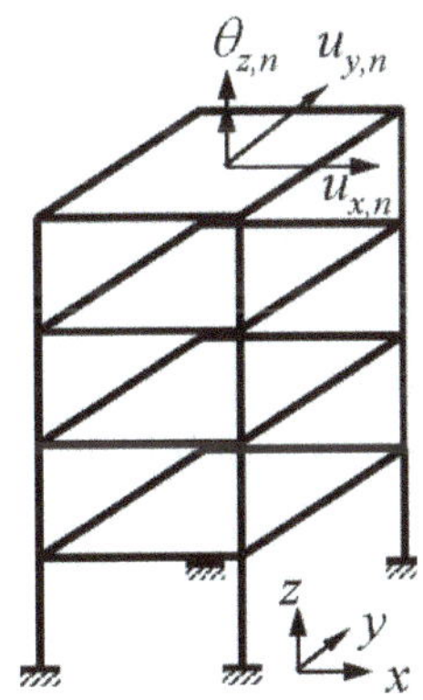

The theoretical derivation of the transformation procedure is as follows (see Belgasmia. 2017). The equations of motion of a MDOF building subjected to base ground motion model is:

$$M\ddot{U} + F(U) = -M\,R\,a \tag{1}$$

Where damping is neglected, M is the mass matrix (assumed diagonal in the original derivation of the N2 method), U and F are vectors representing displacements and internal forces, respectively, R is the influence vector and a is the ground acceleration as function of time, i.e. a = a(t). a is given in one direction only. In the linear elastic case F=KU (where K is the structure stiffness matrix), in the nonlinear case F depends on the displacement history. For uni-directional ground motion, for example in the direction x, the influence vector R consists of 1 in correspondence to the degree of freedom (dofs) in the x direction, and 0 for all other dofs. For example, for the frame of figure 8, considering only the 12 dofs of the 4 stories, $\mathbf{R}^T = \{\mathbf{1} \quad \mathbf{1} \quad \mathbf{1} \quad \mathbf{1}\}$, with $\mathbf{1} = \{1 \quad 0 \quad 0\}$.

The first assumption (and approximation) of the N2 method is that, the displacement U has a constant shape that does not change during the response to the ground motion:

$$U = \phi \mathrm{D}t \tag{2}$$

Or $U(x,t)=\phi(\mathrm{x})\mathrm{D}t_{(}\mathrm{t})$, where D$t$ is the intensity at the pseudo-time t of the displacement shape f, where x indicates that the displacement shape depends on the degree of freedom location. For convenience, f is normalized in such a way that the top-storey displacement is equal to 1, i.e. fn=1. This way Dt gives the top floor displacement at time t.

In the pushover analysis of frames with rigid floors, lateral loads are applied at the center of mass of each storey. The vector of the lateral loads P is

$$P = p\psi = p\,M\,\varphi \tag{3}$$

The magnitude of the lateral load is p, i.e. p = p(t). The distribution of lateral loads is related to the assumed displacement shape f. This is the second assumption of the procedure.

In more complex models; with deformable slabs and with distributed masses at each node, the load vector **P** is applied to all degrees of freedoms with mass in the direction of the applied ground motion.

Note that the displacement shape f is needed only for the transformation from the MDOF (Multi Degree Of Freedom) system to the equivalent SDOF (Single Degree Of Freedom) system of the nonlinear pushover procedure. In the general case of a 3D building, f has nonzero components in the six dofs of each node.

From equation (3), it follows that in a shear frame the lateral force in the i-th storey is proportional to the component f_i of the assumed displacement f, weighted by the storey mass m_i. If the ground motion is applied in the x direction

$$P_i = pm_i\varphi_{x,i} \tag{4}$$

From statics, it follows that

$$P = F \tag{5}$$

That is the internal forces F are equal to the pseudo-static external forces P.

By introducing Equations (2), (3) and (5) into Equation (1), and by pre-multiplying by f', we obtain

$$\varphi^t\,M\,\varphi\,\ddot{D}_t + \varphi^t\,M\,\varphi\;p = \varphi^t\,M\,Ra \tag{6}$$

The left term of Equation (6) is then divided and multiplied by Φ^T **M R** to obtain

$$\underbrace{\varphi^t M R}_{m^*}\underbrace{\frac{\varphi^t M \varphi}{\varphi^t M R}}_{\frac{1}{\Gamma}}\ddot{D}_t + \underbrace{\frac{\varphi^t M \varphi}{\varphi^t M R}}_{\frac{1}{\Gamma}}\underbrace{\varphi^t M R}_{V_d}\, p = \underbrace{\varphi^t M R}_{m^*} a \tag{7}$$

Where $m*$ is the mass of the SDOF (Single Degree Of Freedom) equivalent to the MDOF building

$$m^* = \mathbf{\Phi}^{\mathrm{T}}\,\mathbf{M}\,\mathbf{R} \tag{8}$$

For a shear, building and ground motion applied in the x direction:

$$m^* = \sum m_i \Phi_{x,i} \tag{9}$$

The constant G controls the transformation from MDOF to SDOF and back:

$$\Gamma = \frac{\mathbf{\Phi}^T\,\mathbf{M}\,\mathbf{R}}{\mathbf{\Phi}^T\,\mathbf{M}\,\mathbf{\Phi}} \tag{10}$$

For a shear building and ground motion in the x direction:

$$\Gamma = \frac{\sum m_i \Phi_{x,i}}{\sum m_i \Phi_{x,i}^2} \tag{11}$$

G is a factor that, for Φ equal to the mode shape of one the building modes corresponds to the mode participation factor. In the development of the N2 method, Φ can be any reasonable deformed shape.

V_b is the base shear of the MODF building in the direction of the ground motion. From Equation (3), it follows that:

$$V = \mathbf{\Phi}^T\,\mathbf{M}\,\mathbf{R}\;\; p \tag{12}$$

For a shear, building and ground motion applied in the x direction:

$$V_x = p\sum m_i \Phi_{x,i} = \sum P_{x,i} \tag{13}$$

Equation (7) thus becomes the equation of the SDOF equivalent to the MODF building

$$m^*\ddot{D}^* + F^* = m^* a \tag{14}$$

Figure 9. Capacity curve: transformation from response of MDOF to equivalent SDOF

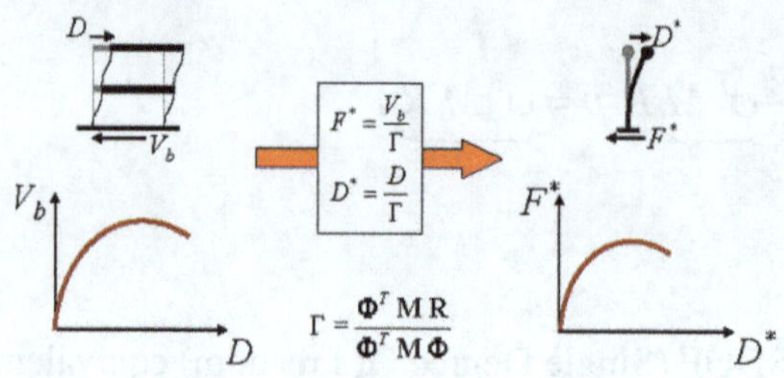

Where:

$$D^* = \frac{D_t}{\Gamma} \qquad F^* = \frac{V_b}{\Gamma} \tag{15}$$

The above derivation allows the transformation of the MDOF pushover capacity curves of figure 8. Into pushover curves for the equivalent SDOF system, as shown in figure 9.

In order to use the capacity curves as a representation of the nonlinear response of the structure, together with a spectrum as a representation of the seismic action on the structure. Some reduction steps are necessary, since the spectra represent the response of SDOF systems, while the capacity curve $V_{b\text{-}d}$ (Base shear - Control point displacement) is related to a MDOF (Mutli Degree Of Freedom) structure. This reduction takes place according to the scheme of figure 9. The displacement of the control point d and the cut at the base V_b are transformed into the corresponding quantities of the system equivalent to SDOF.

Both force and displacement axes are scaled by the same factor G. The stiffness of the system remains the same. Note that the transformation factor G depends on the shape of the assumed displacement shape Φ, and is thus different for different choices of Φ. As shown in figure6, two forms of Φ are suggested in EC8:

1. $F = F_1$, thus $\Gamma = \dfrac{\varphi_1^T MR}{\varphi_1^T M \varphi_1}$
2. $F = R$, thus $\Gamma = 1$.

LINEARIZATION OF THE CAPACITY CURVE AND COMPARISON TO DEMAND SPECTRUM

In order to compare the capacity curve to the demand curve given by the design spectrum, the nonlinear pushover curves of the SDOF are approximated by elastic-perfectly plastic (or bilinear) curves. According to Annex B if the draft EC8 [1] this transformation can be based on the equal energy principle. A target displacement D_m^* is assumed, and equal energy is assumed between bilinear and nonlinear pushover curves. This simple procedure is illustrated in figure10.

Figure 10. Bilinearization of the capacity curve of SDOF

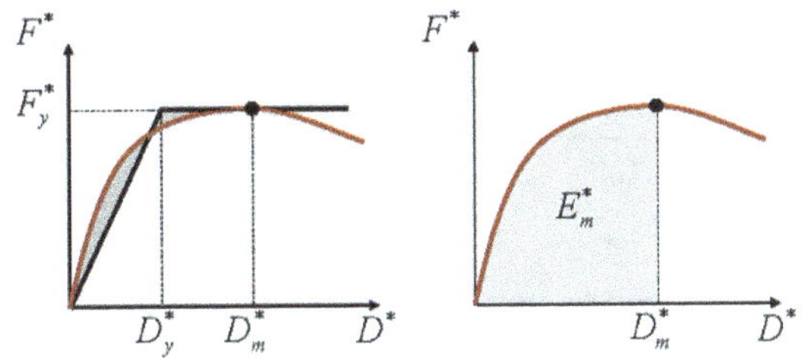

The bilinearization of Figure 10 gives the yield force and the yield displacement

$$D_y^* = 2\left(D_m^* - \frac{E_m^*}{F_y^*}\right),$$

which allow the initial elastic period to be computed as:

$$T^* = 2\pi\sqrt{\frac{m^* D_y^*}{F_y^*}} \tag{16}$$

Secondly, the capacity curve is transformed into capacity spectrum by normalizing the force F^* with respect to the SDOF weight m^*g. The resulting capacity spectrum s is shown in figure11.

Figure 11. SDOF capacity spectrum

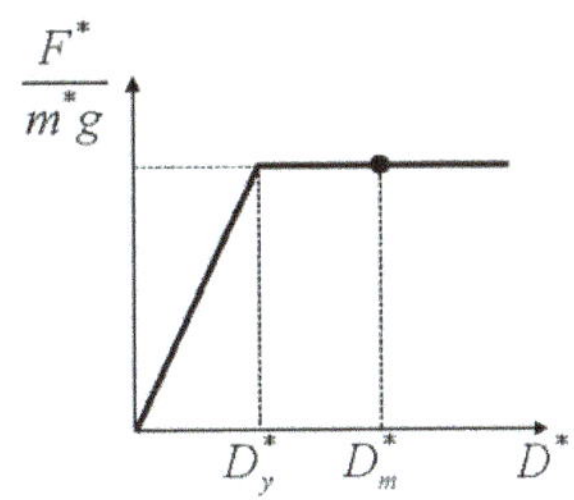

The demand on the building is given by the design spectrum provided by the design codes. In order to compare capacity and demand, the first step is to transform the format of the design spectrum from the classical Acceleration[1] A vs Period T format to the ADRS format, i.e. Acceleration A vs. Displacement D. The procedure is rather simple, as Acceleration and Displacement are related by:

$$S_D = \left(\frac{T}{2\pi}\right)^2 S_A \qquad (17)$$

The transformation to the ADRS spectrum is shown in figure 12. Lines from the origin represent constant period.

Figure 12. Transformation to ADRS linear spectrum

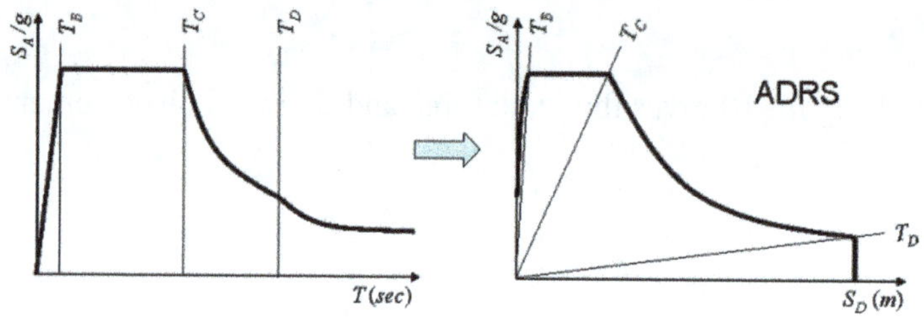

The capacity spectrum of figure 11 is now compared to the ADRS demand spectrum of figure 12. The comparison is not immediate, because the capacity spectrum is nonlinear, while the ADRS spectrum given by the design codes is linear.

For a SDOF system with a bilinear plastic behavior, the acceleration spectrum S_A and the displacement spectrum S_D can be determined as (Vidic, Fajfar & Fischinger 1994).

$$S_A = \frac{S_{Ae}}{R_\mu} \qquad (18)$$

$$S_D = \frac{\mu}{R_\mu} S_{De} = \frac{\mu}{R_\mu} \frac{T^2}{4\pi^2} S_{Ae} = \mu \frac{T^2}{4\pi^2} S_{Ae} \qquad (19)$$

Where subscript *e* indicates *elastic*, m is the ductility factor = maximum inelastic displacement/yield displacement and R_m is the reduction factor due to ductility[2]. The reduction factor R_m can be found in different ways, some analytical, other approximated. In the simple version of the N2 method, the following approximated expressions are given:

$$R_\mu = (\mu - 1)\frac{T}{T_C} + 1 \, T < T_C \qquad 2.20$$

$R_{\mu=}\mu$ T<TC ${}_2.21$

Where TC is a characteristic period of the ground motion that depends on the soil type and is given by Eurocode 8 (2003). It typically corresponds to the transition from the constant acceleration range (short-period range) to the constant velocity range (medium-period range) in the response spectrum. The above equations suggest that in the short-period range the equal displacement principle is applied (elastic and inelastic SDOFs have the same maximum displacement), while in the medium- and long-period range the equal energy principle is applied. These principales are shown in figure 13.

Figure 13. Transformation elastic response – bilinear response: equal maximum displacement and equal energy assumptions

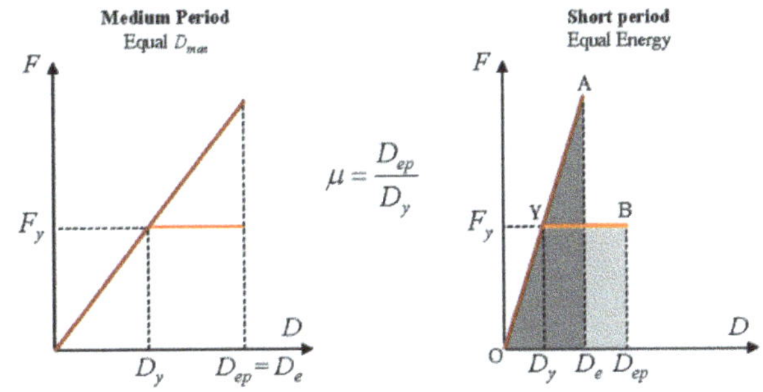

Equation (18), together with Equations (20) and (21) are used to obtain inelastic demand spectra of constant ductility, as shown in figure 14. Note that this procedure is approximate and that inelastic demand spectrum can be determined by a rigorous (but more complex) procedure by using nonlinear dynamic analysis.

Figure 14. Demand spectra for constant ductilities in AD format (based on EC8 spectrum for Zone 1, Soil type A)

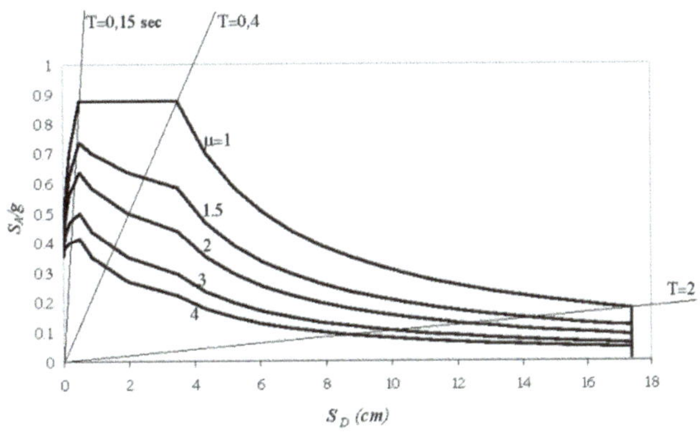

Using the procedures illustrated in the section, the seismic demand on the equivalent SDOF can be determined. The steps are schematically illustrated in figure15 for a bilinear oscillator with medium or long elastic period T*. Given the elastic demand spectrum and the bilinear capacity spectrum, from a theoretical point of view the target displacement is determined by finding the inelastic demand spectrum of ductility $\bar{\mu}$ that intersects the capacity spectrum in a point corresponding to a capacity ductility $\bar{\mu}$. In another words, the point with equal demand and capacity ductility gives the design point.

Figure 15. Capacity and Demand spectra for medium and long period T:determination of the target displacement for SDOF*

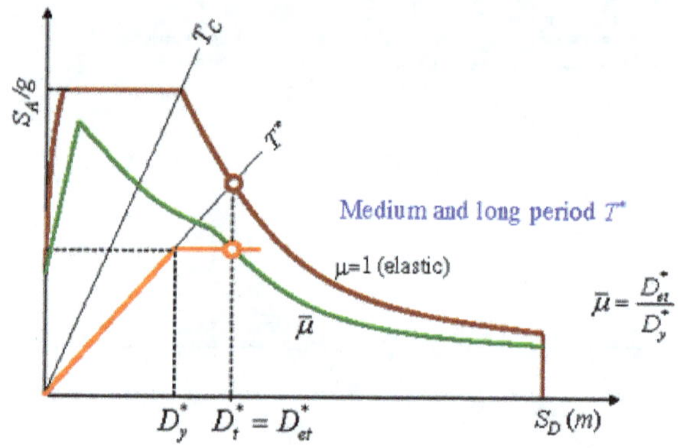

In practice, this is done very easily using the approximate procedure given by Equations (18), (20) and (21):

a) $T^*<T_C$ (Short periods)

a1) $\frac{F_y^*}{m^*} \geq S_A\left(T^*\right)$ → the response remains linear elastic (left of figure 12) $D_t^* = D_{et}^*$

a2) $\frac{F_y^*}{m^*} < S_A\left(T^*\right)$ → the response enters the nonlinear plateau (right of figure16)

$$D_t^* = \frac{D_{et}^*}{q_u}\left(1+\left(q_u-1\right)\frac{T_C}{T^*}\right) \geq D_{et}^*$$

Where $q_u = \frac{S_{Ae}\left(T^*\right)}{F_y^*/m^*}$ is the reduction factor

b) $T^{*3}T_C$ (Medium and long periods) see figure15 $D_t^* = D_{et}^*$

*Figure 16. Capacity and Demand spectra for short period T**

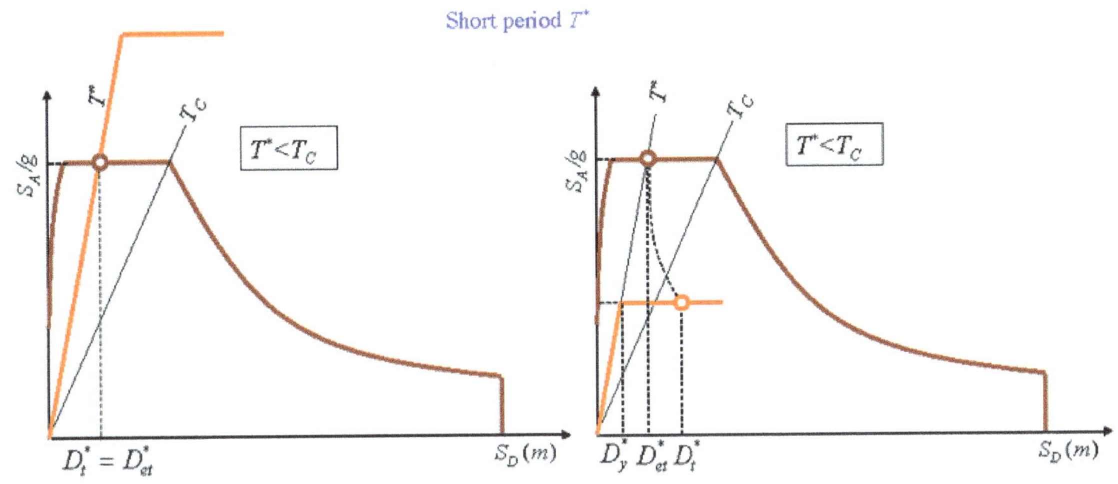

Note that the above procedure is based on the assumption of a tentative target displacement D_m^* (from which the equal energy principle is used to obtain the bilinear capacity curve in figure 10). If the target displacement D_t^* is very different from the assumed value D_m^*, then the procedure must be repeated, setting for instance $D_m^* = D_t^*$. This is a simple iterative procedure that converges very quickly.

The target displacement at the top of the building is obtained by inverting the transformation in figure5 i.e.

$$D_t = \Gamma D_t^* \tag{22}$$

One minor issue concerns the point at which the pushover curve is stopped. While it is not necessary to continue the pushover analysis to unreasonable values of the top floor displacement (this would mean long computational times and convergence problems at large displacement values), there is no general rule on when to stop the pushover curve. This means that it may happen that the pushover curve is stopped at a displacement level smaller that the computed target displacement. In this case, the pushover analysis must be repeated and it must be stopped at higher top displacement values. It is suggested to push the structure to top-displacements of the order of 2%-3% *h*, where *h* is the entire height of the building, for pushover analysis at the Ultimate and Collapse Limit States.

Finally, note that in general for each seismic input direction (*x* and *y*), four different pushover analysis should be performed. Besides considering two different load shapes (figure 6 and figure7) it is in general necessary to consider the forces applied both with the positive and the negative sign, as the irregularity of the building may lead to different responses in the positive and negative directions.

SUMMARY OF NONLINEAR STATIC PUSHOVER ANALYSIS IN EUROCODE 8

The steps necessary for nonlinear static pushover of a structure subject to the design ground motion in a single direction are as follows

1. Create the nonlinear structural model
2. Apply constant gravity loads
3. For each load shape ($F= F_1$ and $F=R$) eventually taken with the positive and negative signs:
   a. Compute the capacity curve for MDOF
   b. Compute the capacity curve for SDOF.
   c. Get the bilinear capacity spectrum for SDOF based on assumed D_m^*
   d. Compute the target displacement D_t^* for SDOF. If D_t^* is very different from D_m^*, go back to 3c with $D_m^* = D_t^*$
   e. Convert D_t^* to the target displacement D_t for MDOF building
   f. Check structural performance corresponding to limit state under consideration

ACCIDENTAL TORSIONAL EFFECTS IN EUROCODE 8

According to Eurocode 8 (2003), the calculated center of mass at each floor, i shall be considered as being displaced from its nominal location in each direction by an accidental eccentricity:

$$e_{ai} = \pm 0.05 L_i \quad (23)$$

Where e_{ai} is the accidental eccentricity of storey mass i from its nominal location, applied in the same direction at all floors and L_i is the floor dimension perpendicular to the seismic action direction.

The procedure is schematically shown in figure 17. The procedure is clearly cumbersome is applied to nonlinear analysis. It is routinely applied to linear analyses.

Figure 17. Application of loads to take into account accidental torsional effects

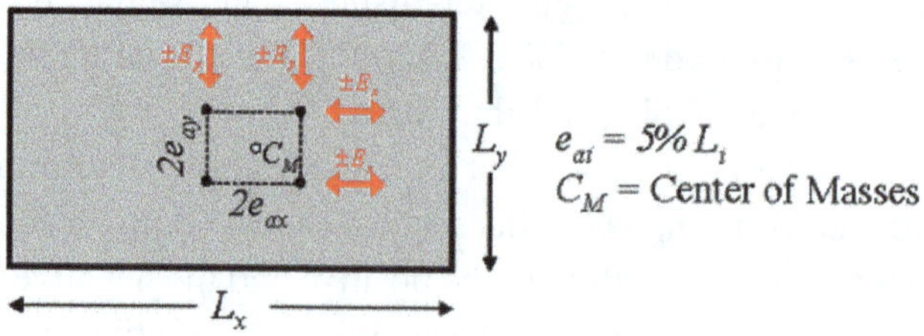

Eurocode 8 (2003) gives an alternative approach, for nonlinear pushover analyses of structures, which are torsionally flexible, where the first or second mode is torsional. It is suggested to consider torsional effects by multiplying the target displacements obtained from the pushover analyses by an amplification factor that results from an elastic modal analysis of the spatial modal. No more information is provided by recent drafts of EC8. This approach refers to recent improvements by Fajfar to the original N2 method applied to asymmetrical building (Modified N2 method, by Fajfar, Kilar, Marusic & Perus (2005)). The procedure proposed in by Fajfar, Kilar, Marusic & Perus (2005) is schematically shown in Figure 14 through Figure 17. After performing the pushover analyses with applied loads in the *x* and *y* direction, two different pushover curves are obtained for the two analyses, and two target displacements at the top floor, D_{tx} and D_{ty}, are obtained (Figure 18).

Figure 18. Pushover analyses in the x and y directions with loads applied at the center of mass

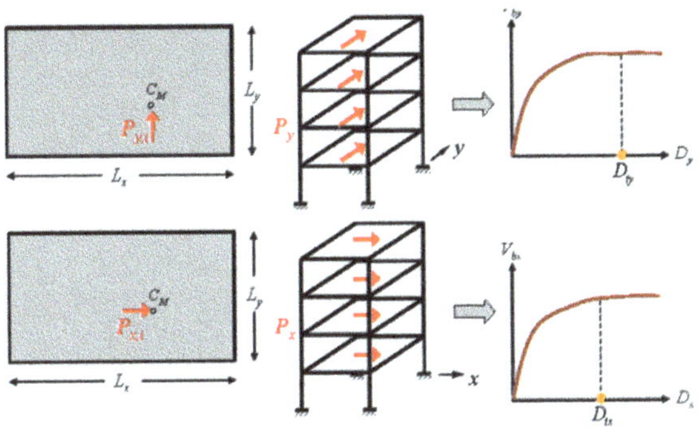

The target displacements D_{tx} and D_{ty} are multiplied by and amplification factor obtained from the modal analysis of the building. This is schematically shown in Figure 19. Given for example a column passing through point *P*, amplification factors are computed for this point. The amplification factors $c_{P,x}$ and $c_{P,y}$, different for the *x* and *y* directions, are given by the ratios between the displacements of point P (($u_{P,x}$ and $u_{P,y}$ for the *x* and *y* directions, respectively) and the displacements of top floor center of mass ($u_{CM,x}$ and $u_{CM,y}$ for the *x* and *y* directions, respectively):

$$c_{P,x} = \frac{u_{P,x}}{u_{C_M,x}} \qquad c_{P,y} = \frac{u_{P,y}}{u_{C_M,y}} \tag{24}$$

Figure 19. Top floor displacements from modal analysis and amplification factors at given point P

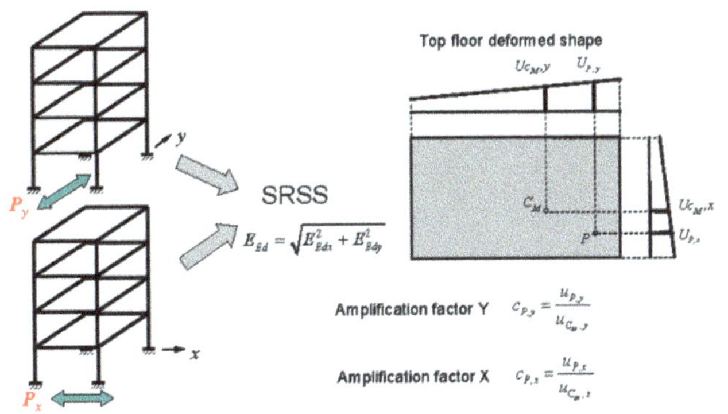

The target displacements for which the column passing through point *P* is checked are then modified to the values $c_{P,x}D_{tx}$ and $c_{P,y}D_{ty}$. This is schematically shown in figure 20. The actions (forces and/or deformations) used for the design checks of the column are those corresponding to the target displacements $c_{P,x}D_{tx}$ and $c_{P,y}D_{ty}$. There can be no de-amplification, thus $c_{P,x} \geq 1$ and $c_{P,y} \geq 1$. The actions in the *x* and *y* directions are then combined using the combination rules discussed in section above.

Figure 20. Amplification of the target displacements computed at the center of mass for a column with projection P on the top floor.

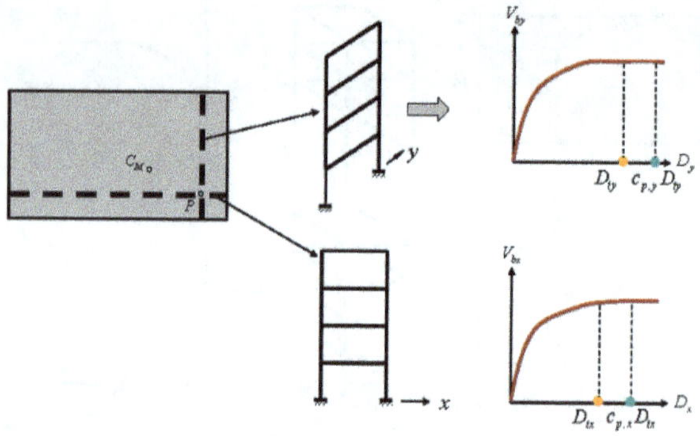

If two planar models are used for the pushover analysis, i.e. the building has plan regularity; the torsional effects may be estimated using the following approach given in Eurocode 8 (2003). The action effects (forces and/or deformations) in the individual load resisting elements resulting from the application of the pushover loads are multiplied by a factor d

$$\delta = 1 + 0,6\frac{x}{L_e} \tag{25}$$

Where x is the distance of the planar frame under consideration from the center of mass of the building plan, measured perpendicularly to the seismic action direction considered, and L_e is the distance between the two outmost lateral loads resisting elements, measured perpendicularly to the seismic action direction considered.

Recent Eurocode 8 (2003), drafts also indicate that for planar models, another procedure, outlined in Eurocode 8 (2003), can be applied, but this procedure, which implies the application of additional torsional moments to a spatial model of the structures, needs further studies before it can be used in regular practice.

COMBINATION OF THE HORIZONTAL SEISMIC ACTION EFFECTS ACCORDING TO EUROCODE 8

The pushover analysis described in sections above refers to the application of the ground motion in a single direction. When the ground motion is applied in the two horizontal directions x and y, combination rules are given by (Eurocode 8 2003). EC8 gives two alternatives for the combination, as shown in Figure 21.

Figure 21. EC8 Combination rules for horizontal seismic forces

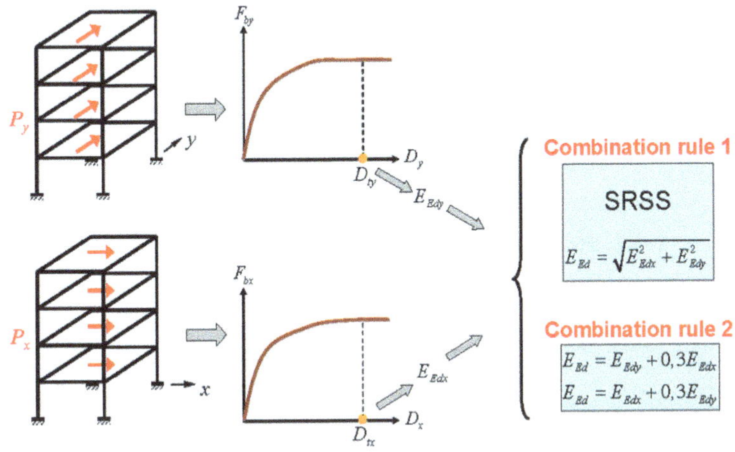

The structural response to each component shall be evaluated separately. If E_{Edx} and E_{Edy} represent the action effects due to the application of the seismic action along the x and y directions, respectively, then the design value of each action may then be estimated by:

1. the SRSS rule $E_{Ed} = \sqrt{E_{Edx}^2 + E_{Edy}^2}$
2. the 100/30 rule $E_{Ed} = E_{Edy}\,"+"\,0,3E_{Edx}$; $E_{Ed} = E_{Edx}\,"+"\,0,3E_{Edy}$

In the second rule, "+" implies "to be combined with" according to EC8.

In nonlinear analysis, the action effects (as they are termed in EC8) may be either forces or deformations. For ductile mechanisms (such as flexure in beams), the action effects are the deformations (for flexure, plastic hinge rotations or curvatures), while for brittle mechanisms (such as shear) the action effects are forces (shear forces in the structural elements). Eurocode 8 (2003) gives some indications on how to check the seismic performance of a building based on nonlinear analysis results, but studies are still under way and a clear checking procedure is still missing.

OTHER NONLINEAR PUSHOVER PROCEDURES

Other nonlinear pushover procedures have been proposed in the published literature. The main drawback of the N2 method included in Eurocode 8 (2003) is the consideration of a fixed load shape, independently of the load intensity.

One alternative pushover method is the so-called adaptive pushover. One such example is the method developed by (Antoniou & Pinho, 2004) and (Elnashai, 2002). The method adapts the form of the applied load to the inelastic evolution of the structure, given by the instantaneous stiffness matrix of the structure. In other words, at each time step, or at given points along the pushover curve, the applied load shape is updated based on the instantaneous mode shapes. This is typically done using either total or incremental updating. In total updating, the total load vector is changed based on the new mode shapes. In the incremental updating form, only the increment of lateral loads includes the change in stiffness and thus the instantaneous mode shapes.

Figure 22. Adaptive pushover procedure (Antoniou & Pinho 2004)

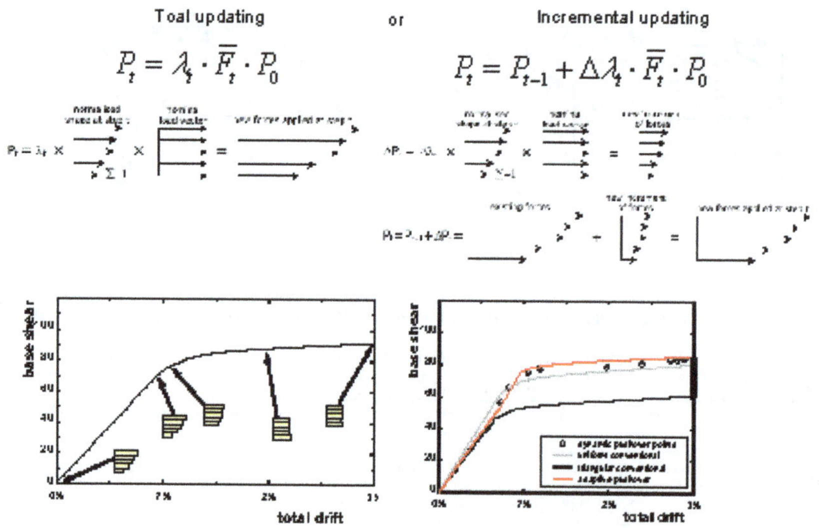

This is schematically shown in figure 22. A comparison between the results of the EC8 procedures (uniform and triangular-first mode load shapes), of the adaptive pushover and of dynamic pushover points (obtained by using accelerograms of increasing intensity) is also shown in figure18. It also appears that the interstorey drift, results obtained with the adaptive pushover are better than time history analyses results. The main criticism to this adaptive method is that the instantaneous stiffness and mode shapes do not necessarily represent the actual damage in the structure at the corresponding level of ground motion intensity, as inelasticity develops as a function of the applied pseudo-static load vector.

Another method is the so-called *modal pushover* developed by (Chopra & Goel 2002). The idea is to extend the modal analysis from the linear to the inelastic range. Let us recall that for a linear elastic system the equations of motion for a ground motion in a single direction are (Chopra 2001).

$$M\ddot{U}+C\dot{U}+KU=-MR\ddot{u}_g \tag{26}$$

The load vector *MR* can be written as expansion of the modal shapes

$$MR=s=\sum s=\sum \Gamma_i M\Phi_i \tag{27}$$

where i indicates the mode, Γ_i is the modal participation factor and f_i the mode shape of mode i. The n-coupled equations of motion (26) in the unknown vector **U** are transformed into the n uncoupled equations of motion in the unkowns Z_i

$$\ddot{Z}_i+2\zeta_i\omega_i\dot{Z}_i+\omega_i^2 Z_i=-\Gamma_i\ddot{u}_g(t) \tag{28}$$

With $U=\sum Z_i\,\Phi_i$. For the linear elastic case, the modal pushover analysis is equivalent to the modal analysis with response spectrum. Each mode is treated as a separate linear elastic SDOF, and the maximum response for each mode is computed based on the modal period (see figure 23). The modal quantities are then combined using the SRSS or the CQC rule.

The nonlinear modal pushover analysis extends the concept to the nonlinear range. The nonlinear equations of motion are

$$M\ddot{U}+C\dot{U}+F(U)=-MR\ddot{u}_g \tag{29}$$

Where F(U) expresses the nonlinearity of the element forces. The decomposition of equation (27) is applied using the initial modal shapes. It is assumed that the response of the system can still be expressed in terms of n uncoupled nonlinear equations of motion. If the excitation is proportional to the i mode, it is assumed that the response is still dominated by the same mode. That is, if the load is proportional to Φ_i, the displacement response is dominated (and thus proportional) to Φ_i. This allows modal decomposition:

$$\ddot{Z}_i+2\zeta_i\omega_i\dot{Z}_i+\frac{F_{si}}{M_i}=-\Gamma_i\ddot{u}_g(t) \tag{30}$$

The global response can be obtained combining the modal responses. The resulting modal pushover procedure is shown in figure 20. The pushover procedure is repeated for each mode, and the resulting target displacements are found for each mode. The maximum seismic effects found using the SRSS or the CQC modal combination rules. The procedure is obviously approximate, but the results published so far show improved results when compared with the N2 method of (Eurocode 8 2003). (Chopra & Goel 2002), studied the nonlinear response of a 9-storey building and showed that if the building is excited with forces proportional to a mode, the nonlinear response is dominated by the same mode. The main criticism to this method is that it combines response of totally independent structures (the single bilinear SDOF systems) and thus lacks a sound physical interpretation. Another problem, common to other nonlinear pushover procedures, is the modal combination based on SRSS or CQC, where the sign of the action is lost.

Figure 23. Spectral modal analysis

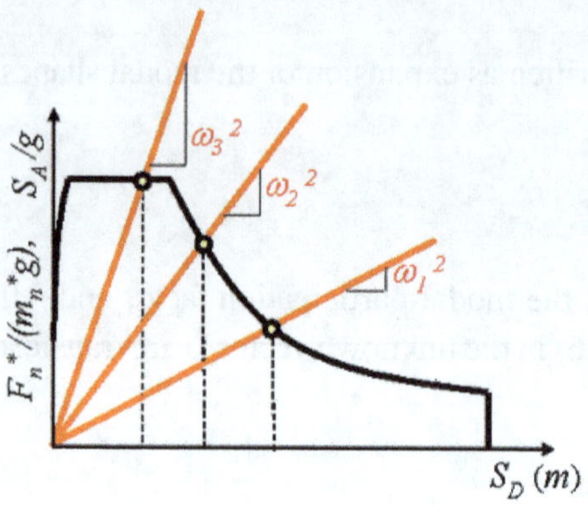

Figure 24. Modal Pushover Analysis

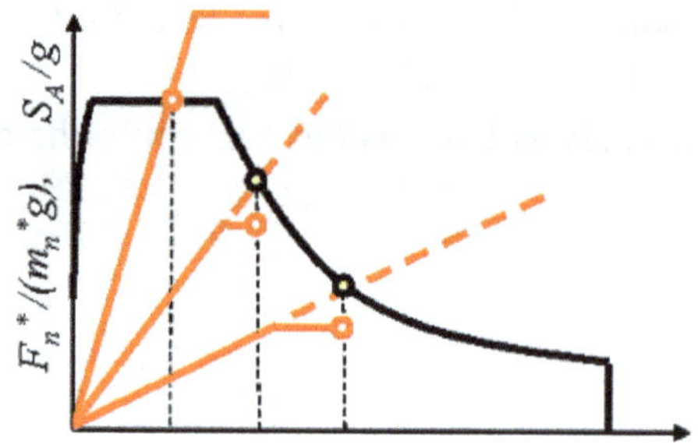

An alternative nonlinear static procedure was proposed by (Antoniou &Pinho 2004) and (Aydinoglu 2003). The procedure, called Incremental Response Spectrum Analysis (IRSA) does not properly belong to the pushover-based based. The procedure is displacement-based, it uses the equal displacement rule and the structure nonlinear behavior is modeled as piece-wise linear. The initial modes are computed and the elastic response spectrum for the initial structure is carried out. In the first stop the spectral ordinates are scaled to the formation of the first plastic hinge, which corresponds in the piece-wise linear capacity curve, to the first change of stiffness. The updated modal quantities are computed and additional spectral ordinated are computed up to the formation of the second plastic hinge. The procedure continues until the entire spectral ordinates are applied. The method does not require any transformation to an equivalent SDOF system. No pushover curve is generated. The method is similar to the procedure presented by (Gupta & Kunnath 2000).

Other pushover procedures have been proposed in the published literature, but no general consensus has been reached on the best procedure. It appears that at this stage the N2 method represents a good compromise between simplicity and precision. A number of studies are however under way in this field.

NONLINEAR TIME-HISTORY ANALYSIS IN EUROCODE 8

The nonlinear time-history analysis is the most general nonlinear method of analysis. It is deemed unpractical by some because of the computational lengths of the analysis, but as computer speeds increase, time-history analysis becomes increasingly feasible. Time-history analysis is a general dynamic method of analysis and there are no computational peculiarities related to earthquake engineering analysis. The input force in earthquake engineering is the input ground motion, expressed in the form of one or more accelerograms. The dynamic equations of motion are:

$$M\ddot{U} + C\dot{U} + F(U) = -MR\ddot{u}_g \tag{31}$$

Where C is the damping matrix, $\ddot{u}_g$ is a vector containing accelerograms and depends on time, i.e $\ddot{u}_g(t)$. In the most general case, $\ddot{u}_g = \{\ddot{u}_{g,x} \quad \ddot{u}_{g,y} \quad \ddot{u}_{g,Z}\}$, that is the input ground motion consists of three simultaneously acting accelerograms in three orthogonal directions. For buildings, the z component of the ground motion is typically neglected or considered only for some structural elements. Thus, for a 3D structural model, only the horizontal components are considered and $\ddot{u}_g = \{\ddot{u}_{g,x} \quad \ddot{u}_{g,y}\}$. The same accelerogram may not be used simultaneously along both horizontal directions, (Eurocode 8 2003).

The only remaining question concerns the selection of the ground motions (accelerograms) for the time-history analysis. First of all, EC8 state that:

1. Minimum of 3 different pairs of accelerograms $\ddot{u}_g = \{\ddot{u}_{g,x} \quad \ddot{u}_{g,y}\}$ should be used. If 3 to 6 ground motions are used, then the maximum of the response quantities from all the analyses should be used as the design value of the action effect E_d in the relevant verifications.
2. If 7 or more different ground motions are used, i.e. if 7 different pairs of accelerograms $\ddot{u}_g = \{\ddot{u}_{g,x} \quad \ddot{u}_{g,y}\}$ are used, then the average of the response quantities from all the analyses should be used as the design value of the action effect E_d in the relevant verifications.

As for the selection of the accelerograms, (Eurocode 8 2003) states that depending on the nature of the application and on the information actually available, the description of the seismic motion may be made by using artificial, recorded or simulated accelerograms.

For all accelerograms, the following requirements hold:

1. As previously stated, at least 3 pairs of accelerograms $\ddot{u}_g = \{\ddot{u}_{g,x} \quad \ddot{u}_{g,y}\}$ should be used;
2. The mean of the zero period spectral response acceleration values (calculated from the individual time histories) should be [3] $a_g S$ for the site in question;
3. In the range of periods between $0{,}2T_1$ and $2T_1$ (where T_1 is the fundamental period of the structure in the direction where the accelerogram will be applied) no value of the mean 5 damping elastic spectrum, calculated from all time histories, should be less the 90% of the corresponding value of the 5% damping elastic spectrum.

For artificial accelerograms, the duration of the acclerograms should be consistent with the magnitude and the other relevant features of the seismic event underlying the establishment of a_g. When site-specific data are not available, the minimum duration T_s of the stationary part of the accelerograms should be equal to 10 sec. (Eurocode 8 2003)

Recorded accelerograms, generated through a physical simulation of source, and travel path mechanisms. It may be used, provided that the samples used are adequately qualified with regard to the seism genetic features of the sources and to the soil conditions appropriate to the site and their values are scaled to the value of $a_g S$ for the zone under construction.

However, no general study is yet available and discussions are under way in those countries that have started employing nonlinear methods of analysis. It is generally recognized that recorded accelerograms, if well selected, are less demanding on the structure, because each single accelerogram does not match well the elastic response spectrum, and the frequency content of each single accelerogram is limited with respect to the elastic response spectrum. On the other hand, generated accelerograms are more demanding, because each single accelerogram has a response spectrum close to the elastic response spectrum. Some claim that generated or simulated accelerograms should be used at the research level for the refinement of the nonlinear methods of analysis, while recorded accelerograms should be used for the actual design of structures. No general consensus has yet been reached, though.

As for recorded accelerograms, there are databases available (such as the European Strong Motion Database[3], the PEER Strong Motion Database[4], the Strong Motion Database of SeismoLinks[5], etc.). The elaboration of these databases of recorded ground motions and the selection of the appropriate accelerograms that satisfy the Eurocode 8 2003, requirements previously mentioned is not trivial. Groups of accelerograms compatible (according to EC8 criteria) with the elastic response spectra for different a_g and soil type are available at http://www.reluis.unina.it/ (in Italian). Figure 25 shows the results of the selection of 7 accelerograms (for analysis in one direction only) compatible with the elastic response spectrum for Zone 3, Soil C. The red line is the mean response spectrum of the 7 accelerograms. In order to limit the strength of the selected accelerograms, one additional conditions was used for the selection: the mean response spectrum should not exceed at any period 110% of the elastic response spectrum of Eurocode 8 2003.

Figure 25. Recorded accelerograms, mean response spectrum and elastic response spectrum for Zone 3, Soil C.

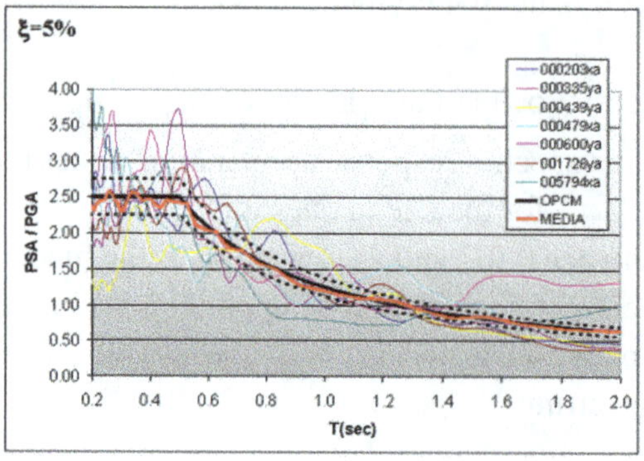

As for generated and simulated accelerograms, a few computer programs are freely available on the internet, notably (Fajfar, Kilar, Marusic & Perus 2005) and other useful software available at http://www.seismolinks.com/Software/Strong-motion_data_processing/.

Finally, Eurocode 8 2003, does not contain any specific procedure for the estimation of the torsional effects. In principle, the general procedure contained in Eurocode 8 2003, for accidental torsional effects should be applied. This is the method shown in figure 21. It would imply applying the ground motions at four different locations, with a considerable increase in computational costs. At this stage, it appears reasonable to neglect the torsional effects when performing nonlinear dynamic analyses.

NONLINEAR MODELING IN EUROCODE 8

Eurocode 8 does not any specific guidelines on how to consider *material* nonlinearities. Only general statements are made in Eurocode 8 2003. The model shall include the strength of structural elements and their post-elastic behavior. As a minimum, a bilinear force-deformation relationship should be used at the element level. For bilinear force-deformation relationship (typical of ductile mechanisms such as the bending moment – curvature in reinforced concrete beams) the elastic stiffness should be that of the cracked section and can be computed as the secant stiffness to the yield point. Zero post-yield stiffness may be assumed. Strength degradation may be included using more refined constitutive laws. These are minimum requirements for basic constitutive laws. For a fiber section model, for example, the section behavior derives from the fibers' behavior. Element properties should be based on mean values of the material properties. For new structures, mean values of the material properties may be estimated from the corresponding characteristic values on the basis of specific Eurocode (for example Eurocode 2 for concrete).

Eurocode 8, which deals with resistance conditions at the Ultimate Limit State, provides guidelines on *geometric* nonlinearities. Second-order, or P-D effects, need not be taken into account if the following condition is satisfied:

$$\theta = \frac{P_{tot}\ d_r}{V_{tot}\ h} \leq 0{,}10 \tag{32}$$

Where θ, is the interstorey drift sensitivity coefficient, Pt_{ot} is the total gravity load at and above the sotrey considered in the seismic design situation. Where dr is the design interstorey drift, evaluated as the average lateral displacement ds at the bottom of the storey under consideration and calculated in accordance with Eurocode 8 2003, Vt_{ot} is the total seismic storey shear, h is the interstorey height. If

$$0.1 < \theta\ £\ 0.2 \tag{33}$$

The second order effects may approximately be taken into account by multiplying the relevant seismic action effects by a factor equal to $1/(1 - \theta)$. Finally, the value of the interstorey drift sensitivity coefficient is limited by

$$\theta = 0.30 \tag{34}$$

The above conditions relate to the ultimate limit state. It appears that these rules apply mainly to linear methods of analysis. For nonlinear analysis, no specifics are given on geometric nonlinearities and the published literature lacks studies on the importance of accounting for geometric nonlinearities in either pushover or time-history analyses. From limited published data and common experience it appears that for most flexible buildings, in zones of high seismicity, geometric nonlinearities will have an important effect on the overall response at the limit state. For analyses at the collapse limit state, geometric nonlinearities should always be considered.

DYNAMICS IN Z_SOIL SOFTWARE

This section briefly reviews the most important features of the procedures used in program Z_Soil (Zimmermann, Truty, Urbanski & Podles 2006) for dynamic problems. No special procedures are needed at the computational levels for earthquake engineering analysis. A simple procedure was added for the transformations involved in the pushover analysis.

NONLINEARITIES FOR FRAME ANALYSIS IN Z_SOIL

Recent versions of Z_Soil include advanced models for frame analysis. In particular, nonlinear beams with fiber section models are available. The cross section is divided into fibers and the constitutive laws of each fiber is assigned from the constitutive law library available in Z_Soil. Both displacement-based and force-based formulations are available. Force-based elements (Spacone, Filippou & Taucer 1996) are exact within the classical Euler-Bernoulli beam theory. As for geometric linearities; these are considered in the general framework of the program and thus follow a corotational approach (Zimmermann, Truty, Urbanski & Podles 2006).

EQUATIONS OF MOTION

The following equations of motion are used for nonlinear dynamics:

$$\mathbf{Ma} + \mathbf{Cv} + \mathbf{N}(\mathbf{d}) = \mathbf{F} \tag{35}$$

Where **a**, **v** and **d** represent approximations of $\ddot{\mathbf{U}}$, $\dot{\mathbf{U}}$ and **U**, respectively, in Equation (29). In order to be consistent with the notation of (Zimmermann, Truty, Urbanski & Podles 2006), **N** indicates the internal forces and **F** the time depend externally applied forces. For dynamic analysis, Rayleigh damping is used

$$\mathbf{C} = \alpha\mathbf{M} + \beta\mathbf{K} \tag{36}$$

Where a and b are obtained by imposing damping rates at 2 different modal frequencies (i,j):

$$\begin{Bmatrix} \xi_i \\ \xi_j \end{Bmatrix} = 0,5 \begin{bmatrix} 1/\omega_i & \omega_j \\ 1/\omega_j & \omega_j \end{bmatrix} \begin{Bmatrix} \alpha \\ \beta \end{Bmatrix} \tag{37}$$

Remarks:

1. **C** takes into account minor nonlinearities ignored in **N**(**d**)
2. **K** is typically approximed by the initial stiffness, thus $\mathbf{K}=\mathbf{K}_0$

SEIMIC INPUT

Seismic input is given as imposed displacements or accelerations of the ground. Independently of the input format, the solving algorithms can find the solution in terms of accelerations, displacements, or velocities.

NONLINEAR NEWMARK ALGORITHM FOR ARBITRARY DYNAMIC INPUT

Consider the nonlinear equations in their static form. Equilibrium must be satisfied at any time $t_{n+1} = (n+1)\cdot\Delta t$.

The algorithm consists of the following 3 equations:

$$\mathbf{M}\,\mathbf{a}_{n+1} + \mathbf{C}\,\mathbf{v}_{n+1} + \mathbf{N}(\mathbf{d}_{n+1}) = \mathbf{F}_{n+1} \tag{38}$$

With

$$\begin{aligned} \mathbf{d}_{n+1} &= \mathbf{d}_n + \Delta t\,\mathbf{v}_n + \frac{\Delta t^2}{2}[(1-2\beta)\mathbf{a}_n + 2\beta\,\mathbf{a}_{n+1}] \\ \mathbf{v}_{n+1} &= \mathbf{v}_n + \Delta t[(1-\gamma)\mathbf{a}_n + \gamma\,\mathbf{a}_{n+1}] \end{aligned} \tag{39}$$

Where subscripts n and n+1 indicate values at instants t_n and t_{n+1}, respectively. The algorithm is completed by the following predictors

$$\begin{aligned} \tilde{\mathbf{d}}_{n+1} &= \mathbf{d}_n + \Delta t\,\mathbf{v}_n + \frac{\Delta t^2}{2}(1-2\beta)\mathbf{a}_n \\ \tilde{\mathbf{v}}_{n+1} &= \mathbf{v}_n + (1-\gamma)\Delta t\,\mathbf{a}_n \end{aligned} \tag{40}$$

And correctors

$$
\begin{aligned}
\mathbf{d}_{n+1} &= \tilde{\mathbf{d}}_{n+1} + \beta\,\Delta t^2\,\mathbf{a}_{n+1} \\
\mathbf{v}_{n+1} &= \tilde{\mathbf{v}}_{n+1} + \gamma\,\Delta t\;\mathbf{a}_{n+1}
\end{aligned}
\tag{41}
$$

NEWTON-RAPHSON NONLINEAR IMPLICIT NEWMARK ALGORITHM

Let's consider the static procedure first. If superscript i+1 indicates iterations, in the vicinity of the displacement $\mathbf{d}_{n+1}^{i+1}$

$$
\mathbf{N}\left(\mathbf{d}_{n+1}^{i+1}\right) \cong \mathbf{N}\left(\mathbf{d}_{n+1}^{i}\right) + \frac{d\mathbf{N}}{d\mathbf{d}}\,\Delta\mathbf{d} = \mathbf{N}\left(\mathbf{d}_{n+1}^{i}\right) + \mathbf{K}_T\,\Delta\mathbf{d}
\tag{42}
$$

hence:

$$
\begin{cases}
\mathbf{K}_T \cdot \Delta\mathbf{d} = \mathbf{F}_{n+1} - \mathbf{N}\left(\mathbf{d}_{n+1}^{i}\right) & \left(\text{with } \mathbf{d}_{n+1}^{0} = \mathbf{d}_n\right) \\
\mathbf{d}_{n+1}^{i+1} = \mathbf{d}_{n+1}^{i} + \Delta\mathbf{d} &
\end{cases}
\tag{43}
$$

Convergence is reached when:

$$
\left\| \mathbf{F}_{n+1} - \mathbf{N}\left(\mathbf{d}_{n+1}^{i+1}\right) \right\| < TOLERANCE
\tag{44}
$$

Remarks:

The following algorithmic alternative are possible

- Update **K** at each step and iteration: full Newton-Raphson
- Opportunistic update: modified Newton-Raphson
- $\mathbf{K}_T = \mathbf{K}_0$: constant stiffness algorithm

Let's now consider the dyamics procedure in terms of incremental displacements $\Delta\mathbf{d}$
Let's define

$$
\begin{aligned}
\mathbf{N}^*(\mathbf{d}) &= \mathbf{Ma} + \mathbf{Cv} + \mathbf{N}(\mathbf{d}) \\
\mathbf{K}^* &= \left(\frac{d\mathbf{N}^*}{d\mathbf{d}}\right) = \frac{\mathbf{M}}{\Delta t^2 \beta} + \frac{\gamma}{\Delta t \beta}\mathbf{C} + \mathbf{K}
\end{aligned}
\tag{45}
$$

Initialization

$$\mathbf{d}_0 = \mathbf{d}(0)$$
$$\mathbf{v}_0 = \mathbf{v}(0) \tag{46}$$

For every Δt
Compute predictors

$$\tilde{\mathbf{d}}_{n+1} = \mathbf{d}_n + \Delta t\ \mathbf{v}_n + (1-2\beta)\frac{\Delta t^2}{2}\ \mathbf{a}_n$$
$$\tilde{\mathbf{v}}_{n+1} = \mathbf{v}_n + (1-\gamma)\Delta t\ \mathbf{a}_n$$
$$\mathbf{d}_{n+1}^{i=0} = \tilde{\mathbf{d}}_{n+1} \tag{47}$$
$$\mathbf{v}_{n+1}^{i=0} = \tilde{\mathbf{v}}_{n+1}$$
$$\mathbf{a}_{n+1}^{i=0} = \mathbf{0}$$

For every $\Delta\mathbf{d}$
Solve

$$\Delta\mathbf{F}_{n+1}^{i} = \mathbf{F}_{n+1} - \mathbf{N}^{*}\left(\mathbf{d}_{n+1}^{i}\right)$$
$$\mathbf{K}_{T}^{*i}\ \Delta\mathbf{d} = \Delta\mathbf{F}_{n+1}^{i} \tag{48}$$
$$\mathbf{d}_{n+1}^{i+1} = \mathbf{d}_{n+1}^{i} + \Delta\mathbf{d}$$

Compute correctors

$$\mathbf{a}_{n+1}^{i+1} = \frac{\mathbf{d}_{n+1}^{i+1} - \tilde{\mathbf{d}}_{n+1}}{\Delta t^{2\ 2}} \tag{49}$$
$$\mathbf{v}_{n+1}^{i+1} = \tilde{\mathbf{v}}_{n+1} + \Delta t^{3}\ \mathbf{a}_{n+1}^{i+1}$$

Next iteration

$$\text{if} \quad \Delta\mathbf{F}_{n+1}^{i} > TOLERANCE \quad \Rightarrow \quad i = i+1$$

Next time step

$$\text{if} \quad t < T_{max} \quad \Rightarrow \quad n = n+1$$

Remarks: The following algorithmic alternative are possible

1 Update **K** at each step and iteration: full Newton-Raphson

2 Opportunistic update: modified Newton-Raphson
3 $\mathbf{K}_T = \mathbf{K}_0$: constant stiffness algorithm

The algorithm does not work with b=0. A similar algorithm, expressed in terms of incremental accelerations **Da,** can be derived which supports b=0.

NONLINEAR HHT-A ALGORITHM FOR ARBITRARY DYNAMIC INPUT

The original linear form of the algorithm is written:

$$\mathbf{M}\,\mathbf{a}_{n+1} + (1+\alpha)\mathbf{C}\,\mathbf{v}_{n+1} - \alpha\,\mathbf{C}\,\mathbf{v}_n + (1+\alpha)\mathbf{K}\,\mathbf{d}_{n+1} - \alpha\,\mathbf{K}\,\mathbf{d}_n = \mathbf{F}_{n+\alpha} \tag{50}$$

With

$$[*]_{n+\alpha} = (1+\alpha)[*]_{n+1} - \alpha[*]_n \tag{51}$$

Thus

$$\mathbf{F}_{n+\alpha} = (1+\alpha)\mathbf{F}_{n+1} \quad - \alpha\,\mathbf{F}_n \tag{52}$$

The corresponding nonlinear statement can be written as

$$\mathbf{M}\,\mathbf{a}_{n+1} + \mathbf{C}\,\mathbf{v}_{n+\alpha} + \mathbf{N}\left(\mathbf{d}_{n+\alpha}\right) = \mathbf{F}_{n+\alpha} \tag{53}$$

With

$$\begin{aligned} \mathbf{d}_{n+1} &= \mathbf{d}_n + \Delta t\,\mathbf{v}_n + \frac{\Delta t^2}{2}\left[(1-2\beta)\mathbf{a}_n + 2\beta\,\mathbf{a}_{n+1}\right] \\ \mathbf{v}_{n+1} &= \mathbf{v}_n + \Delta t\left[(1-\gamma)\mathbf{a}_n + \gamma\,\mathbf{a}_{n+1}\right] \end{aligned} \tag{54}$$

Remarks:

1 Hilber proposes: $-0.3<\alpha<0$, $\gamma=\dfrac{(1\text{-}2\alpha)}{2}$, $\beta=\dfrac{(1\text{-}\alpha)^2}{4}$
2 α=0 corresponds to Newmark's algorithm

MODELIZATION OF SOIL STRUCTURE INTERACTION

Soil-Structure Interaction (SSI) is a term used essentially to describe the forces that occur between different linked systems ; the structure, the foundation and the soil surrounding the foundation. It is used also when these forces are able to change the dynamic response of the structure as well as the response of the soil when it is subjected to a free field ground motion (Kausel 2010).

Soil-structure interaction principles varies widely across both the structural and geotechnical engineering disciplines. Most structural engineers can appreciate that soil-structure interaction effects are more pronounced in soft soils, and many are aware that foundation input motions can differ from free-field ground motions.

Current methods of Soil-Structure Interaction analysis can be classifiés according to the choice of surface surrounding the basement in the soil that divides the soil-structure model into two parts: the external part and the internal part.

Nowadays, there exist two main methods for Soil-Structure Interaction problems: Direct and Substructure approchées, to understand them, we start with basic definitions.

DIRECT METHOD

In the "Direct" method, the model is analyzed as a complete system (the soil and the structure are usually discretized in a one finite element model with foundation and structural elements) Figure 26, and probably iut is the easiest way to analyze soil-structure interaction for seismic excitation, from a conception point of view.

However, from a computational stand point, it is difficult to obtain a direct solution of the SSI problem, décause the number of dynamic degrees of freedom in the soil region is high, and systems that contains significant nonlinearities in the soil or structural materials.

Figure 26. Finite element model illustrating the direct method (Nist 2012)

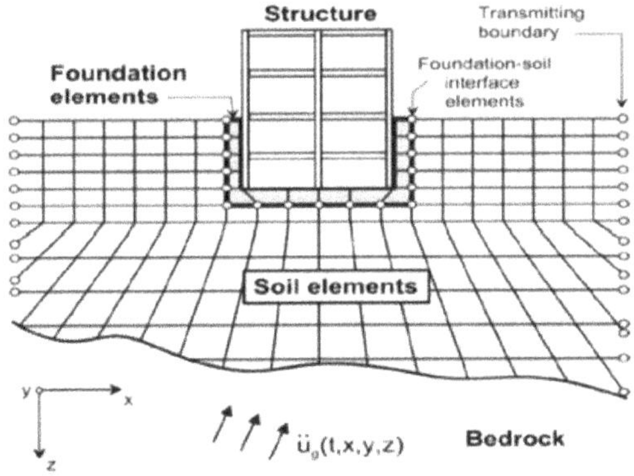

SUBSTRUCTURE METHOD

The "Substructure" method, allows the complicated soil-structure system to be broken down into manageable parts that are combined in final to formulate the complete solution see figure 27 .

In this approach, two important steps are taken in consideration, the first one is to evaluate free-field accelerogram which is computed without considering the presence of the structure, then we proceed to the calculation of soil dynamic stiffness coefficients that are interpreted by the spring-dashpot system. In the second step, the free-field accelerogram is used to analyze the structure whose base is coupled with a discrete translational and rotational spring and dashpot elements.

Figure 27. Substructure method in soil structure interaction (Wolf 1985)

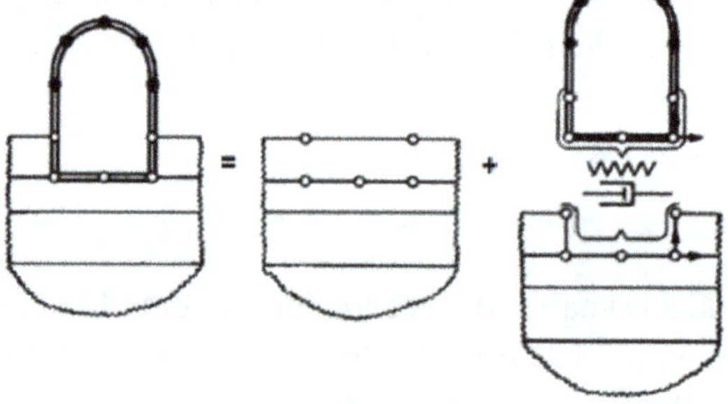

The behavior of the soil is accounted for by simple mechanical elements such as springs and dashpots. Different configurations can be taken into account by connecting several springs and dashpots whose parameters are described in the next paragraphs.

The stiffness and damping characteristics of the soil-foundation interaction are characterized using relatively simple impedance function models or a series of distributed springs and dashpots as schematized in figure 28 and figure 29.

Figure 28. Springs Model (Nist 2012)

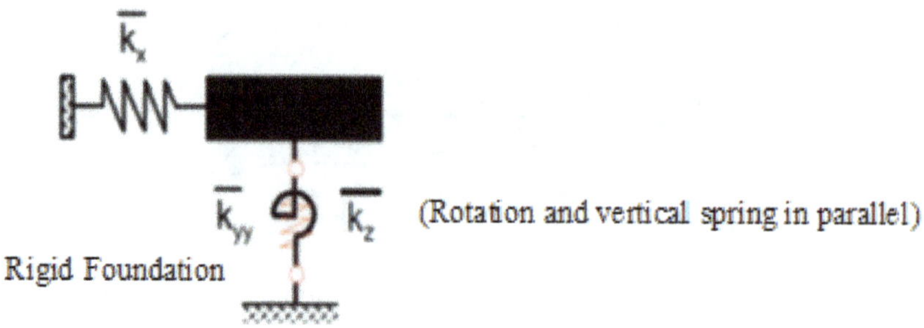

Figure 29. Model with springs and dashpots (Nist 2012)

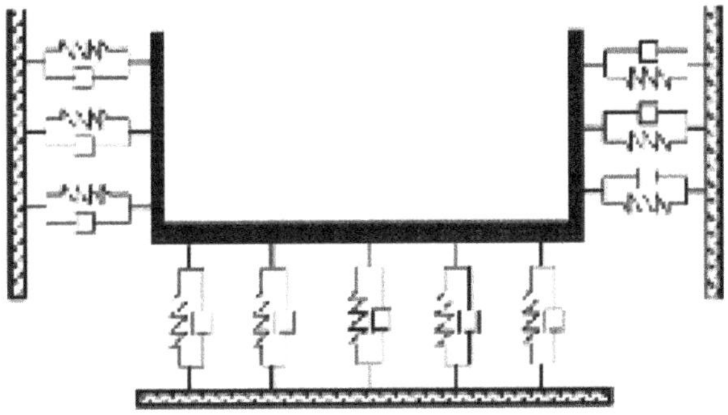

IMPEDANCE

Historically, the first problems for impedances considering soil inertia were solved semi analytically, for homogeneous half-space without the internal damping and for circular surface stamps. It turned out that horizontal impedances behaved more or less like pairs of springs and viscous dashpots. This fact created a base for using the above-mentioned "soil springs and dashpots" elements.

The foundation dashpot elements consider two different sources of damping ; the hysteric (strain dependent) damping caused by the inelastic response of the soil supporting the foundation and the radiation (geometry dépendent) damping caused by the stress waves that travel away from the foundation as the soil deforms. Radiation damping is much greater than material damping for most foundation configurations.

Many impedance function solutions are available for rigid circular or rectangular foundations located on the surface of, or embedded within, a uniform, elastic, or visco-elastic half-space and the two tables below were provided by (Gazetas 1990).

Table 2 lists stiffness and damping coefficients named K and C respectively for the six degrees of freedom (three translational and three rotational) for an arbitrarily shaped foundation on surface of a homogeneous half-space, that include the approximate effects of radiation damping in the soil. Table 3 contains the same parameters but this time for a foundation embedded in a half space with an arbitrary basement shape.

APPLICATIONS

Example 1

An existing three-storey reinforced concrete building is studied using the nonlinear frame analysis capabilities of Z_Soil. The building is in Bonefro, Italy, and is a good example of residential buildings of the 70's and 80's in Italy, prior to the introduction of the seismic code in the early 80's (figure 30). Foundation, floor and roof plans are shown in figure 27 through figure 29, the model built in Z_Soil is shown in figure 35.

Table 2. Spring and dashpot coefficients for a foundation on surface of homogeneous half space

Motion		Spring constant K	Dashpot constant C
Vertical		$K_Z = \frac{2GL}{1-\upsilon}\left(0.73+154\left(\frac{A_b}{4L^2}\right)^{0.75}\right)$	$C_Z = \rho V_{La} A_b \tilde{C}_Z$
Horizontal	Lateral direction	$K_y = \frac{2GL}{2-\upsilon}\left(2+2.5\left(\frac{A_b}{4L^2}\right)^{0.95}\right)$	$C_y = \rho V_S A_b \tilde{C}_y$
	Longitudinal direction	$K_y = K_y - \left[\frac{0.2}{0.75-\upsilon} GL\left(1-\left(\frac{B}{L}\right)\right)\right]$	$C_x = \rho V_S A_b$
Rocking	Longitudinal x axis	$K_{rx} = \frac{G}{1-\upsilon} I_{bx}^{0.75}\left(\frac{L}{B}\right)^{0.25}\left(2.4+\frac{B}{2L}\right)$	$C_{rx} = \rho V_{La} I_{bx} \tilde{C}_{rx}$
	Lateral y axis	$K_{ry} = \frac{3G}{1-\upsilon} I_{by}^{0.75}\left(\frac{L}{B}\right)^{0.15}$	$C_{ry} = \rho V_{La} I_{by} \tilde{C}_{ry}$
Torsion		$K_t = 3GI_{bZ}^{0.75}\left(\frac{B}{L}\right)^{0.4}\left(\frac{I_{bZ}}{B^4}\right)^{0.2}$	$C_t = \rho V_s I_{bZ} \tilde{C}_t$

BONEFRO BUILDING MODELING

Two models were created for the building. A general 3D model (figure 35) and a 2D model of a single frame (discussed later in the report). The following material properties are used for the fiber section in Z_Soil:

Concrete: E=2.7e7 KN/m^2
Steel E=2.1e8 KN/m^2
u=0.16 u=0.2
f_t=300 KN/m^2
f_t = f_c =500000 KN/m^2
f_c=3000KN/m^2

The following reinforcement details are used:

For the floor beams geometry, an equivalent rectangular beam is used. The beam dimensions are found imposing the same inertia of the T-beam. The results are:

In our case b/d = 6.67m & h/t = 2.5m thus b'/b = 0.336m so b'= 0.672m.

The nodal masses computed based on the loads given by EC8 are shown in the following page for the three floors.

Table 3. Spring and dashpot coefficients for embedded foundations in a half space

Motion		Spring constant	Dashpot constant
Vertical		$K_{Z,e}=K_z\left[1+\frac{1}{21}\left(\frac{D}{B}\right)\left(1+1.3\left(\frac{A_b}{4L^2}\right)\right)\right]\left(1+0.2\left(\frac{A_w}{A_b}\right)^{\frac{2}{3}}\right)$	$C_{Z,e}=C_Z+\rho V_{La}A_w$
Horizontal	Lateral direction	$K_{y,e}=K_y\left[1+0.15\left(\frac{D}{B}\right)^{0.5}\right]\left[1+0.52\left(\frac{hA_w}{BL^2}\right)^{0.4}\right]$	$C_{y,e}=C_y+4\rho V_s Bd+4\rho V_{La}Ld$
Horizontal	Longitudinal direction	$K_{x,e}=K_x\left(\frac{K_{y,e}}{K_y}\right)$	$C_{x,e}=C_x+\rho V_{La}Bd+4\rho V_s Ld$
Rocking	Longitudinal x axis	$K_{rx,e}=K_{rx}\left\{1+1.26\left(\frac{d}{B}\right)\left[1+\left(\frac{d}{B}\right)\left(\frac{d}{D}\right)^{-0.2}\left(\frac{B}{L}\right)^{0.5}\right]\right\}$	$C_{rx,e}=C_{rx}+\rho I_{bx}\left(\frac{d}{B}\right)$ $C_{rx,e}=C_{rx}+\rho I_{bx}\left(\frac{d}{B}\right)\left\{V_{La}\left(\frac{d^2}{B^2}\right)+3V_s+V_S\left(\frac{B}{L}\right)\left[1+\left(\frac{d^2}{B^2}\right)\right]\right\}\eta_r$ $\eta_{rx}=0.25+0.65\sqrt{a_0}\left(\frac{d}{D}\right)^{-\frac{a_0}{2}}\left(\frac{D}{B}\right)^{-\frac{1}{4}}$
Rocking	Lateral y axis	$K_{ry,e}=K_{ry}\left\{1+0.92\left(\frac{d}{L}\right)\left[1.5+\left(\frac{d}{L}\right)^{1.9}\left(\frac{d}{L}\right)^{-0.6}\right]\right\}$	$C_{ry,e}=C_{ry}$ by replacing x by y and interchanging B by L with $a_0=w\frac{B}{V_s}$
Torsion		$K_{t,e}=K_t\Gamma_W\Gamma_{tre}$ $\Gamma_W=1+0.4\left(\frac{D}{d}\right)^{0.5}\left(\frac{j_s}{j_r}\right)\left(\frac{B}{D}\right)^{0.6}$ $\Gamma_{tre}=1+0.5\left(\frac{D}{B}\right)^{0.1}\left(\frac{B^4}{I_{bx}}\right)^{0.13}$ $j_s=\left(\frac{4}{3}\right)d\left(B^3+L^3\right)+4BLd\left(L+B\right)$ $j_r=\left(\frac{4}{3}\right)BL\left(B^2+L^2\right)$	$C_{t,e}=C_t+4\rho d\left[\left(\frac{1}{3}\right)V_{aL}B^3+L^3\right]+V_s BL(L+B)\eta_t$ $\eta_t=\left(\frac{d}{D}\right)^{-0.5}\frac{{a_0}^2}{\left[{a_0}^2+\left(\frac{1}{2}\right)\left(\frac{L}{B}\right)^{-1.5}\right]}$

Notes: D: embedded depth of foundation in soil;

d: contact depth between foundation and soil; h=D-d/2;

I_{bx}, I_{by}, I_{bz}: are Polar moment of inertia of the bottom surface of foundation about the x, y and z axes ; B, L: half-width and half-length of the foundation.

G,υ,ρ Are the shear modulus, Poisson's ratio and mass density of the soil respetively ;

A_w: area of foundation side contact with soil;

A_b: actual area of bottom surface of foundation.

$\tilde{C}_Z, \tilde{C}_y, \tilde{C}_{rx}, \tilde{C}_{ry}, \tilde{C}_t$ Are the damping adjustment coefficients of foundation surface, which is determined by the table of $\left(a_0, \frac{L}{B}\right)$; V_s: shear wave velocity of the soil and finally V_{La} is Lymer's analog wave velocity that equals: $V_{La}=\frac{3.4}{\pi(1-\nu)}V_s$

Figure 30. Bonefro reinforced concrete 3-storey building

Figure 31. Foundation plan of Bonefro building

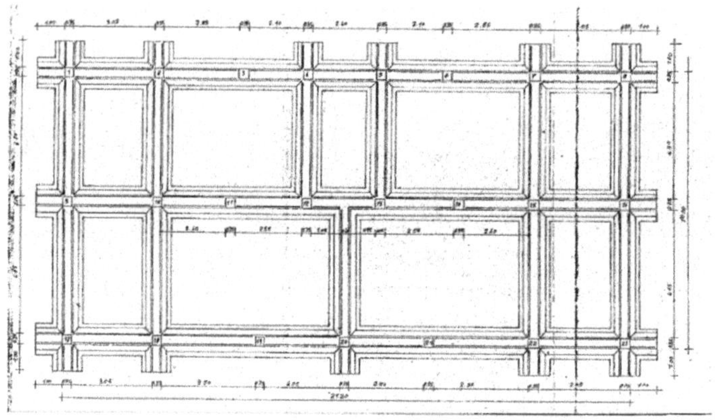

Figure 32. First floor plan of Bonefro building

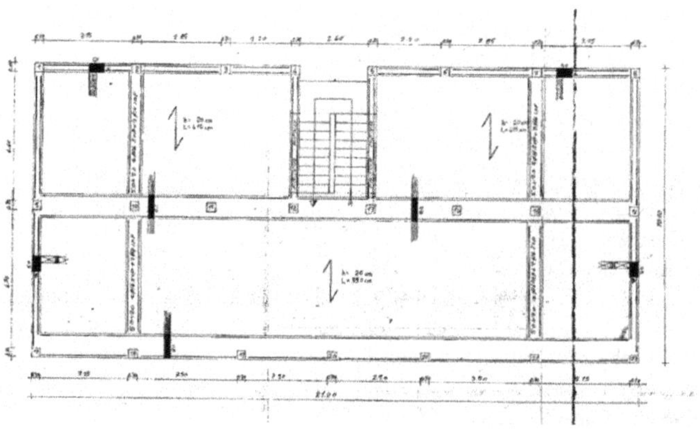

Figure 33. Second floor plan of Bonefro building

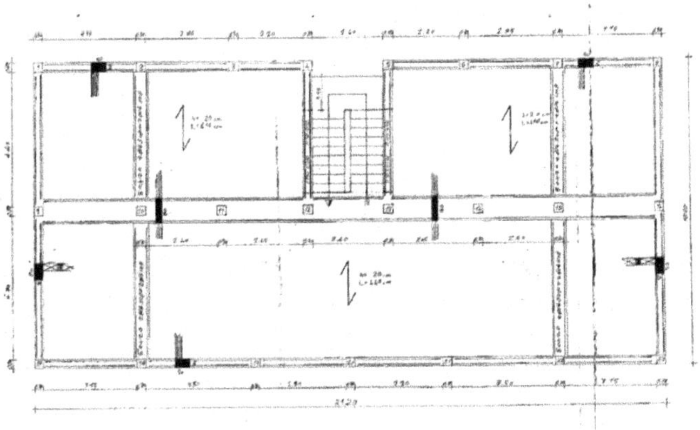

Figure 34. Third level + roof plan of Bonefro building

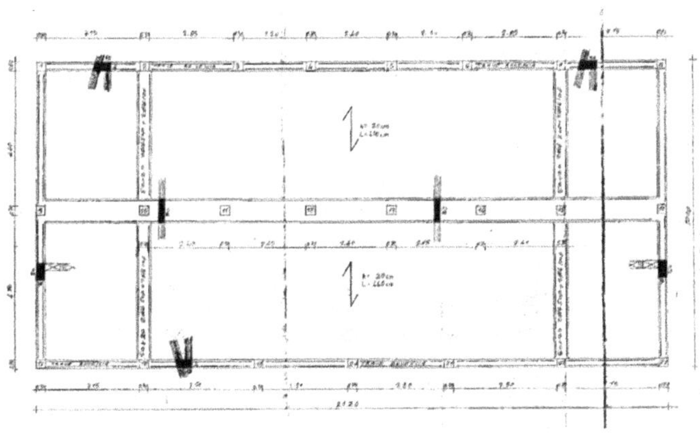

Table 4. Reinforcement and characteristics elements details

elements & characteristics	b(m)	h(m)	Reinforcement
element 1	0.3	0.3	2Φ16 top; 2Φ16 bottom; 2 Φ 14 middle
element 2	0.3	0.3	2Φ16 top ; 2Φ16 bottom
element 3	0.3	0.5	2Φ14 top ; 4Φ14 bottom
element 4	0.8	0.2	6Φ14 top ; 6Φ14 bottom
element 5	0.5	0.2	2Φ14 top ; 2Φ14 bottom
element 6	0.3	0.2	2Φ14 top ; 2Φ14 bottom

Figure 35. Bonefro building model in Z_Soil

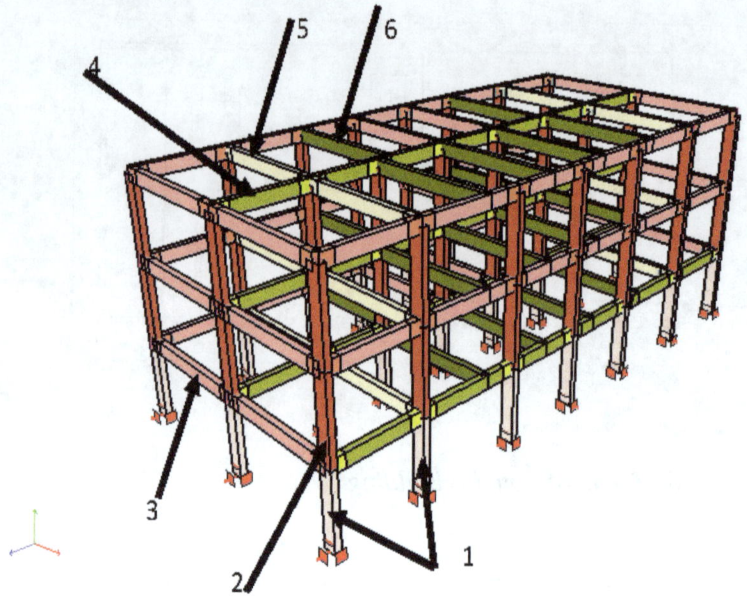

Figure 36. Rectangular section for floor beams

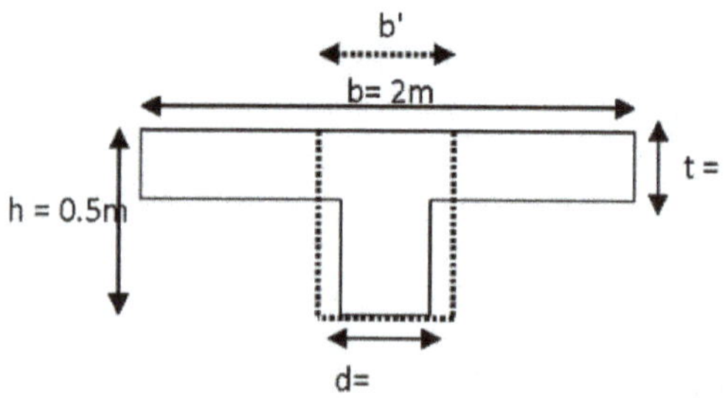

NONLINEAR PUSHOVER RESPONSE ON 2D FRAME WITH Z_SOIL (SMALL DISPLACEMENTS)

First, the response of a single 2D frame is studies. The side frame analyzed is shown in figure 33. Its model with force-based elements in Z_Soil is shown in figure 39.

The Eurocode 8 pushover procedure is applied to the 2D model of figure33 and figure 34. The frame is modeled with force-based elements. The load distributions are shown in figure 40. The "uniform" load distribution corresponds to P=MR, while the "modal" load distribution corresponds to $P=MF_1$. The elastic response response spectrum refers to a type 1 spectrum of Eurocode 8, ground type A (S=1, T_B= 0,15 sec, T_C = 0,4 sec, T_D = 2 sec) 5% damping (h=1) and a_g = 0.15g.

Figure 41 and Figure 42 show the result of the pushover analyses for modal and uniform load distributions, respectively. In both cases, the structure remains elastic.

Figure 37. Nodal masses for Bonefro building

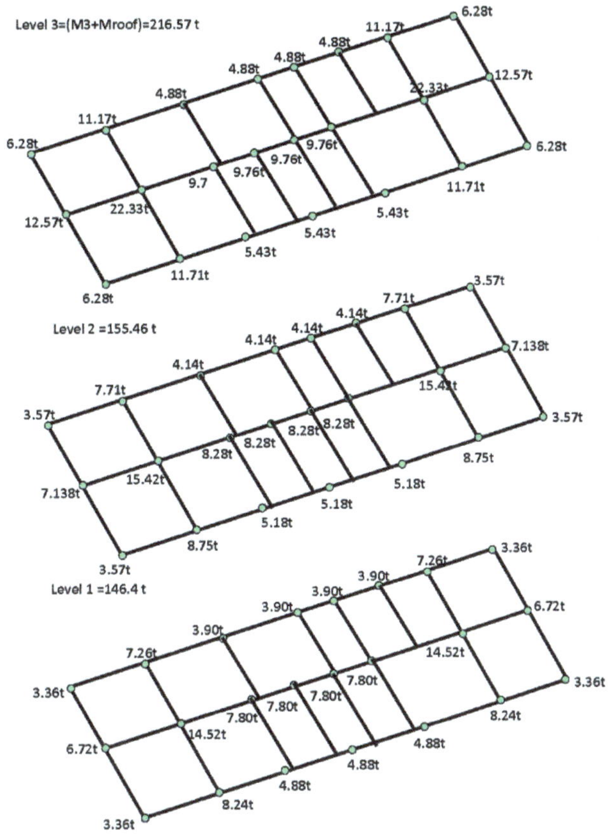

Figure 38. 2D frame Z-soil model

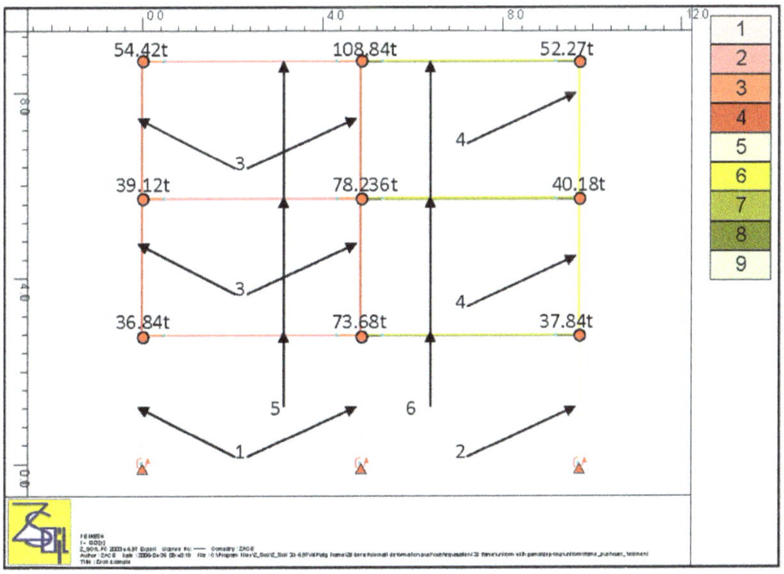

Figure 39. 2D frame studied

Figure 40. Load distributions used in the pushover analyses.

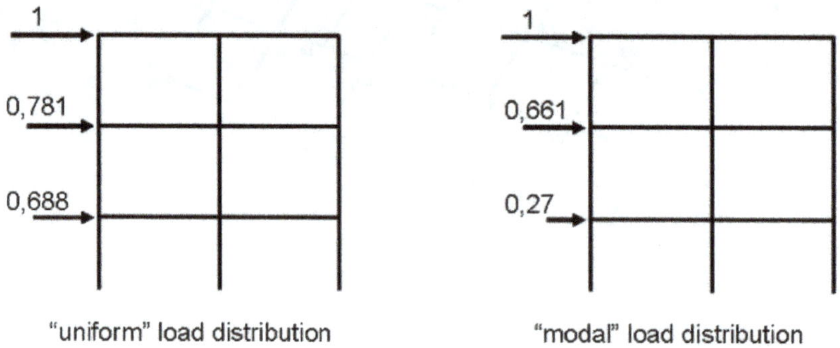

Figure 41. Determination of target displacement for modal load distribution.

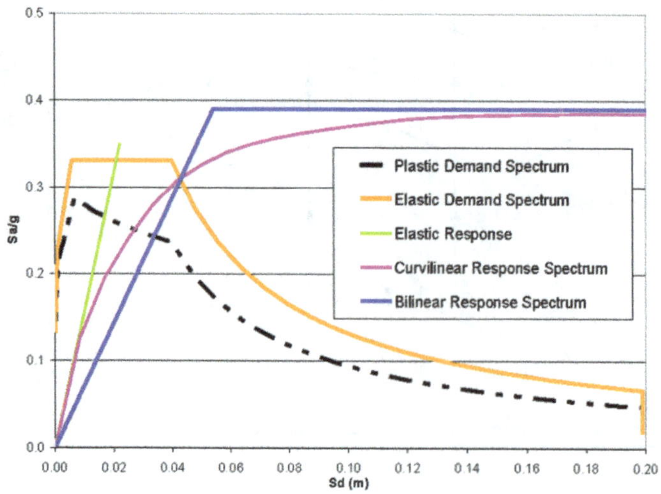

Figure 42. Determination of target displacement for uniform load distribution.

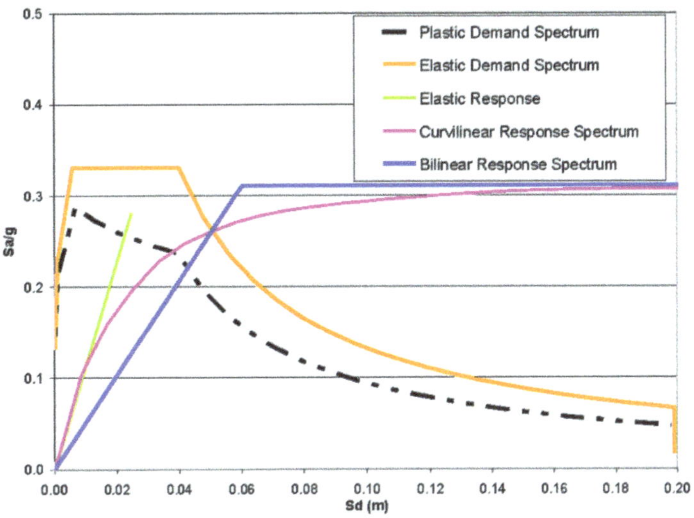

NONLINEAR PUSHOVER ON 3D FRAME WITH Z_ SOIL (SMALL DISPLACEMENTS)

Figure 43, shows the building model and the top floor point at which displacements are measured Point P). Force-based elements are used. Modal and uniform load distributions are considered.

Figure 43. Determination of target displacement for modal load distribution.

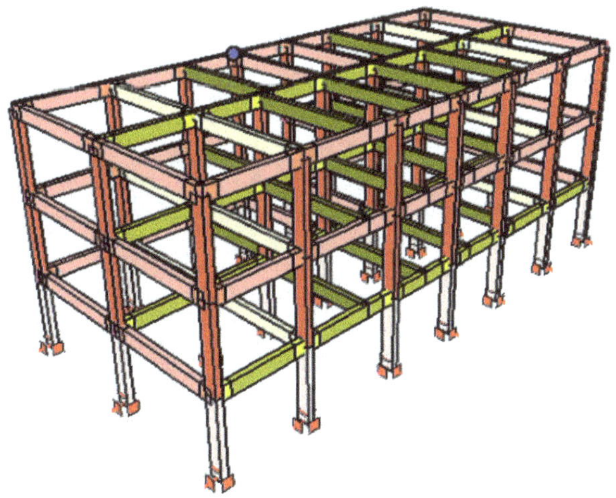

Figure 44. Determination of target displacement for modal loading in x direction load distribution

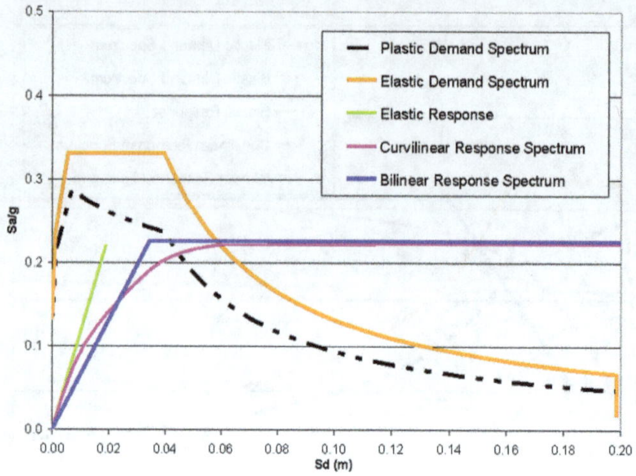

Figure 45. Determination of target displacement for modal loading100% in x & 30% in z direction load distribution (flexibility based)

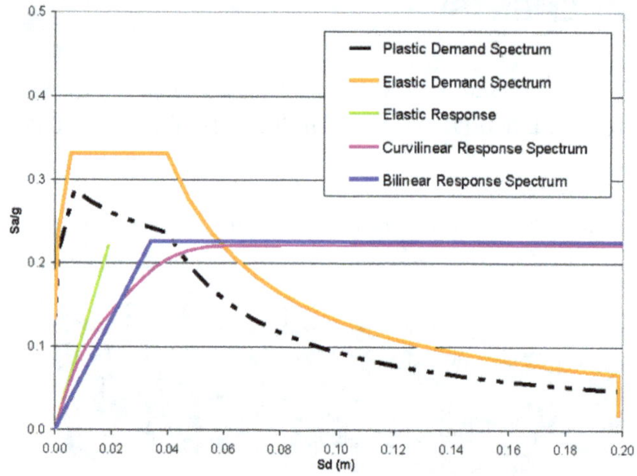

The target displacements for 2D & 3Dmodel of Bonefro building in the case of two loads distribution are around 0.051, 0.042, 0.052 and 0.06m for uniform and modal load distribution respectively these results are for SDOF in order to have the displacement of MDOF we must multiply the results of SDOF by G which is equal for 3D modal to 1.28 and for 2D modal to 1.2, thus the MDOF results are 0.0612, 0.0504, 0.067and 0.077m . for 2D and 3D model in the case of uniform and modal load distribution respectively.

Figure 46. Determination of target displacement for uniform load distribution

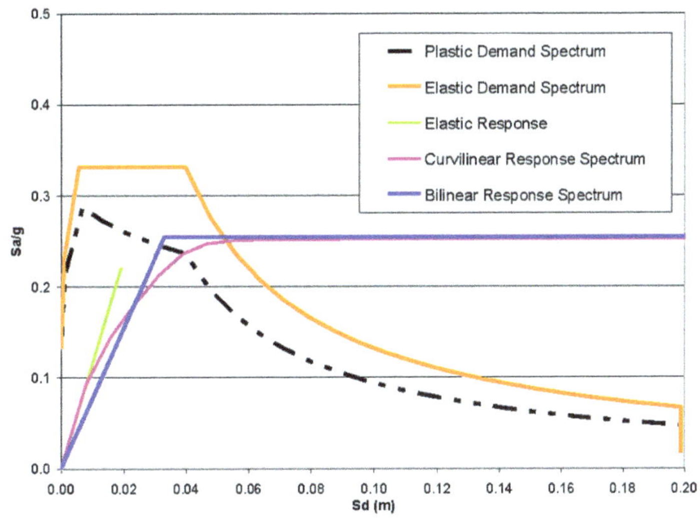

Figure 47. Comparison between 3D model pushover curve under modal 100x and modal 100%x &30% z load distribution

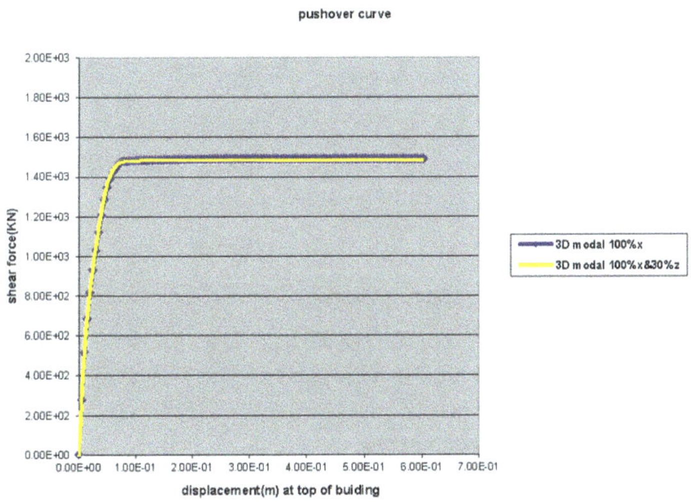

Table 5. Target displacement for 2D and 3D unser uniforme and modal load pattern

	Target displacement under Uniform load (SDOF)	Target displacement under Modal load (SDOF)	Target displacement Uniform load (MDOF)	Target displacement under Modal load (MDOF)
2D	0.051m	0.042m	0.0612m	0.0504m
3D	0.052m	0.06m	0.067m	0.077m

SIMULATED ACCELEROGRAMS

Three accelerograms compatible with the linear response spectrum are applied see figure 48. The three accelerograms (applied in one direction only – the x direction indicated in Figure 49) are simulated accelerograms generated with a computer program by (Sabetta & Pugliese 1996). The input parameters for this program are the epicentral distance, the earthquake magnitude and the type of soil). Sabetta program allows the generation of non-stationary artificial earthquakes according to an empirical method based on the regression of the relations of attenuation of a collection of earthquakes measured in Italy (95 accelerogram of 17 earthquakes magnitude 4.6 to 6.8). It generates a significant number of earthquakes and it takes the average of these earthquakes. The input parameters that yield an acceptable mean spectrum are magnitude = 6.02, epicentral distance = 22.8 km, type soil = shallow. The relevant mean spectrum is shown in Figure 3.34. The three accelerograms are shown in Figure 49. The three earthquakes are scaled to 0,33g PGA.

Figure 48. Mean spectrum of three simulated ground motions (Belgasmia 2015)

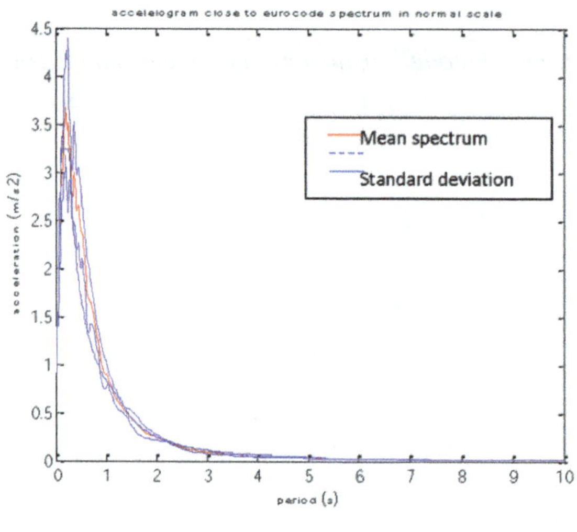

3D MODEL NONLINEAR TIME HISTORY

A time history analysis of the 3D model is presented with the three accelerograms of the previous section. The results for input ground motion applied in one direction only, the x direction of Figure 3.7, are compared for pushover and time-history analysis. Stiffness proportional Rayleigh damping is prescribed, with 5% damping at 2 Hertz. The resulting values for Rayleigh damping are α=0 and β=0.008. The top-floor response to the three accelerograms of in figure 49 are shown in figures 50 to figure 52.

Because three accelerograms are used, Eurocide 8 2003, prescribes that the maximum of the three maximum actions must be used for verifications. The maximum top floor displacement for the three earthquakes is 0.072m. The target displacement for the pushover analysi is 0,077m.

Figure 49. Three generated accelerograms

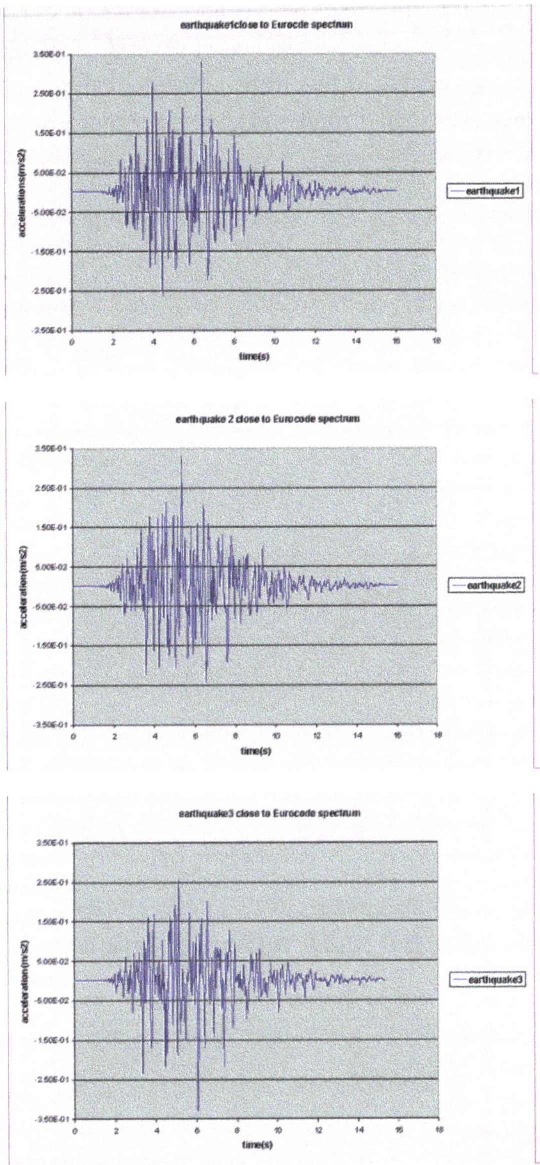

Figure 53 shows on the same plot the pushover curve and the based shear measured during the time history analysis at given displacements, 0.01, 0.02 and 0.04 m. Because during a cyclic analysis the top displacement reaches at several instances a given value, several points correspond to a given displacement. This is also illustrated for the time history response to earthquake 3 in Figure 49. The displacement of 0,04 m is attained at several instances. It is interesting to note from figure 48, that in the time history analysis the maximum based shear is reached even for small displacements due to the cylcic nature of the response. Figure 53 points out how the pushover curve gives different information compare to the three time history. Pushover analysis provides the maximum base shear one can expect for a given maximum target displacement (corresponding to a given earthquake intensity). The time histories provide not only the maximum values, but the entire history. For the example at hand, the max displacement of the time history is 0.072m, and the maximum base shear is approximately equal to 1500 kN, but during the time history this value of the base shear can be reached at several instances and for different values of the top displacement. On the pushover curve this translates into a single point that provides maximum displacement and maximum base shear that can be expected for that given earthquake.

Figure 50. Top floor response (at P point) of 3D building to earthquake 1(Belgasmia 2015)

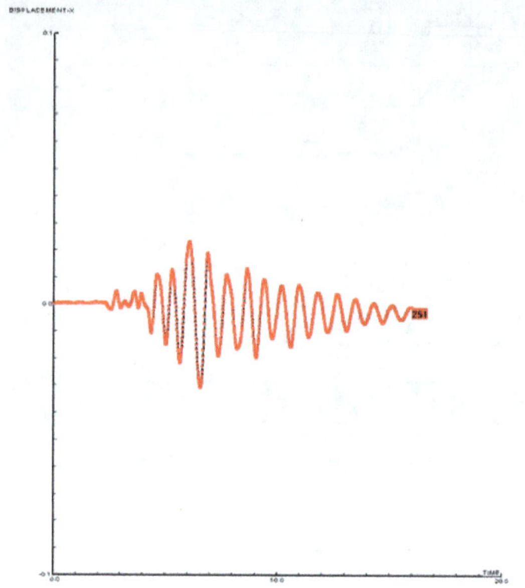

Figure 51. Top floor response (at P point) of 3D building to earthquake 2 (Belgasmia 2015)

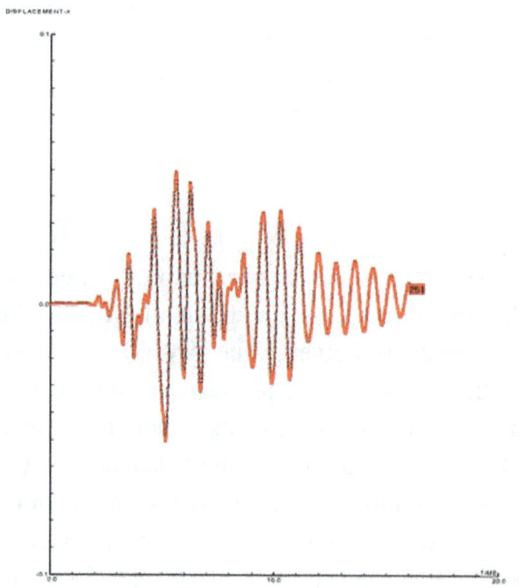

Figure 52. Top floor response (at P point) of 3D building to earthquake 3 (Belgasmia 2015)

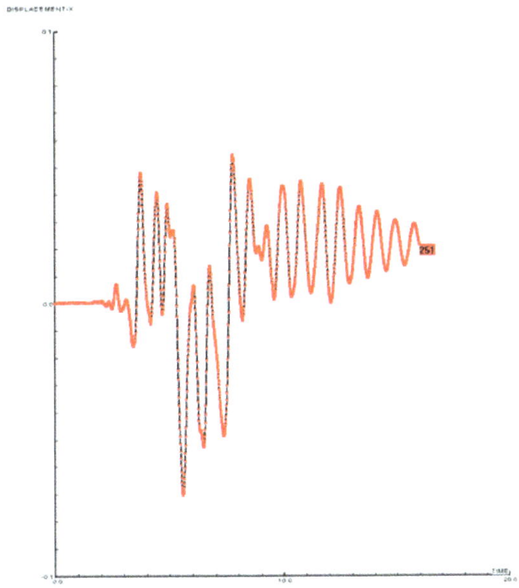

Figure 53. Base shear at selected top-floor displacements from TH analyses: comparison with pushover curve (Belgasmia 2015)

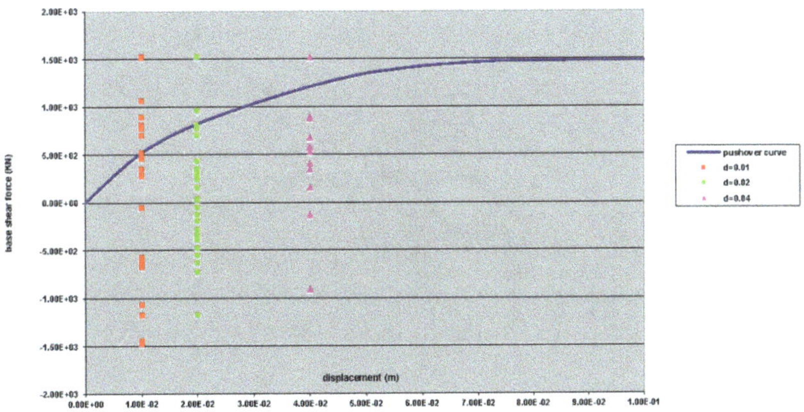

SENSITIVITY TO SEISMIC PARAMETERS

The sensitivity of the time history analyses to the seismic paramaters used in generating the accelerograms with the program by (Sabetta & Pugliese 1996) is presented here. The three parameters magnitude, epicenter distance and soil type were changed. The results are shown in the following tables. The reference values are magnitude = 6.02, epicenter distance = 22.8 km, shallow soil type. These are the values used for generating the three earthquakes used for the time history analyses.

Figure 54. Instances at which the top displacement due to earthquake 3 is equale to 0,04 m (Belgasmia 2015)

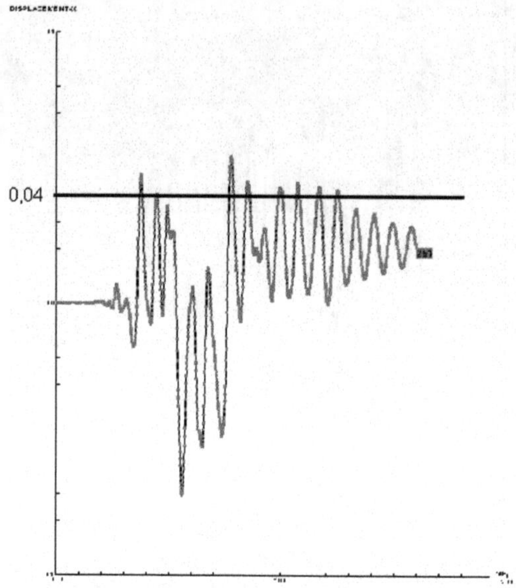

Figure 55. Deformed shape of 3D model for earthquake 3, d= 0.04 m t=7.64 s V_b =1520 kN

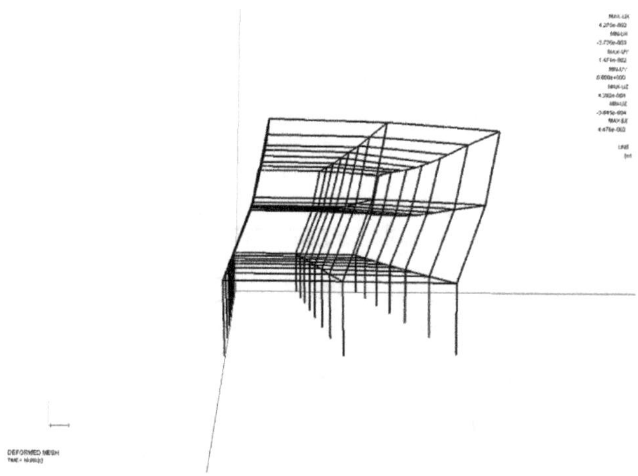

Table 6. Sensitivity to magnitude

	Magnitude= 5.72	Magnitude= 6.02	Magnitude= 6.32
Maximum response under modal load in x direction for static pushover analysis		Max = 1.28* 0.068m Max= 0.077m	
Maximum response under earthquakes in x direction for Time history analysis	Max = 0.06m	Max = 0.072m	Max = 0.106 m

Figure 56. Deformed shape of 3D model for earthquake 3, d=0.04 m t=10 s V_b =882 kN

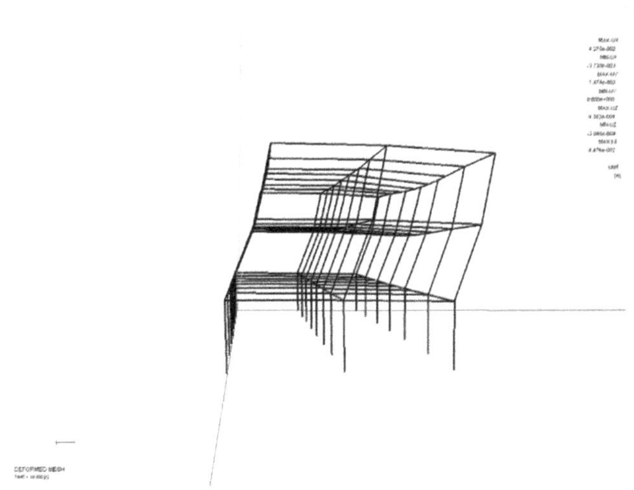

Figure 57. Deformed shape of 3D model for earthquake 3, d=0.04 m t=10.9 s, V_b=675 kN

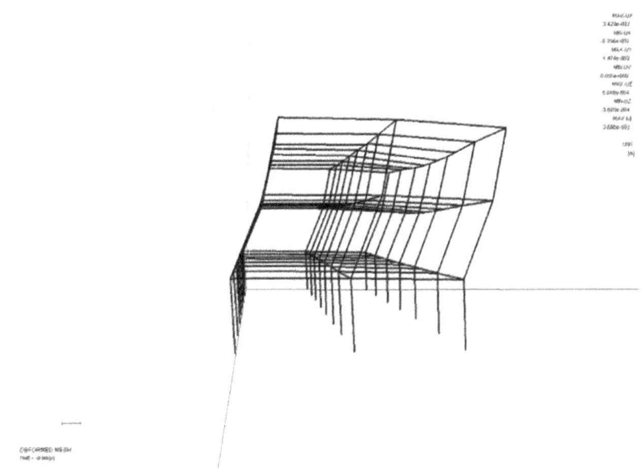

Table 7. Sensitivity to epicenter distance

	Epicenter distance=11.4Km	**Epicenter distance=22.8Km**	**Epicenter distance=45.6Km**
Maximum response under modal load in x direction for static pushover analysis		Max = 1.28* 0.068m Max= 0.077m	
Maximum response under earthquakes in x direction for Time history analysis	Max = 0.065m	Max = 0.072m	Max = 0.095m

The results are summarized in compact form in Figure 58. The magnitude and epicenter results are normalized with respect to the reference values indicated in bold in the above tables. There seems to be a very high sensitivity to the earthquake magnitude, but further studies are needed.

Table 8. Sensitivity to soil type

	Soil type shallow	Soil type deep
Maximum response under modal load in x direction for static pushover analysis	Max = 1.28* 0.068m Max= 0.077m	
Maximum response under earthquakes in x direction for Time history analysis	Max = 0.072m	Max = 0.065m

Figure 58. Sensitivity of response (expressed in terms of d_{max} at top floor)

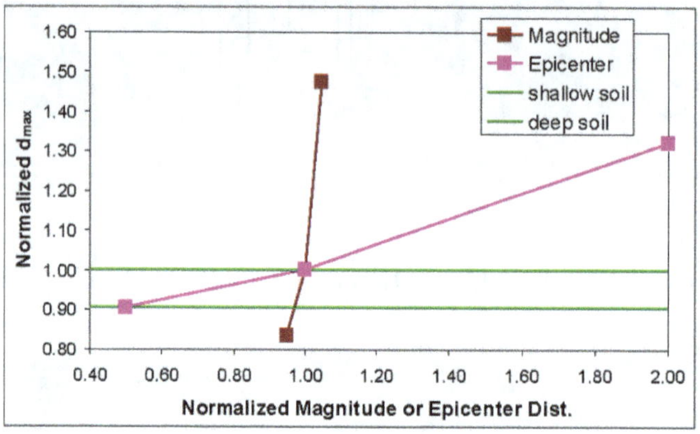

RESPONSE SENSITIVITY TO LARGE DEFORMATION

The pushover analyses are repeated using the large displacement analysis capabilities in Z_Soil. The results are shown in the following figures and are summarized in the end table. The pushover curve become strain softening after yielding, due to the geometric nonlinearities. The increase in target displacement is approx. 3.3% for modal distribution and 11.5% for uniform distribution.

Figure 59. Pushover curve with modal load distribution of 3D model with large deformations

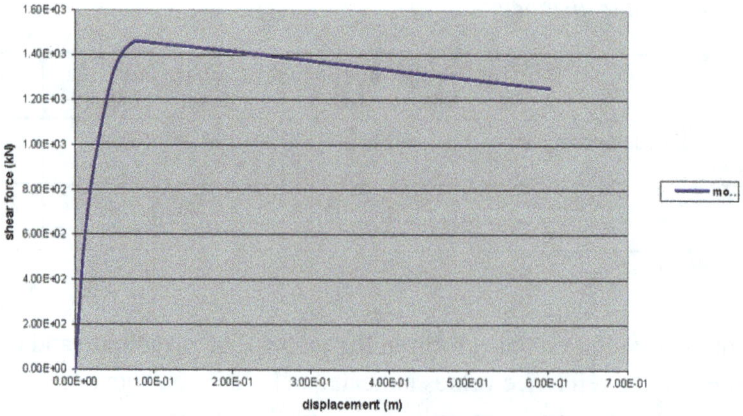

Figure 60. Target displacement for pushover curve

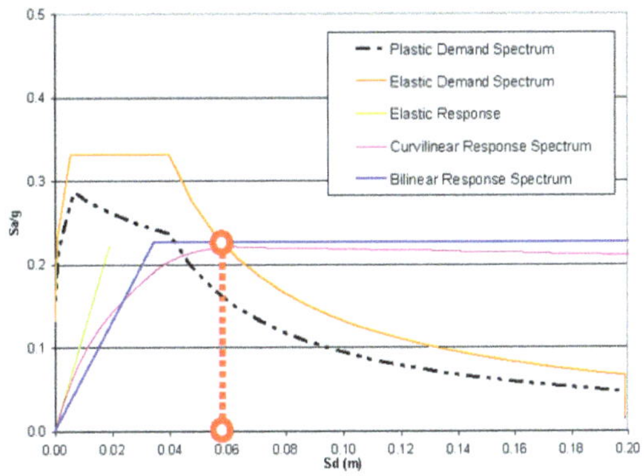

Figure 61. Pushover curve with uniform load distribution of 3D model with large deformations

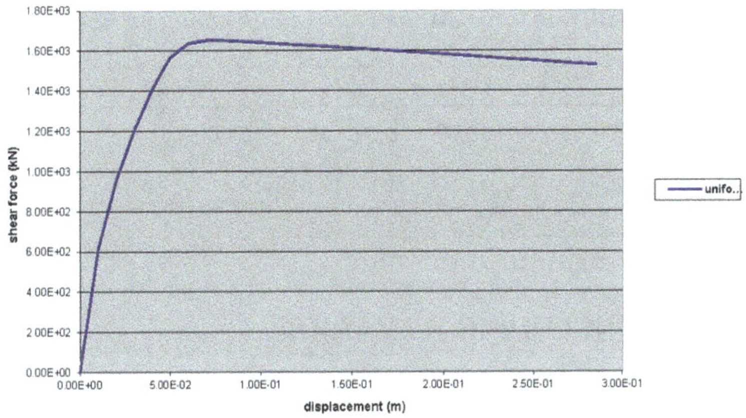

Figure 62. Target displacement for pushover curve

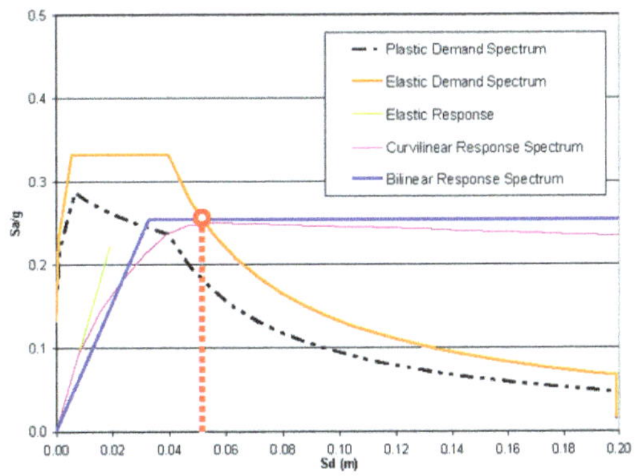

Figure 63. Pushover curves for modal and uniform load distributions

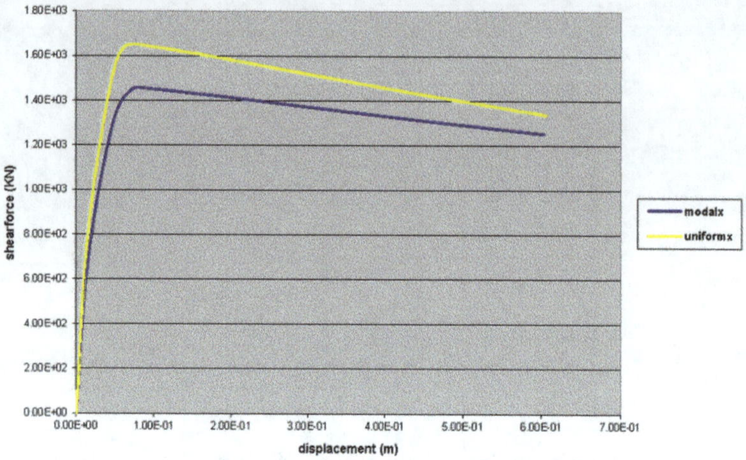

Table 9. Difference between small and large displacement static nonlinear results

	Modal load	Uniform load
Target displacement of SDOF for static pushover analysis (small displacements)	0.06m	0.052m
Target displacement of SDOF for static pushover analysis (large displacements)	0.062m (+3.3%)	0.058m (+11.5)

OBSERVATION AND CONCLUDING REMARKS

After satisfactory comparison of flexibility and displacement based formulation, only the first one was used in this study.

- 2D analysis indicates a larger target displacement for uniform loading vs modal loading (figures 41 & figure 42) 0.051m against 0.042m. (reference point P structural displacements 0.061m vs 0.05m).
- 3D analysis, uniform loading vs modal loading, indicates the opposite trend, (figures 44 & figure 46), 0.052m against 0.06m. (at P 0.067m vs 0.077m)
- Combining loading, 100% in x direction and 30% in z direction shows no influence on the push-over curve and hence on the target displacement (figure 45 & figure 47).
- The comparison between pushover and dynamics gives difference of 7% (d_{max} modal from Pushover/ d_{max} dynamics) =0.077/0.072.
- The sensitivity to Magnitude, in dynamics a (+-5%) variation in EQ magnitude yields a variation in max displacement of (+47% and –17%).
- The sensitivity to epicenter distance, in dynamics (via Sabetta program) a (+100%, -50%) variation yields a variation in max displacement of (+31%, -10%).
- The sensitivity to soil, in dynamics (via Sabetta program): shallow/deep = 0.072/0.065=1.1 the difference is about 10%.

- The maximum base shear indicated by the pushover analysis is systematically reached for almost any maximum top displacement, this is probably indicative of significant influence of 2nd 3rd etc modes (figure 54).
- Large displacement induce global softening and increase target displacements in 3D pushover +11% for uniform loading 0.058/0.052, +3% for modal loading 0.062/0.06.

We notice no influence in dynamics.

Example 2

In this exemple we present a study of 3D nonlinear static analysis under seismic loads of a fourteen-story reinforced concrete building designed according to the Algerian code (C.B.A 1993) and (R.P.A 2003), with typical story height of 3.5m, the structure is a beam-column system with a flat slab at different floor levels. The strength of the concrete is 25 MPa and the steel reinforcement is 400 MPa. Plastic hinge property of the confined cross sections is determined by considering transversal and longitudinal reinforcement. In order to consider the influence of different ground soil conditions three wave shear(V_S) velocities, including 400 m/s, 200 m/s, 100 m/s that represents stiff, medium and soft soil are chosen, SAP2000 finite element software is used for three dimensional modeling and analyses of the building. (see figure 64).

In the current analysis, we will proceed as follows:

1. Establish the model of the RC high-rise building.
2. Apply the seismic load increasingly. The structural deformation is growing and plastic hinges are formed in some sections of the structure.
3. A collapse or a target displacement is reached when increasing the horizontal load.
4. Display the base shear-displacement curves and the floor displacement for two different cases:
    a. Case 1: The structure will be fixed at the base.
    b. Case 2: The interaction between the soil and the structure will be modeled with springs and dashpots for different types of soil: stiff, medium and soft soil.

Figure 64. 3D models with fixed base and spring-dashpot elements (Bougherra 2018)

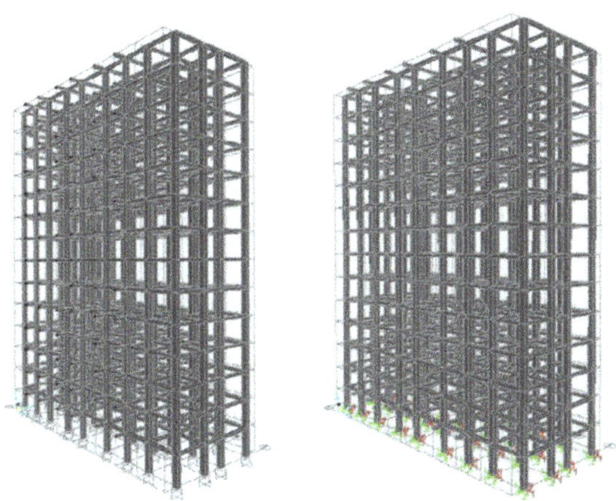

ANALYSIS RESULTS

The results of the analysis given by SAP2000 software for different types of soil and for a fixed base are shown in Table 10 and from figure 65 to figure 67, which represents floor displacement, the Capacity curve, story drift plastic hinge formation respectively:

Table10. Floor displacement for different types of soil

Story	Fixed base(m)	Stiff(m)	Medium(m)	Soft(m)
0	0	0.000015	0.000066	0.000244
1	0.030252	0.03048	0.041728	0.078495
2	0.090437	0.090767	0.114792	0.189983
3	0.170931	0.171317	0.20912	0.322467
4	0.264465	0.264871	0.317478	0.469256
5	0.364684	0.365081	0.433283	0.624119
6	0.466058	0.466426	0.550734	0.781454
7	0.564027	0.564355	0.664971	0.936449
8	0.655134	0.655419	0.772212	1.085215
9	0.737101	0.737344	0.869862	1.224898
10	0.808718	0.808921	0.956585	1.353759
11	0.869725	0.869892	1.032217	1.471218
12	0.920811	0.920946	1.097583	1.577875
13	0.963615	0.963721	1.154445	1.675493
14	1.000739	1.00082	1.205506	1.766946

Table 10 shows an augmentation in the lateral displacement between the model with fixed base and the based on soft soil. The largest differences are obvious on the five top floors.

The capacity curves shown in figure 65 represents the relationship between the values of base shear and the top lateral displacement. As the curves show, effects of soil structure interaction are not obvious in the elastic stage of the curve but it is more apparent in the nonlinear stage. We deduce that with a given soft soil the structure tends to reflect a lower capacity than the structures with higher values of wave shear.

The curves of figure 66 represents the inter story drifts for different types of soil. It is obvious from this curves that SSI has an impact on structural behavior especially in the case of the inter story drift of the bottom floors. We deduce that the relative displacement increases with softer foundation soil.

Figure 67 shows that formation of plastic hinges changes when considering SSI effects, the yellow hinges represents the step of collapse structure, the position changes and the number of hinges increases with lower values of wave shear. In addition, we deduce that the hinges that represents the collapse statue in the model with a fixed base appear in the twelfth step, while in the models with horizontal and rocking springs, collapse hinges are formed in the eleventh step.

Figure 65. Capacity curve for different types of soil (Bougherra 2018)

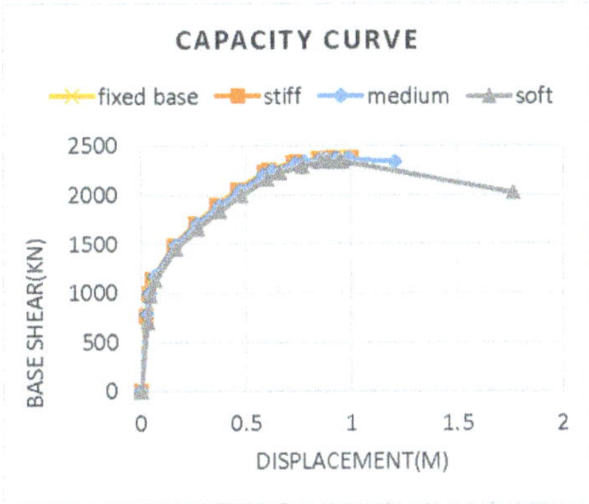

Figure 66. Inter story drift for different types of soil (Bougherra 2018)

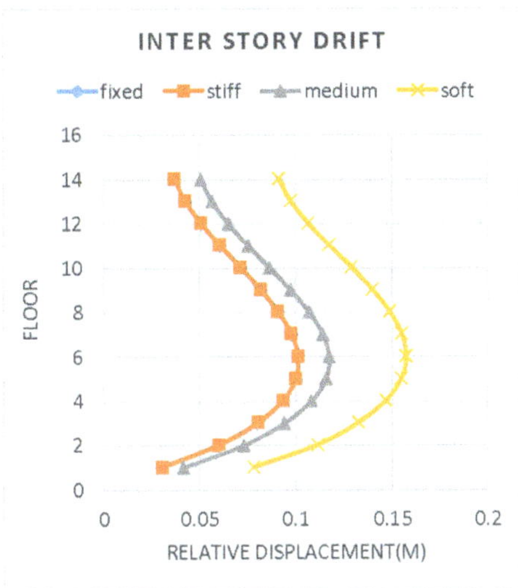

Figure 67. Plastic hinges distribution for fixed, stiff, medium and soft soil (Bougherra 2018)

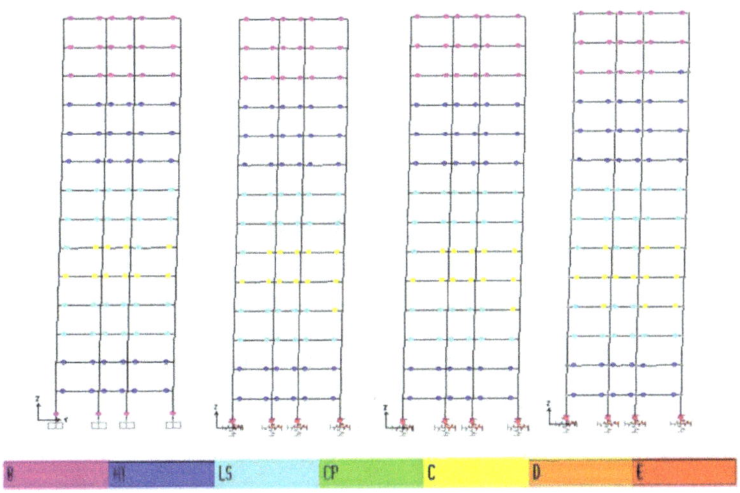

REFERENCES

Antoniou, S., & Pinho, R. (2004). Advantages and Limitations of Adaptive and Non-adaptive Force-Based Pushover Procedures. *Journal of Earthquake Engineering*, *8*(4), 497–522. doi:10.1080/13632460409350498

Aydinoglu, M. N. (2003). An Incremental Response Spectrum Analysis Procedure Based on Inelastic Spectral Deformation for Multi-Mode Seismic Performance Evaluation. *Bulletin of Earthquake Engineering*, *1*(1), 3–36. doi:10.1023/A:1024853326383

Aydinoglu, M. N. (2004). An Improved Pushover Procedure for Engineering Practice: Incremental Response Spectrum Analysis (IRSA). *International Workshop on Performance-based seismic design. Concepts and Implementation*. Pacific Earthquake Research Center.

Belgasmia, M., & Moussaoui, S. (2015). Comparison Between Static Nonlinear and Time History Analysis using Flexibility-Based Model for an Existing Structure and Effect of Taking into Account Soil Using Domain Reduction Method for a Single Media. *KSCE Journal of Civil Engineering*, *19*(3), 651–663. doi:10.100712205-015-0351-y

Belgasmia, M., Spacone, E., & Zimmermann, Th. (2006). *Seismic Evaluation of Constructions: Static Pushover procedure and time history analysis for Nonlinear Frames LSC Internal Report*. Swiss Federal Institute of Technology.

Belgasmia, M. (2017). Definition of Static Nonlinear Procedure and Flexibility-Based model with Application on 2D Model for an Existing Structure and Comparing Results with Time History Analysis. In *Modeling and Simulation Techniques in Structural Engineering*. IGI Global.

Bougherra, S., & Belgasmia, M. (2018). Effect of soil structure interaction on the response of R.C building to seismic loads, a case study. *International Journal of Forensic Engineering*, *4*(2), 133–144. doi:10.1504/IJFE.2018.098504

CBA 93. (1993). *Code du Béton Algérien*. CGS Alger.

Chopra, A. K. (2001). *Dynamics of Structures: Theory and Applications to Earthquake Engineering* (2nd ed.). Prentice Hall.

Chopra, A. K., & Goel, R. K. (2002). A modal pushover analysis procedure for estimating seismic demands for buildings. *Earthquake Engineering & Structural Dynamics*, *31*(3), 561–582. doi:10.1002/eqe.144

Elnashai, A.S. (2002). Do we really need Inelastic Dynamic Analysis. *Journal of Earthquake Engineering*, *6*(S1), 123-130.

Eurocode 8. (2003). *Design of Structures for Earthquake Resistance*. European Committee for Standardization.

Fajfar, P. (1999). Capacity Spectrum Method Based on Inelastic Demand Spectra. *Earthquake Engineering & Structural Dynamics*, *28*(9), 979–993. doi:10.1002/(SICI)1096-9845(199909)28:9<979::AID-EQE850>3.0.CO;2-1

Fajfar, P. (2002). Structural Analysis in Earthquake Engineering – A Breakthrough of Simplified Non-Linear Methods. *Proc. 12th European Conference on Earthquake Engineering*.

Fajfar, P., Kilar, V., Marusic, D., & Perus, I. (2015). The extension of the N2 method to asymmetric buildings. *Proceedings of the 4th European Workshop on the Seismic Behaviour of Irregular and Complex Structures.*

Gazetas, G. (1990). Formulas and Charts for Impedances of Surface and Embedded foundations. *Journal of Geotechnical Engineering*, *117*(9), 1363–1381. doi:10.1061/(ASCE)0733-9410(1991)117:9(1363)

Gupta, B., & Kunnath, S. K. (2000). Adaptive Spectra-Based Pushover Procedure for Seismic Evaluation of Structures. *Earthquake Spectra*, *16*(2), 367–391. doi:10.1193/1.1586117

Kausel, E. (2010). Early history of soil–structure interaction. *Soil Dynamics and Earthquake Engineering*, *30*(9), 822–832. doi:10.1016/j.soildyn.2009.11.001

Nist, G. (2012). Soil-structure interaction for building structures. *US Department of Commerce. GCR*, *12*, 917–921.

R.P.A 99. (2003). *Règles Parasismiques Algériennes 1999.* Algiers: Centre National de Recherche Appliquée en Génie Parasismique.

Sabetta, F., & Pugliese, A. (1996). Estimation of Response Spectra and Simulation of Nonstationary Earthquake Ground Motions. *Bulletin of the Seismological Society of America*, *86*(2), 337–352.

SAP. (2000). Integrated structural analysis and design software. Computers and Structures, Inc.

Spacone, E., Filippou, E. F. C., & Taucer, F. F. (1996). Fiber Beam-Column Model for Nonlinear Analysis of R/C Frames. I: Formulation, II: Applications. *Earthquake Engineering & Structural Dynamics*, *25*(7), 711–742. doi:10.1002/(SICI)1096-9845(199607)25:7<711::AID-EQE576>3.0.CO;2-9

Vidic, T., Fajfar, P., & Fischinger, M. (1994). Consistent Inelastic Design Spectra: Strength and Displacement. *Earthquake Engineering & Structural Dynamics*, *23*(5), 502–521. doi:10.1002/eqe.4290230504

Wolf, J. P., & Obernhuber, P. (1985). Non-linear soil-structure-interaction analysis using dynamic stiffness or flexibility of soil in the time domain. *Earthquake Engineering & Structural Dynamics*, *13*(2), 195–212. doi:10.1002/eqe.4290130205

Zimmermann, T., Truty, A., Urbanski, A., & Podles. K. (2006). *Z-Soil user manual.* Zace Services Ltd.

KEY TERMS AND DEFINITIONS

Capacity Curve: Is a type of graph expressing the relationship between pattern load acting on the structure vs base shear force.

Nonlinear Time History: Is a method in which the action is dynamic and structure behavior is nonlinear.

Pushover Analysis: Is also called static nonlinear analysis; because it is a method in which, the action is static, but the structure behavior is nonlinear.

Small Displacement: Is a mathematical approach to the description of the deformation of a solid body in which the displacements of the material particles are assumed much smaller.

Soil Structure Interaction: Consists of the interaction between soil (ground) and a structure built upon it. And the study of the process in which the response of the soil influences the motion of *the* structure and the motion of the structure influences the response of the soil.

ENDNOTES

1 The design spectra are actually given in terms of pseudo-acceleration rather than acceleration, but for low damping the two are basically identical, thus the "pseudo-" is typically dropped. It is worth recalling that the pseudo-acceleration is defined as.

2 Note that the ductility reduction factor R_m is different from the reduction factor R (called behavior factor q in Eurocode 8). q takes into account both energy dissipation and the overstrength R_s, that is $R= R_m R_s$.

3 http://www.isesd.cv.ic.ac.uk/

4 https://peer.berkeley.edu/smcat/index.html

5 http://www.seismolinks.com/Strong_Motion_Databases/

About the Author

Mourad Belgasmia is a Lecturer and Thesis Adviser joined the team of the Civil Engineering department of university of Bejaia (Algeria) on December 2001 as a lecturer (assistant professor), till 2008, Since 2008 till now with the team of the Civil Engineering department of university of Setif 1 (Algeria). He spent two years (from 2004 to 2006) at the laboratory of structural and continuum Mechanics at Swiss Federal Institute of Technology in Lausanne, Switzerland, working then on seismic evaluation of existing structures and new structures using Pushover Analysis. And at several occasion since for shorter periods the work was mainly on approximate method for seismic evaluation of buildings this research is documented in two substantial internal reports. His responsibilities during this period in the laboratory of structural and continuum Mechanics at Swiss Federal Institute of Technology(EPFL) Switzerland (Lausanne) From July 2004 to July 2006 is Assistant of the implementation of new method in z-soil software. His teaching experience is Strength of materials, Finite element method, Continuum Mechanics, Steel Construction, Computer language programming, Structural Dynamic, Pre-stressed concrete. His research directions are seismic expertise, dynamic of structures, object oriented programming, xfem extend finite element method. Mourad Belgasmia is a book editor, (co) author of 18 papers in refereed international journals and more than 20 international conferences.

Index

R

S

T

U

V

Recommended Reference Books

IGI Global's reference books are available in three unique pricing formats:
Print Only, E-Book Only, or Print + E-Book.

Shipping fees may apply.

www.igi-global.com

ISBN: 978-1-5225-9276-1
© 2020; 612 pp.
List Price: US$ **345**

ISBN: 978-1-5225-8325-7
© 2019; 169 pp.
List Price: US$ **165**

ISBN: 978-1-4666-8693-9
© 2015; 337 pp.
List Price: US$ **215**

ISBN: 978-1-4666-8823-0
© 2016; 824 pp.
List Price: US$ **420**

ISBN: 978-1-4666-8150-7
© 2015; 380 pp.
List Price: US$ **205**

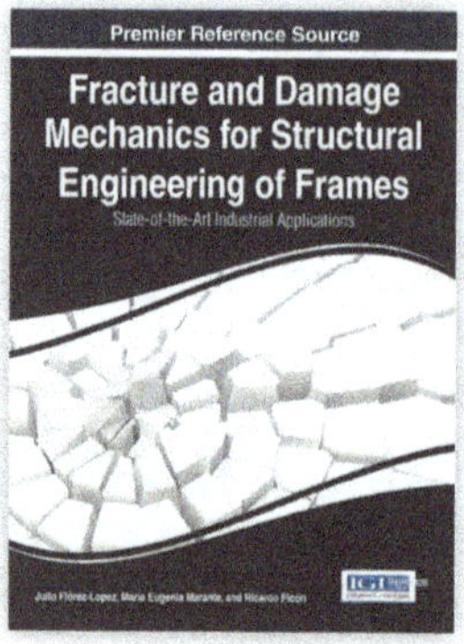

ISBN: 978-1-4666-6379-4
© 2015; 602 pp.
List Price: US$ **225**

Do you want to stay current on the latest research trends, product announcements, news, and special offers?
Join IGI Global's mailing list to receive customized recommendations, exclusive discounts, and more.
Sign up at: **www.igi-global.com/newsletters.**

Publisher of Peer-Reviewed, Timely, and Innovative Academic Research

IGI Global DISSEMINATOR OF KNOWLEDGE

www.igi-global.com Sign up at www.igi-global.com/newsletters facebook.com/igiglobal twitter.com/igiglobal linkedin.com/igiglobal

Ensure Quality Research is Introduced to the Academic Community

Become an Evaluator for IGI Global Authored Book Projects

The overall success of an authored book project is dependent on quality and timely manuscript evaluations.

Applications and Inquiries may be sent to:
development@igi-global.com

Applicants must have a doctorate (or equivalent degree) as well as publishing, research, and reviewing experience. Authored Book Evaluators are appointed for one-year terms and are expected to complete at least three evaluations per term. Upon successful completion of this term, evaluators can be considered for an additional term.

If you have a colleague that may be interested in this opportunity, we encourage you to share this information with them.

IGI Global Author Services

Providing a high-quality, affordable, and expeditious service, IGI Global's Author Services enable authors to streamline their publishing process, increase chance of acceptance, and adhere to IGI Global's publication standards.

Benefits of Author Services:

- **Professional Service:** All our editors, designers, and translators are experts in their field with years of experience and professional certifications.
- **Quality Guarantee & Certificate:** Each order is returned with a quality guarantee and certificate of professional completion.
- **Timeliness:** All editorial orders have a guaranteed return timeframe of 3-5 business days and translation orders are guaranteed in 7-10 business days.
- **Affordable Pricing:** IGI Global Author Services are competitively priced compared to other industry service providers.
- **APC Reimbursement:** IGI Global authors publishing Open Access (OA) will be able to deduct the cost of editing and other IGI Global author services from their OA APC publishing fee.

Author Services Offered:

English Language Copy Editing
Professional, native English language copy editors improve your manuscript's grammar, spelling, punctuation, terminology, semantics, consistency, flow, formatting, and more.

Scientific & Scholarly Editing
A Ph.D. level review for qualities such as originality and significance, interest to researchers, level of methodology and analysis, coverage of literature, organization, quality of writing, and strengths and weaknesses.

Figure, Table, Chart & Equation Conversions
Work with IGI Global's graphic designers before submission to enhance and design all figures and charts to IGI Global's specific standards for clarity.

Translation
Providing 70 language options, including Simplified and Traditional Chinese, Spanish, Arabic, German, French, and more.

Hear What the Experts Are Saying About IGI Global's Author Services

"Publishing with IGI Global has been ***an amazing experience*** *for me for sharing my research. The* ***strong academic production*** *support ensures quality and timely completion."* – **Prof. Margaret Niess, Oregon State University, USA**

"The service was ***very fast, very thorough, and very helpful*** *in ensuring our chapter meets the criteria and requirements of the book's editors. I was* ***quite impressed and happy*** *with your service."* – **Prof. Tom Brinthaupt, Middle Tennessee State University, USA**

Learn More or Get Started Here:

For Questions, Contact IGI Global's Customer Service Team at cust@igi-global.com or 717-533-8845

IGI Global
PUBLISHER of TIMELY KNOWLEDGE
www.igi-global.com

IGI Global
PUBLISHER of TIMELY KNOWLEDGE
www.igi-global.com

Celebrating Over 30 Years of Scholarly Knowledge Creation & Dissemination

InfoSci®-Books

A Database of Nearly 6,000 Reference Books Containing Over 105,000+ Chapters Focusing on Emerging Research

GAIN ACCESS TO **THOUSANDS** OF REFERENCE BOOKS AT **A FRACTION** OF THEIR INDIVIDUAL LIST **PRICE**.

InfoSci®-Books Database

The **InfoSci®-Books** is a database of nearly 6,000 IGI Global single and multi-volume reference books, handbooks of research, and encyclopedias, encompassing groundbreaking research from prominent experts worldwide that spans over 350+ topics in 11 core subject areas including business, computer science, education, science and engineering, social sciences, and more.

Open Access Fee Waiver (Read & Publish) Initiative

For any library that invests in IGI Global's InfoSci-Books and/or InfoSci-Journals (175+ scholarly journals) databases, IGI Global will match the library's investment with a fund of equal value to go toward **subsidizing the OA article processing charges (APCs) for their students, faculty, and staff** at that institution when their work is submitted and accepted under OA into an IGI Global journal.*

INFOSCI® PLATFORM FEATURES

- Unlimited Simultaneous Access
- No DRM
- No Set-Up or Maintenance Fees
- A Guarantee of No More Than a 5% Annual Increase for Subscriptions
- Full-Text HTML and PDF Viewing Options
- Downloadable MARC Records
- COUNTER 5 Compliant Reports
- Formatted Citations With Ability to Export to RefWorks and EasyBib
- No Embargo of Content (Research is Available Months in Advance of the Print Release)

**The fund will be offered on an annual basis and expire at the end of the subscription period. The fund would renew as the subscription is renewed for each year thereafter. The open access fees will be waived after the student, faculty, or staff's paper has been vetted and accepted into an IGI Global journal and the fund can only be used toward publishing OA in an IGI Global journal. Libraries in developing countries will have the match on their investment doubled.*

To Recommend or Request a Free Trial:
www.igi-global.com/infosci-books
eresources@igi-global.com • Toll Free: 1-866-342-6657 ext. 100 • Phone: 717-533-8845 x100

IGI Global
PUBLISHER of TIMELY KNOWLEDGE
www.igi-global.com

www.ingramcontent.com/pod-product-compliance
Lightning Source LLC
LaVergne TN
LVHW060303011026
849161LV00011B/2766
* 9 7 8 1 7 9 9 8 4 3 9 9 3 *